CW01455406

DESIGNERS' GUIDES TO THE EUROCODES

DESIGNERS' GUIDE TO EUROCODE 1: ACTIONS ON BUILDINGS

EN 1991-1-1 AND -1-3 TO -1-7

Eurocode Designers' Guide series

Designers' Guide to EN 1990 Eurocode: Basis of structural design. H. Gulvanessian, J.-A. Calgaro and M. Holický. 978 0 7277 3011 4. Published 2002.

Designers' Guide to Eurocode 8: Design of structures for earthquake resistance. EN 1998-1 and EN 1998-5. General rules, seismic actions, design rules for buildings, foundations and retaining structures. M. Fardis, E. Carvalho, A. Elnashai, E. Faccioli, P. Pinto and A. Plumier. 978 0 7277 3348 1. Published 2005.

Designers' Guide to EN 1994-1-1. Eurocode 4: Design of Composite Steel and Concrete Structures, Part 1-1: General Rules and Rules for Buildings. R.P. Johnson and D. Anderson. 978 0 7277 3151 7. Published 2004.

Designers' Guide to Eurocode 7: Geotechnical design. EN 1997-1 General rules. R. Frank, C. Bauduin, R. Driscoll, M. Kavvadas, N. Krebs Ovesen, T. Orr and B. Schuppener. 978 0 7277 3154 8. Published 2004.

Designers' Guide to Eurocode 3: Design of Steel Structures. EN 1993-1-1 General rules and rules for buildings. L. Gardner and D. Nethercot. 978 0 7277 3163 0. Published 2005.

Designers' Guide to Eurocode 2: Design of Concrete Structures. EN 1992-1-1 and EN 1992-1-2 General rules and rules for buildings and structural fire design. R.S. Narayanan and A.W. Beeby. 978 0 7277 3105 0. Published 2005.

Designers' Guide to EN 1994-2. Eurocode 4: Design of composite steel and concrete structures. Part 2 General rules for bridges. C.R. Hendy and R.P. Johnson. 978 0 7277 3161 6. Published 2006

Designers' Guide to EN 1992-2. Eurocode 2: Design of concrete structures. Part 2: Concrete bridges. C.R. Hendy and D.A. Smith. 978-0-7277-3159-3. Published 2007.

Designers' Guide to EN 1991-1-2, EN 1992-1-2, EN 1993-1-2 and EN 1994-1-2. T. Lennon, D.B. Moore, Y.C. Wang and C.G. Bailey. 978 0 7277 3157 9. Published 2007.

Designers' Guide to EN 1993-2. Eurocode 3: Design of steel structures. Part 2: Steel bridges. C.R. Hendy and C.J. Murphy. 978 0 7277 3160 9. Published 2007.

Designers' Guide to EN 1991-1.4. Eurocode 1: Actions on structures, general actions. Part 1-4 Wind actions. N. Cook. 978 0 7277 3152 4. Published 2007.

Designers' Guide to Eurocode 1: Actions on buildings. EN 1991-1-1 and -1-3 to -1-7. H. Gulvanessian, P. Formichi and J.-A. Calgaro. 978 0 7277 3156 2. Published 2009.

Designers' Guide to Eurocode 1: Actions on Bridges. EN 1991-1-1, -1-3 to -1-7 and EN 1991-2. J.-A. Calgaro, M. Tschumi and H. Gulvanessian. 978 0 7277 3158 6. Forthcoming: 2009.

www.thomastelford.com/books
www.eurocodes.co.uk

DESIGNERS' GUIDES TO THE EUROCODES

DESIGNERS' GUIDE TO EUROCODE 1: ACTIONS ON BUILDINGS

EN 1991-1-1 AND -1-3 TO -1-7

H. Gulvanessian, P. Formichi and J.-A. Calgaro

with contributions to Part 7 from Geoff Harding

Series editor
H. Gulvanessian

Published by Thomas Telford Limited, 40 Marsh Wall, London E14 9TP, UK.
http://www.thomastelford.com

Distributors for Thomas Telford books are
USA: ASCE Press, 1801 Alexander Bell Drive, Reston, VA 20191-4400
Australia: DA Books and Journals, 648 Whitehorse Road, Mitcham 3132, Victoria

First published 2009

A catalogue record for this book is available from the British Library

ISBN: 978-0-7277-3156-2

Typeset by Academic + Technical, Bristol
Printed and bound in Great Britain by CPI Antony Rowe, Chippenham
Index created by Indexing Specialists (UK) Ltd, Hove

Preface

EN 1991: Eurocode 1: Actions on Structures includes ten parts which provide comprehensive information and guidance on all actions that it is normally necessary to consider in the design of building and civil engineering structures. All parts have now been published by the European Committee for Standardisation (CEN) as European Standards (EN).

Aims and objectives of this guide

The principal aim of this book is to provide the user with guidance on the interpretation and use on actions on buildings in the following parts of *EN 1991: Actions on Structures.*

EN 1991-1-1: *Eurocode 1: Actions on Structures: Part 1-1: Densities, self-weight and imposed loads*
EN 1991-1-3: *Eurocode 1: Actions on Structures: Part 1-3: Snow loads*
EN 1991-1-4: *Eurocode 1: Actions on Structures: Part 1-4: Wind actions*
EN 1991-1-5: *Eurocode 1: Actions on Structures: Part 1-5: Thermal actions*
EN 1991-1-6: *Eurocode 1: Actions on Structures: Part 1-6: Actions during execution*
EN 1991-1-7: *Eurocode 1: Actions on Structures: Part 1-7: Accidental actions*

Guidance on the interpretation and use of EN 1991-1-2: *Eurocode 1: Actions on Structures; Part 1-2: Actions on structures exposed to fire* is not provided in this book but the user will find this information in the Thomas Telford *Designers' Guide to the Eurocode Fire Parts (EN 1991-1-2; EN 1992-1-2; EN 1991-1-3 and EN 1991-1-4)* by Colin Bailey, Tom Lennon, David Moore and Yong Wang.

The guidance given on the part in this Designers' Guide on EN 1991-1-4: *Eurocode 1: Actions on Structures: Part 1-4: Wind Actions* is for the design of the everyday general building. For information on matters relating to dynamic sensitive building, vortex shedding etc. reference should be made to the Thomas Telford Designers' Guide on *EN 1991-1-4: Wind Actions* by N. J. Cook.

For guidance on the interpretation use on actions on bridges in the appropriate parts of *EN 1991: Actions on Structures*, the user should use the sister book to this volume, namely the Thomas Telford *Designers' Guide to EN 1991: Actions on Bridges.*

In producing this guide the authors have endeavoured to provide explanations and commentary to the clauses in EN 1991 for all the categories of users identified in the foreword of each Eurocode part. Although the Eurocodes are primarily intended for the design of buildings and civil engineering works, EN 1991 is intended for the consideration of more categories of users who include:

- designers and contractors

- clients
- product manufacturers
- public authorities and other bodies who produce regulations.

Layout of this guide

EN 1991: Eurocode 1: Actions on Structures has ten parts which are described in the Introduction to this Designers' Guide. This guide gives guidance on the parts mentioned above. The Guide is divided into seven parts with for example Part 1 covering EN 1991-1-1; Part 3 covering EN 1991-1-3; Part 5 covering EN 1991-1-5 etc. Each part of the guide has a number of chapters, each numbered as for the sections of each Eurocode part with further chapters for each of the annexes. For example, for EN 1991-1-1 Chapters 1 to 6 correspond to Sections 1 to 6 of EN 1991-1-1, and Chapters 7 to 8 correspond to Annexes A and B of EN 1991-1-1 respectively.

All cross-references in this guide to sections, clauses, sub-clauses, annexes, figures and tables of EN 1991 or its National Annexes are in italic type. Where text from a clause of EN 1991 has been directly reproduced, this is also shown in italics.

Examples and background information are set in shaded boxes.

Some tables given in this guide show the style of the tables of EN 1991 but do not give the complete information. Complete data can be obtained from EN 1991.

Acknowledgements

This book would not have been possible without the successful completion of EN 1991 and the authors would like to thank all those who contributed to its preparation. Those involved included the members of the Project Teams and the National Delegations. The following individuals are especially thanked: Professor Luca Sanpaolesi, Professor Gerhard Sedlacek, Dr Paul Luchinger, Mr Lars Albretkson and Mr Malcolm Greenley.

Mr Geoff Harding deserves a special thank you for his advice on Part 7 of this Designer's guide.

While this book was being prepared, Professor Gulvanessian was involved in producing the Institution of Structural Engineer's manual on EN 1990 and EN 1991. Some of the content of the current guide springs from the IstructE committee producing the manual and the authors acknowledge in particular John Tubman, David Dibbs-Fuller and John Littler.

Professor Gulvanessian would also especially like to thank Rohan Rupasinghe of the Building Research Establishment (BRE) who was his constant sounding board.

This book is dedicated to the following:

- The authors' employers and supporters, BRE Garston and the Department of Communities and Local Government, London; the University of Pisa and the General Council for Environment and Sustainable Ministry of Ecology, Energy, Sustainable Development and Town and Country Planning, Paris.
- The authors' wives, Vera Gulvanessian, Enrica Formichi and Elisabeth Calgaro, for their support and patience over the years.

Contents

Preface		**v**
	Aims and objectives of this guide	v
	Layout of this guide	vi
	Acknowledgements	vi
Introduction		**1**
	Background to the Eurocode programme	1
	Status and field of application of the Eurocodes	3
	National standards implementing the EN Eurocode	3
	Additional information on EN 1991	4
	National Annexes on EN 1991	7
	References	7
PART 1: EN 1991-1-1		**9**
Chapter 1.	**General**	**11**
	1.1. Scope	11
	1.2. Normative references	12
	1.3. Distinction between principles and application rules	12
	1.4. Terms and definitions	13
	1.5. Symbols	14
Chapter 2.	**Classification of actions**	**15**
	2.1. Self-weight	15
	Example 2.1	16
	2.2. Imposed loads	17
Chapter 3.	**Design situations**	**19**
	3.1. General	19
	3.2. Permanent loads	20
	Example 3.1	20
	3.3. Imposed loads	22
	Example 3.2. A cantilever beam against overturning (reference Table A1.2(A) of EN 1990)	23
	Example 3.3. A three-span continuous floor slab (reference Table A1.2(B) of EN 1990)	25

Chapter 4. Densities of construction and stored materials **29**
4.1. General 29

Chapter 5. Self-weight of construction works **31**
5.1. Representation of actions 31
5.2. Characteristic values of self-weight 32

Chapter 6. Imposed loads on buildings **35**
6.1. Representation of actions 35
6.2. Load arrangements 37
6.3. Characteristic values of imposed loads 38

Chapter 7. Annex A (informative)
Tables for nominal density of construction materials, and nominal density and angles of repose for stored materials **55**

Chapter 8. Annex B of EN 1991-1-1: Vehicle barriers and parapets for car parks **57**
Example 58

References **59**

PART 2: EN 1991-1-2 **61**

Chapter 1. Eurocode 1 – Actions on structures: Part 1.2: General Actions – Actions on structures exposed to fire **63**

References **63**

PART 3: EN 1991-1-3 **65**

Chapter 1. General **67**
1.1. Scope 67
1.2. Normative references 68
1.3. Assumptions 68
1.4. Distinction between Principles and Application Rules 69
1.5. Design assisted by testing 69
1.6. Terms and definitions 69
1.7. Symbols 70

Chapter 2. Classification of actions **71**

Chapter 3. Design situations **73**
3.1. General 73
3.2. Normal conditions 74
3.3. Exceptional conditions 74

Chapter 4. Snow load on the ground **77**
4.1. Characteristic values 77
4.2. Other representative values 80
4.3. Treatment of exceptional snow loads on the ground 80
Example 4.1 81

Chapter 5. Snow load on roofs **83**
5.1. Nature of the load 83
5.2. Load arrangements 83
5.3. Roof shape coefficients 89

Example 5.1 92
Example 5.2 93
Example 5.3 93
Example 5.4 97
Example 5.5 98

Chapter 6. Local effects 101
6.1. General 101
6.2. Drifting at projections and obstructions 101
Example 6.1 101
6.3. Snow overhanging the edge of a roof 102
6.4. Snow loads on snowguards and other obstacles 104
Example 6.2 104
Example 6.3 105

Chapter 7. Annex A. Design situations and load arrangements to be used for different locations 107

Chapter 8. Annex B. Snow load shape coefficients for exceptional snow drifts 109

Chapter 9. Annex C. European ground snow load maps 111
Example C.1 114

Chapter 10. Annex D. Adjustment of the ground snow load according to return period 117

Chapter 11. Annex E. Bulk weight density of snow 121

References 123

PART 4: EN 1991-1-4 125

Chapter 1. General 127
1.1. Scope 127
1.2. Definitions and symbols 127

Chapter 2. Design situations 131

Chapter 3. Modelling of wind actions 133

Chapter 4. Wind velocity and velocity pressures 135
4.1. Basis for calculation 135
4.2. Basic values 135
4.3. Mean wind 137
4.4. Wind turbulence 138
UK National Annex 138
4.5. Peak velocity pressure 139
UK National Annex 140
4.6. Explanation of h_{dis} displacement height 144

Chapter 5. Wind actions 145
5.1. General 145
5.2. Wind pressures on surfaces 145
5.3. Wind forces 145

Chapter 6. Structural factor $c_s c_d$ 149

	Example 6.1. Comparing $c_s c_d$ values between EN 1991-1-4 and the BSI NA	150
	Wake buffeting	151
Chapter 7.	**Pressure and force coefficients**	**153**
	General	153
	Choice of aerodynamic coefficient	153
Chapter 8.	**Annexes to EN 1991-1-4**	**155**
References		**156**
PART 5: EN 1991-1-5		**157**
Chapter 1.	**General**	**159**
	1.1. Scope	159
	1.2. Normative references	159
	Introductory advice for using this EN 1991-1-5 for the design of buildings	160
	1.3. Assumptions	161
	1.4. Distinction between Principles and Application Rules	161
	1.5. Terms and definitions	161
	1.6. Symbols	161
Chapter 2.	**Classification of actions**	**163**
	Example 2.1	164
Chapter 3.	**Design situations**	**165**
Chapter 4.	**Representation of actions**	**167**
Chapter 5.	**Temperature changes in buildings**	**169**
	5.1. General	169
	5.2. Determination of temperatures	169
	5.3. Determination of temperature profiles	170
Chapter 6.	**Annex A to Thermal Actions Part of the Manual**	**175**
Chapter 7.	**Annex B to Thermal Actions Part of the Manual**	**177**
	Temperature Profiles in Buildings and other Construction Works covered in Annex D of EN 1991-1-5	177
References		**178**
PART 6: EN 1991-1-6		**179**
Chapter 1.	**General**	**181**
	1.1. Scope	181
	1.2. Normative references	182
	1.3. Assumptions	182
	1.4. Distinction between Principles and Application Rules	183
	1.5. Terms and definitions	183
	1.6. Symbols	183
	Introductory advice for using this EN 1991-1-6 for the design of buildings	183

Chapter 2. Classification of actions **185**
2.1. General 185
2.2. Construction loads 185

Chapter 3. Design situations and limit states **189**
3.1. General – Identification of design situations 189
Choice of characteristic values of variable actions for transient design situations 190
Example 3.1 193
3.2. Ultimate limit states 194
3.3. Serviceability limit states 194

Chapter 4. Representation of actions **197**
4.1. General 197
Determination of γ and ψ values for construction loads Q_c 198
4.2. Actions on structural and non-structural members during handling 199
4.3. Geotechnical actions 200
Casting in in-situ concrete 200
4.4. Actions due to prestressing 201
4.5. Predeformations 201
4.6. Temperature, shrinkage, hydration effects 202
4.7. Wind actions 202
Treatment of wind actions during execution 202
4.8. Snow loads 203
4.9. Actions caused by water 203
4.10. Actions due to atmospheric icing 204
4.11. Construction loads 204
Defining construction loads 204
Background on construction loads Q_c for buildings 204
4.12. Accidental actions 210
4.13. Seismic actions 211

Annex A1 (normative) Supplementary rules for buildings **213**
A1.1. Ultimate limit states 213
A1.2. Serviceability limit states 213
A1.3. Horizontal actions 214

Annex A2 (normative) Supplementary rules for bridges **215**

Annex B (informative) Actions on structures during alteration, reconstruction or demolition **217**

References **217**

PART 7: EN 1991-1-7 **219**

Chapter 1. General **221**
1.1. Scope 221
EN 1990 requirements affecting EN 1991-1-7 222
1.2. Normative references 223
1.3. Assumptions 223
1.4. Distinction between Principles and Application Rules 223
1.5. Terms and definitions 223
1.6. Symbols 225

Chapter 2. Classification of actions **227**

Chapter 3. Design situations **229**
3.1. General 229
Design for accidental actions and the acceptance of localized damage 230
3.2. Accidental design situations – strategies for identified accidental actions 232
3.3. Accidental design situations – strategies for limiting the extent of localized failure 234
Example 3.1 235
3.4. Accidental design situations – use of consequence classes 236

Chapter 4. Impact **239**
4.1. Field of application 239
4.2. Representation of actions 240
4.3. Accidental actions caused by road vehicles 241
4.4. Accidental actions caused by forklift trucks 245
4.5. Accidental actions caused by derailed rail traffic under or adjacent to structures 245
4.6. Accidental actions caused by ship traffic 248
4.7. Accidental actions caused by helicopters 249

Chapter 5. Internal explosions **251**
5.1. Field of application 251
Gas explosions 251
5.2. Representation of actions 252
5.3. Principles of design 253

Chapter 6. Annex A (informative): Design of consequences of localized failures in buildings from an unspecified cause **255**
Introduction 255
A.1. Scope 256
A.2. Introduction 256
Background to the UK requirements for achieving robustness in buildings 256
A.3. Consequence classes of buildings 257
A.4. Recommended strategies 259
A.5. Horizontal ties 259
Example 6.1. Horizontal ties for framed structures 261
Example 6.2. Horizontal ties for load-bearing structures 262
A.6. Vertical ties 263
Example 6.3. Vertical ties for framed structures 263
A.7. Nominal section of load-bearing wall 264
A.8. Key elements 264
Example 6.4. Vertical ties for load-bearing construction 264

Chapter 7. Annex B (informative): Information on Risk Analysis **265**
B.1. Introduction 265
Application of risk assessment to Class 3 buildings 265
B.2. Definitions 266
B.3. Description of the scope of risk analysis 266
B.4. Methods of risk analysis 267

	B.5. Risk acceptance and mitigating measures	269
	B.6. Risk mitigating measures	270
	B.7. Reconsideration	271
	B.8. Communication of results and conclusions	271
	B.9. Application to buildings and civil engineering structures	271
	Introduction to B.9	271
Chapter 8.	**Annex C (informative): Dynamic design for impact**	**277**
	C.1. General	277
	Introduction to this chapter	277
	C.2. Impact dynamics	278
	C.3. Impact from aberrant road vehicles	279
	C.4. Impact by ships	279
Chapter 9.	**Annex D (informative): Internal explosions**	**281**
	D.1. Dust explosions in rooms, vessels and bunkers	281
	Introduction to Annex D	281
	D.2. Natural gas explosions	283
	D.3. Explosions in road and rail tunnels	283
References		**284**
Index		**285**

Introduction

The material in this Introduction relates to the forwards of the parts of the European Standard EN 1991: *Actions on Structures*. The forewords include the following aspects:

- The background to the Eurocode programme
- Status and field of application of the Eurocodes
- National Standards implementing Eurocodes
- Additional information on EN 1991
- National Annexes on EN 1991.

The forewords of all of the Eurocode parts have common text. Guidance on the common text is provided in the Introduction of the *Designers' Guide to EN 1990: Eurocode: Basis of Structural Design*[1] to which reference should be made. Only information essential to the user of EN 1991 is given here.

The European Commission (EC), in close cooperation with representatives of Member States (the Eurocode National Correspondents (ENC)) has prepared a document *Application and Use of the Eurocodes*.[2] The reader is recommended to study this document, guidance to which is given in reference 1.

Background to the Eurocode programme

The objectives of the Eurocodes and their status

In 1975, the Commission of the European Community decided on an action programme in the field of construction based on article 95 of the Treaty of Rome. The objective of the programme was the elimination of technical obstacles to trade and the harmonisation of technical specifications.

Within this action programme the Commission took the initiative to establish a set of harmonised technical rules for the structural design of construction works, with the following European Commission objective: '*The Eurocodes to establish a set of common technical rules for the design of buildings and civil engineering works which will ultimately replace the differing rules in the various Member States.*'

For almost 15 years, the Commission, with the help of a Steering Committee containing representatives of Member States, conducted the development of the Eurocode programme, which led to the publication of a first-generation set of European codes in the 1980s.

In 1989 the Special Agreement between CEN and the European Commission (BC/CEN/03/89) transferred the preparation and the publication of the Eurocodes to CEN, thus providing the Eurocodes with a future status of European EN Standards.

This links, *de facto*, the Eurocodes with the provisions of all the Council's directives and/or the Commission's decisions dealing with European standards; for example:

- The Construction Products Directive (N.B. it is proposed to convert this to a Regulation, the Construction Products Regulations (CPR) in 2010)
- Public Procurement Directives, on public works and services for execution, design etc. of buildings and civil engineering works.

Eurocode programme

The structural Eurocodes, as shown in Table 0.1, each generally consisting of a number of parts, some of which relate to bridges, have all now been released in EN form.

Table 0.1. The structural Eurocodes

EN Number	The structural Eurocodes
EN 1990	*Eurocode: Basis of Structural Design*
EN 1991	*Eurocode 1: Actions on Structures*
EN 1992	*Eurocode 2: Design of Concrete Structures*
EN 1993	*Eurocode 3: Design of Steel Structures*
EN 1994	*Eurocode 4: Design of Composite Steel and Concrete Structures*
EN 1995	*Eurocode 5: Design of Timber Structures*
EN 1996	*Eurocode 6: Design of Masonry Structures*
EN 1997	*Eurocode 7: Geotechnical Design*
EN 1998	*Eurocode 8: Design of Structures for Earthquake Resistance*
EN 1999	*Eurocode 9: Design of Aluminium Structures*

Each of the ten Eurocodes listed above are made up of separate parts, which cover the technical aspects of the structural and fire design of buildings and civil engineering structures. The Eurocodes are a harmonised set of documents that have to be used together. Figure 0.1 shows the links among the various Eurocodes.

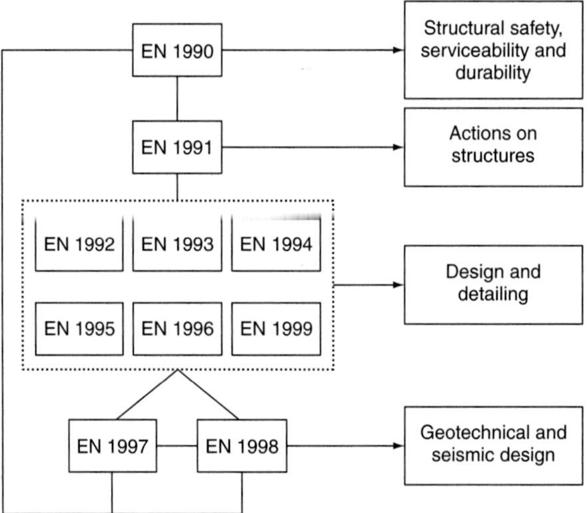

Fig 0.1. Links between the Eurocodes

Potential benefits of the use of Eurocodes

The intended benefits of the Eurocodes are as follows:

- Provide a common understanding regarding the design of structure between owners, operators and users, designers, contractors and manufacturers of construction products.
- Provide common design criteria and methods to fulfil the specified requirements for mechanical resistance, stability and resistance to fire, including aspects of durability and economy.
- Facilitate the marketing and use of structural components and kits in Member States.

- Facilitate the marketing and use of materials and constituent products, the properties of which enter into design calculations, in Member States.
- Be a common basis for research and development.
- Allow the preparation of common design aids and software.
- Benefit the European civil engineering enterprises, contractors, designers and product manufacturers in their worldwide activities, and increasing their competitiveness.

Status and field of application of the Eurocodes

The status of the Eurocodes and their relationship with the Construction Product Directive's Interpretive Documents are comprehensively described in the *Designers' Guide to EN 1990*.[1]

National Standards implementing the EN Eurocode

It is the responsibility of each National Standards Body (e.g. British Standards Institution (BSI) in the UK) to implement Eurocodes as National Standards.

The National Standard implementing each Eurocode part, will comprise, without any alterations, the full text of the Eurocode part and its annexes as published by the CEN. This may be preceded by a national title page and national foreword, and may be followed by a National Annex (see Fig. 0.2).

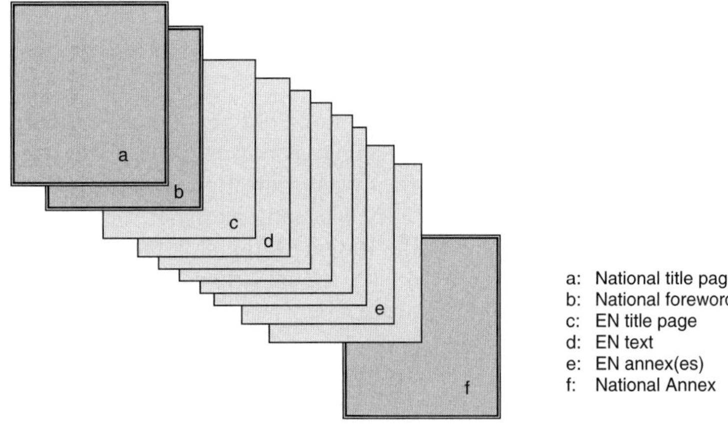

Fig 0.2. National implementation of Eurocodes

Rules and contents of National Annexes for Eurocodes

The European Commission recognises the responsibility of regulatory authorities (e.g. the Building Regulations Division of the Office of the Deputy Prime Minister in the UK) or national competent authorities (e.g. the Highways Agency or Railway Safety in the UK) in each EU Member State. It has safeguarded their right to determine values related to safety matters at national level through a National Annex.

The National Annex may only contain information on those parameters, which are left open in the Eurocode for national choice, known as Nationally Determined Parameters, to be used for the design of buildings and civil engineering works to be constructed in the country concerned.

Nationally Determined Parameters (NDPs)

Possible differences in geographical or climatic conditions (e.g. wind or snow maps) or in ways of life, as well as different levels of protection that may prevail at national, regional or local level, can be taken into account by choices left open about values, classes or alternative methods that are identified in the EN Eurocodes to be determined nationally.

The values, classes or methods to be chosen or determined at national level, called Nationally Determined Parameters, will allow the EU Member States to choose the level of safety, including aspects of durability and economy, applicable to works in their territory. They include:

- values and/or classes where alternatives are given in the Eurocodes
- values to be used where only a symbol is given in the Eurocodes
- country-specific data (geographical, climatic, etc.), e.g. snow maps
- procedures to be used where alternative procedures are given in the Eurocodes.

National Annexes

The National Standards bodies should publish the parameters in a National Annex on behalf of and with the agreement of the national competent authorities.

A National Annex is not required if the EN Eurocode Part is not relevant for the Member State (e.g. seismic design for some countries).

The Annex may also contain

- decisions on the application of informative annexes
- references to non-contradictory complementary information to assist the user in applying the Eurocode.

A National Annex cannot change or modify the content of the EN Eurocode text in any way other than where it indicates that national choices may be made by means of Nationally Determined Parameters.

Annexes not transferable

Each EU Member State will have a different National Annex; the Annex used must be the one applicable to where the building or civil engineering work is being constructed. For example, a UK designer will have to use the appropriate Eurocode with the UK National Annex when designing a building in the UK. The same designer, designing a building in Italy, will have to use the Eurocode with the Italian National Annex.

When using Eurocodes in countries where a National Annex is not available (e.g. in some overseas countries) in accordance with Guidance paper L[2] the recommended values may be used with the designer ensuring correct use of climatic etc. information.

Additional information on EN 1991

Background and current programme

EN 1991: *Eurocode 1: Actions on Structures* provides comprehensive information and guidance on all actions that are normally necessary to consider in the design of building and civil engineering structures. All parts have now been published by the European Committee for Standardisation (CEN) as European Standards (EN).

EN 1991 comprises ten EN Parts (see Table 0.2). The parts are referred to in this Introduction by their proposed EN numbers. These parts provide the actions for use with EN 1990: *Eurocode: Basis of Structural Design*, and EN 1992 to EN 1999 as appropriate, for design and verification on the basis of overall principles which are given in EN 1990.

Definitions of actions in EN 1990: *Eurocode Basis of Structural Design*

EN 1990 classifies actions according to their:

- variation in time: permanent, variable, accidental
- nature and/or structural response: static or dynamic
- origin: direct or indirect
- spatial variation: fixed or free.

Table 0.2. The parts of Eurocode 1: Actions on structures

EN number	Title
1991-1-1	*Densities, Self-weight and Imposed loads*
1991-1-2	*Actions on Structures Exposed to Fire*
1991-1-3	*Snow Loads*
1991-1-4	*Wind Actions*
1991-1-5	*Thermal Actions*
1991-1-6	*Actions During Execution*
1991-1-7	*Accidental Actions*
1991-2	*Traffic Loads on Bridges*
1991-3	*Actions Induced by Cranes and Machinery*
1991-4	*Actions in Silos and Tanks*

The classification of actions by their variation in time is shown in tabulated form in Table 0.3.

A comprehensive table showing classification of actions according to the above criteria is included in Part 6 of this Designers' Guide.

The term *single action* is also used to define an action which is statistically independent in time and space from any other action acting on the structure.

The self-weight of a structure can be represented by a single characteristic value (G_k), provided the variability of G is small, and it can be calculated on the basis of the nominal dimensions and the mean unit mass. If the variability of G is not small and the statistical distribution is known, two values are used: an upper value ($G_{k,sup}$) and a lower value ($G_{k,inf}$). More information on this subject has been given in Part 1 of this Designers' Guide.

A variable action has the following representative values (see Fig. 0.3):

- the characteristic value (Q_k)
- the combination value ($\psi_0 Q_k$)
- the frequent value ($\psi_1 Q_k$)
- the quasi-permanent value ($\psi_2 Q_k$).

The combination value ($\psi_0 Q_k$) takes account of the reduced probability of simultaneous occurrence of the most unfavourable values of several independent variable actions. It is used for the verification of ultimate limit states and irreversible serviceability limit states. The frequent value ($\psi_1 Q_k$) is used for verification of ultimate limit states involving accidental actions and reversible limit states. The quasi-permanent value ($\psi_2 Q_k$) is also used for ultimate limit state verification involving accidental actions and for reversible serviceability limit states. The recommended values of ψ_0, ψ_1, ψ_2 for buildings are shown, reproduced from EN 1990,[3] in Table 0.4.

Information on combining actions for particular design situations is given in Part 1 of this paper for both ultimate and serviceability limit states.

Table 0.3. Classification of actions

Permanent action	Variable action	Accidental action
(a) Self-weight of structures, fittings and fixed equipment	(a) Imposed floor loads	(a) Explosions
(b) Prestressing force	(b) Snow loads	(b) Fire
(c) Water and soil pressures	(c) Wind loads	(c) Impact from vehicles
(d) Indirect action, e.g. settlement of supports	(d) Indirect action, e.g. temperature effects	
	(e) Actions due to traffic	

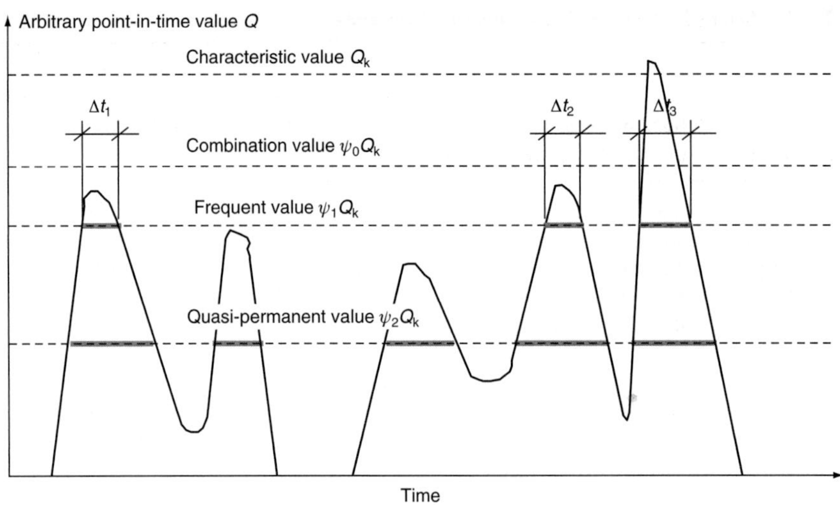

Fig. 0.3. Representative values

Values of actions depending on design working lives

The characteristic values of actions which are either given directly or obtained from basic information supplied in EN 1991, are normally for a 50-year return period.

For the design of buildings and civil engineering works for a shorter or a longer design working life than 50 years, the value of the action determined from the appropriate Eurocode Part should be multiplied by a coefficient k (see Table 0.5 for climatic actions).

The expressions for determining k are given in Part 6 of this Designers' Guide. For determining k for the climatic actions the recommended values of the appropriate parameters have been used. For snow, a coefficient of variation $v = 0.6$ has been used.

Table 0.5 covers both the design working life of the construction works together with execution phases (ref: EN 1991-1-6), and the value for $Q_{k,R}$ where R is the design working life can be obtained from the following expression:

$$Q_{k,R} = kQ_{k,50}$$

Table 0.4. ψ factors for buildings (Table A1.1 of EN 1990)

Action	ψ_0	ψ_1	ψ_2
Imposed loads on buildings, category (see Part 1)			
Category A: domestic, residential areas	0.7	0.5	0.3
Category B: office areas	0.7	0.5	0.3
Category C: congregation areas	0.7	0.7	0.6
Category D: shopping areas	0.7	0.7	0.6
Category E: storage areas	1.0	0.9	0.8
Category F: traffic area, vehicle weight $\leq$ 30 kN	0.7	0.7	0.6
Category G: traffic area, 30 kN $<$ vehicle weight $\leq$ 160 kN	0.7	0.5	0.3
Category H: roofs	0	0	0
Snow loads on buildings (see Part 3)			
Finland, Iceland, Norway, Sweden	0.7	0.5	0.2
Remainder of CEN Member States, for sites located at altitude $H > 1000$ m a.s.l.	0.7	0.5	0.2
Remainder of CEN Member States, for sites located at altitude $H \leq 1000$ m a.s.l.	0.5	0.2	0
Wind loads on buildings (see Part 4)	0.6	0.2	0
Temperature (non-fire) in buildings (see Part 5)	0.6	0.5	0

Table 0.5. Coefficient k of actions $Q_{k,R}$ for different return periods R

			Coefficient k			
Design working life	Return period, R	Probability of exeedence, P	for thermal, $T_{max,R}$	for thermal, $T_{min,R}$	for snow, $s_{n,R}$	for wind, $v_{b,R}$
≤ 3 days	2 years	0.5	0.8	0.45	0.64	0.77
3 days $< t \leq 3$ months	5 years	0.2	0.86	0.63	0.75	0.85
3 months $< t \leq 1$ year	10 years	0.1	0.91	0.74	0.83	0.90
1 year $< t \leq 50$ years	50 years	0.02	1	1	1	1
80 years	80 years	0.0125	1.03	1.08	1.09	1.03
120 years	120 years	0.0083	1.05	1.14	1.16	1.05
500 years	500 years	0.002	1.13	1.36	1.42	1.12
1000 years	1000 years	0.001	1.17	1.47	1.55	1.16

Note: The expressions for determining k are given in Part 6 of this Designers' Guide. For determining k for the climatic actions, the recommended values of the appropriate parameters have been used. For snow, a coefficient of variation $v = 0.6$ has been used.

National Annexes on EN 1991

The ten parts of EN 1991 allow national choice through a number of clauses. A National Standard implementing a part of EN 1991 should have a National Annex containing all NDPs to be used for the design of construction works in the relevant country.

The clauses in the parts of EN 1991 where national choice is allowed are listed in the foreword of each EN 1991 part.

References

1. Gulvanessian, H., Calgaro, J.-A. and Holický, M. *Designers' Guide to EN 1990: Eurocode: Basis of Structural Design*. Thomas Telford, London, 2002.
2. Guidance Paper L (concerning the Construction Products Directive – 89/106/EEC) – *Application and Use of the Eurocodes*: European Commission, Enterprise Directorate-General, 2004.
3. European Committee for Standardisation. EN 1990: *Eurocode: Basis of Structural Design*. CEN, Brussels, 2002.

PART 1: EN 1991-1-1: Eurocode 1: Part 1.1 – Densities, self-weight and imposed loads

CHAPTER 1

General

This chapter is concerned with the general aspects of EN 1991-1-1: *Eurocode 1 – Actions on structures: Part 1.1: General Actions – Densities, self-weight, imposed loads for buildings.*[1] The material described in this chapter is covered in the following clauses:

- Scope — *Clause 1.1: 1991-1-1*
- Normative references — *Clause 1.2: 1991-1-1*
- Distinction between Principles and Application Rules — *Clause 1.3: 1991-1-1*
- Terms and definitions — *Clause 1.4: 1991-1-1*
- Symbols — *Clause 1.5: 1991-1-1*

1.1. Scope

1.1.1. Primary scope

EN 1991-1-1: *Eurocode 1 – Actions on structures: Part 1.1: General Actions – Densities, self-weight, imposed loads for buildings* is one of the ten Parts of EN 1991 (see Table 0.2).

It gives design guidance and specifies actions for the structural design of buildings and civil engineering works for the following topics: *cl.1.1(1): 1991-1-1*

- Densities of construction materials and stored materials, for the design of buildings and civil engineering works.
- Self-weight of construction works for the design of buildings and civil engineering works.
- Imposed loads for buildings.

1.1.2. Scope in relation to densities of construction materials and stored materials

With regard to densities of construction materials and stored materials, Chapter 4 of EN 1991-1-1 gives nominal values for densities of specific building, bridge and stored materials. Where relevant, for appropriate materials EN 1991-1-1 also provides the angle of repose. *cl.1.1(2): 1991-1-1*

1.1.3. Scope in relation to self-weight of construction works

Section 5 of EN 1991-1-1 gives methods for the determination of the characteristic values of self-weight of construction works. *cl.1.1(3): 1991-1-1*

1.1.4. Scope in relation to imposed loads for buildings

Section 6 of EN 1991-1-1 gives characteristic values of imposed loads for floors and roofs according to category of use in the following areas in buildings: *cl.1.1(4): 1991-1-1*

- residential, social, commercial and administration areas
- garage and vehicle traffic areas
- areas for storage and industrial activities
- roofs
- helicopter landing areas.

EN 1991-1-1 also gives the loads:

cl.1.1(5): 1991-1-1
- on traffic areas for vehicles up to a gross vehicle weight of 160 kN, and
- for barriers or walls, which have the function of barriers, horizontal forces are given.

cl.1.1(6): 1991-1-1 An informative Annex B provides additional guidance for vehicle barriers in car parks.
cl.1.1(7): 1991-1-1 No comment is necessary for ***Clause 1.1(7): 1991-1-1***.

1.2. Normative references

cl.1.2: 1991-1-1 No comment is necessary for ***Clause 1.2: 1991-1-1***.

1.3. Distinction between principles and application rules

cl.1.3(1): 1991-1-1 The clauses in EN 1991-1-1 are set out as either Principles or Application Rules.

cl.1.3(2): 1991-1-1
- Principles are defined as '*general statements and definitions for which there is no alternative, as well as requirements and analytical models for which no alternative is permitted unless specifically stated*'.

cl.1.3(3): 1991-1-1
- '*The Principles are identified by the letter P following the paragraph number*'. The word **shall** is always used in the Principle clauses.

cl.1.3(4): 1991-1-1
- '*The Application Rules are generally recognised rules which comply with the Principles and satisfy their requirements*'.

cl.1.3(5): 1991-1-1
- '*It is permissible to use alternative design rules different from the Application Rules given in EN 1991-1-1 for works, provided that it is shown that the alternative rules accord with the relevant Principles and are at least equivalent with regard to the structural safety, serviceability and durability which would be expected when using the Eurocodes*'.

cl.1.3(5): 1991-1-1 EN 1991-1-1, through a note to ***Clause 1.3(5): 1991-1-1*** states:

If an alternative design rule is substituted for an Application Rule, the resulting design cannot be claimed to be wholly in accordance with EN 1991-1-1 although the design will remain in accordance with the Principles of EN 1991-1-1. When EN 1991-1-1 is used in respect of a property listed in an Annex Z of a product standard or an ETAG, the use of an alternative design rule may not be acceptable for CE marking.

cl.1.3(5): 1991-1-1 With regard to the note to ***Clause 1.3(5): 1991-1-1***, the European Commission guidance paper L, Application and Use of the Eurocodes[2] states:

National Provisions should avoid replacing any EN Eurocode provisions, e.g. Application Rules, by national rules (codes, standards, regulatory provisions, etc).
When, however, National Provisions do provide that the designer may – even after the end of the co-existence period – deviate from or not apply the EN Eurocodes or certain provisions thereof (e.g. Application Rules), then the design ***will not*** *be called 'a design according to EN Eurocodes'.*

cl.1.3(6): 1991-1-1
- '*Application Rules are identified by a number in brackets*' (only). The word **should** is normally used for application rules. The word **may** is also used for example as an alternative application rule. The words **is** and **can** are used for a definitive statement or as an assumption.

1.4. Terms and definitions

Most of the definitions given in EN 1991-1-1 derive from ISO 2394,[3] ISO 3898[4] and ISO 8930[5]. In addition reference should be made to EN 1990 which provides a basic list of terms and definitions which are applicable to EN 1990 to EN 1999, thus ensuring a common basis for the Eurocode suite. *cl.1.4: 1991-1-1*

For the structural Eurocode suite, attention is drawn to the following key definitions, which may be different from current national practices:

- '*action*' means a load, or an imposed deformation (e.g. temperature effects or settlement)
- '*effects of actions*' or '*action effects*' are internal moments and forces, bending moments, shear forces and deformations caused by actions.

From the many definitions provided in EN 1990, the following apply for use with EN 1991.

(1) From common terms used in EN 1990. The definitions for: *cl.1.5.1: 1990*

- construction works *cl.1.5.1.1: 1990*
- structure *cl.1.5.1.6: 1990*
- structural member *cl.1.5.1.7: 1990*
- execution. *cl.1.5.1.8: 1990*

(2) From special terms relating to design in general. The definitions for: *cl.1.5.2: 1990*

- design situations *cl.1.5.2.2: 1990*
- transient design situation *cl.1.5.2.3: 1990*
- persistent design situation *cl.1.5.2.4: 1990*
- accidental design situation *cl.1.5.2.5: 1990*
- design working life *cl.1.5.2.8: 1990*
- hazard *cl.1.5.2.9: 1990*
- load arrangement *cl.1.5.2.10: 1990*
- load case *cl.1.5.2.11: 1990*
- limit states *cl.1.5.2.12: 1990*
- ultimate limit states *cl.1.5.2.13: 1990*
- serviceability limit states *cl.1.5.2.14: 1990*
- resistance *cl.1.5.2.15: 1990*
- strength *cl.1.5.2.16: 1990*
- reliability *cl.1.5.2.17: 1990*
- basic variable *cl.1.5.2.19: 1990*
- nominal value. *cl.1.5.2.22: 1990*

(3) From terms relating to actions. The definitions for:

- action (F) *cl.1.5.3.1: 1990*
- effect of action (E) *cl.1.5.3.2: 1990*
- permanent action (G) *cl.1.5.3.3: 1990*
- variable action (Q) *cl.1.5.3.4: 1990*
- accidental action (A) *cl.1.5.3.5: 1990*
- seismic action (A_E) *cl.1.5.3.6: 1990*
- geotechnical action *cl.1.5.3.7: 1990*
- fixed action *cl.1.5.3.8: 1990*
- free action *cl.1.5.3.9: 1990*
- single action *cl.1.5.3.10: 1990*
- static action *cl.1.5.3.11: 1990*
- dynamic action *cl.1.5.3.12: 1990*
- quasi-static action *cl.1.5.3.13: 1990*
- characteristic value of an action *cl.1.5.3.14: 1990*
- reference period *cl.1.5.3.15: 1990*
- combination value of a variable action ($\psi_0 Q_k$) *cl.1.5.3.16: 1990*
- frequent value of a variable action ($\psi_1 Q_k$) *cl.1.5.3.17: 1990*

cl. 1.5.3.18: 1990 • quasi-permanent value of a variable action ($\psi_2 Q_k$)
cl. 1.5.3.20: 1990 • representative value of an action (F_{rep})
cl. 1.5.3.21: 1990 • design value of an action (F_d)
cl. 1.5.3.22: 1990 • combination of actions.

cl. 1.5.5: 1990 (4) From terms relating to geometrical data. The definitions for:

cl. 1.5.5.2: 1990 • design value of a geometrical property (a_d).

Gulvanessian *et al.*[6] provide an explanation of the definitions in EN 1990.

The following comments are made to help the understanding of particular definitions in EN 1991-1-1.

cl. 1.4.1: 1991-1-1 • 'bulk weight density'. The bulk weight density is defined as '*the overall weight per unit volume of a material, including a normal distribution of micro-voids, voids and pores*'. In everyday usage 'bulk weight density' is frequently abbreviated to 'density'. The definition of density is mass per unit volume. The term density in EN 1991-1-1 is used for weight per unit volume, area or length.

cl. 1.4.2: 1991-1-1 • 'angle of repose'. According to ISO 9194,[7] which is the base document for the parts of EN 1991-1-1 referring to actions due to self-weight of structures, for non-structural elements and stored materials it should be considered a deviation of $\pm 30\%$ for the angle of repose. Annex B of EN 1991-1-1 takes this into account for some materials.

cl. 1.4.3: 1991-1-1 • 'gross weight of vehicle'. No further comment necessary.

cl. 1.4.4: 1991-1-1 • 'structural elements'. It should be emphasised that *structural elements* are referred to as *structural members* in the EN 1990 definitions. The definition in EN 1991-1-1 differs from that given in ISO 8930,[5] which defines structural elements as '*Physically distinguishable part of a structure, for instance a column, a beam a slab, a shell element.*'

cl. 1.4.5: 1991-1-1 • 'non structural elements'. No further comment necessary.

cl. 1.4.6: 1991-1-1 • 'partitions'. No further comment necessary.

cl. 1.4.7: 1991-1-1 • 'movable partitions'. No further comment necessary.

1.5. Symbols

cl. 1.5(1): 1991-1-1 The notation in *Clause 1.5: EN 1991-1-1* is based on ISO 3898[4].

cl. 1.5(2): 1991-1-1 EN 1990 *Clause 1.6* provides a comprehensive list of symbols, some of which may be appropriate for use with EN 1991-1-1. The symbols given in *Clause 1.5(2): 1991-1-1* are additional notations specific to this part of EN 1991-1-1.

CHAPTER 2

Classification of actions

This chapter is concerned with the classification of the actions in EN 1991-1-1: *Eurocode 1 – Actions on structures: Part 1.1: General Actions – Densities, self-weight, imposed loads for buildings*. The material described in this chapter is covered in the following clauses.

- Self-weight *Clause 2.1: 1991-1-1*
- Imposed loads *Clause 2.2: 1991-1-1*

2.1. Self-weight

EN 1991-1-1 classifies self-weight as a permanent fixed action in most cases. *cl.2.1(1): 1991-1-1*

Permanent actions (G), which normally are likely to act throughout the design life of a structure, include self-weight of structures, fixed equipment etc.

A fixed action has a fixed distribution in space such that its magnitude and direction are determined unambiguously for the whole structure when determined at one point on it. Static water pressure represents an example of a fixed action (see Clause 1.5.3.8 of EN 1990).

Normally with self-weight the variability with time is negligible. Thus the variability of G may be neglected as G does not vary significantly during the design working life of the structure and its coefficient of variation is small. Therefore the characteristic value (G_k) can be taken equal to the mean value (μ) of the statistical distribution for G, which may be assumed to be normal (Gaussian). See Fig. 2.1.

However when self-weight can vary in time, it should be taken into account by the upper ($G_{k,sup}$) and lower ($G_{k,inf}$) characteristic values in accordance with *EN 1990: Clause 4.1.2*. According to EN 1990, $G_{k,inf}$ is the 5% fractile and $G_{k,sup}$ is the 95% fractile of the statistical distribution for G, which may be assumed to be normal (Gaussian) as shown in Fig. 2.1. *cl.2.1(2): 1991-1-1*

For example: the self-weight of items such as parapets, waterproofing, coatings, screeds, etc. is taken into account using an upper and lower characteristic value because the variability with time may not be small. *cl.2.1(2): 1991-1-1*

The following relationships (see *Designers' Guide to EN 1990*[6]) can be used to determine the lower value $G_{k,inf}$ and the upper value $G_{k,sup}$:

$$G_{k,inf} = \mu_G - 1.64\sigma_G = \mu_G(1 - 1.64V_G) \qquad \text{(D 2.1)}$$

$$G_{k,sup} = \mu_G + 1.64\sigma_G = \mu_G(1 + 1.64V_G) \qquad \text{(D 2.2)}$$

where V_G is the coefficient of variation of G.

Normally the variation of G could be considered significant when $V_G \geq 0.10$. In this case the variability of G should be taken into account for all structures and members.

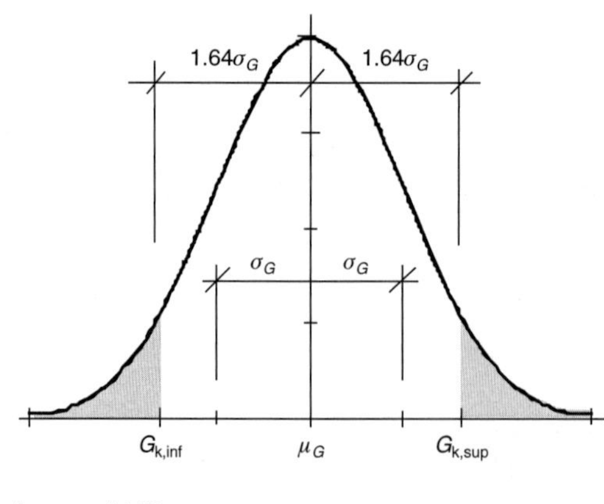

Fig. 2.1. Definition of mean value (μ_G) and lower ($G_{k,inf}$) and upper ($G_{k,sup}$) characteristic values of permanent actions based on normal (Gaussian) distribution

For special cases for particular design situations (e.g. when considering overturning and strength of retaining walls), both the lower value $G_{k,inf}$ and the upper value $G_{k,sup}$ need to be used in the design when the coefficient of variation is greater than 0.05.

Two values, namely an upper value $G_{k,sup}$ and a lower value $G_{k,inf}$, may have to be used when considering the stabilising and de-stabilising effects of self-weight on for example a multi-span floor where the quality control during the construction of the floor is below that expected. This can occur for example when casting in in situ concrete, where there may not be site supervision on behalf of the client. However, for quality-controlled constructions (e.g. for a steel frame or precast concrete member, where the members are manufactured under factory conditions), the variability of G may be neglected as G does not vary significantly during the design working life of the structure and its coefficient of variation is small. Therefore the characteristic value (G_k) can be taken equal to the mean value (μ) of the statistical distribution for G, which may be assumed to be normal (Gaussian). See Fig. 2.1.

Example 2.1

Assume the design specifies a 200 mm thick (i.e. for determining the mean value μ_G of G) reinforced concrete slab, to be cast in situ. If the designer considers a coefficient of variation $V = 0.10$, due to deviations from the design due to construction errors, then using expressions (D 2.1) and (D 2.2), $G_{k,inf}$ and $G_{k,sup}$ will therefore be 16.4% less than or greater than the mean value μ_G.

Thus for a specified 200 mm slab the self-weight should be determined for a thickness of 167.2 mm for $G_{k,inf}$ and 232.8 mm for $G_{k,sup}$.

When the self-weight is classified as a free action (e.g. for movable partitions, or floor coverings), it should be treated as an additional imposed load. This applies in particular
cl.2.1(2): 1991-1-1 when the 'permanent' free action is favourable. A free action may have various spatial distributions over the structure.

cl.2.1(3)P: 1991-1-1 No comment is necessary in the case of buildings for Clause 2.1(3)P: 1991-1-1. For comments regarding ballast on railway bridges, see Designers' Guide to EN 1991 for bridges.[8]

cl.2.1(4)P: 1991-1-1 EN 1991-1-1 stipulates that earth loads on roofs and terraces (e.g. for roof gardens) be considered as permanent actions.

cl.2.1(3)P: 1991-1-1 For both ballast and earth loads (e.g. for roof gardens) the self-weight will vary with time
cl.2.1(4)P: 1991-1-1 '*due to variations in moisture content and variation in depth, that may be caused by uncontrolled accumulation during the design life of the structure*' and these should be considered in the
cl.2.1(5)P: 1991-1-1 design.

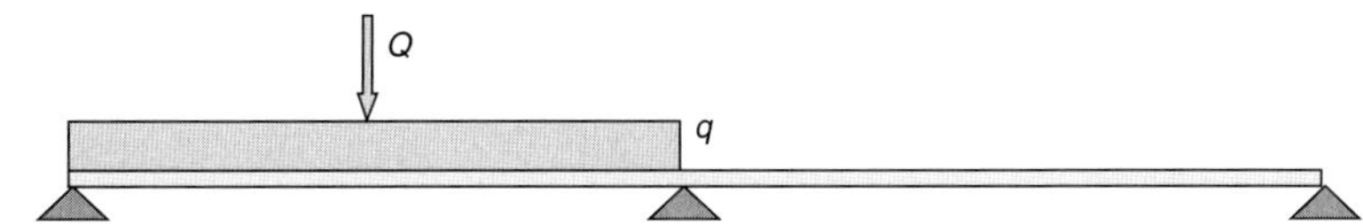

Fig. 2.2. The position, magnitude and direction of a free action

Information on earth pressures is given in EN 1997 and the Designers' Guide to EN 1997[9]. *cl.2.1(5)P: 1991-1-1*

2.2. Imposed loads

EN 1991-1-1 classifies imposed loads as free actions. See Section 1.1.4 of this Guide for the imposed loads covered by EN 1991-1-1. *cl.2.2(1)P: 1991-1-1*

Imposed loads for bridges are covered by EN 1991-2 and its appropriate Designers' Guide.[8]

A free action may have various spatial distributions over the structure. An action that is caused by people on a floor is an example of a free action. See Fig. 2.2.

No comment is necessary on ***Clause 2.2(2): 1991-1-1***. *cl.2.2(2): 1991-1-1*

The characteristic load models specified in EN 1991-1-1 include the effects of acceleration (dynamic effects) caused by the actions, which are taken into account as quasi-static actions (*see EN 1990 1.5.3.13 and 5.1.3(3)*), when there is no risk of resonance or other significant dynamic response of the structure. For structures susceptible to dynamic excitation from synchronised rhythmical movement of people or dancing or jumping, the load model should be determined from a special dynamic analysis. *cl.2.2(3): 1991-1-1*

For structures that are susceptible to dynamic excitation, the design should take account of the load–structure interaction and dynamic characteristics of the structure, e.g. natural frequency, mass, damping and mode shapes.

When oscillation or vibration is a serviceability criterion, separate consideration of their influence is necessary, e.g. on the operation of equipment in the building and/or on the comfort of the users and occupiers of the building.

Some particular cases of dynamic loading conditions and structural types are described in the three examples below.

Example of 'synchronised rhythmical movements'

Structures with members which may be subject to dancing and jumping are liable to inadvertent or deliberate synchronised movement of occupants. This may be to music with a strong beat, such as occurring at pop concerts and aerobics events. These activities generate dynamic effects that can result in enhanced vertical and horizontal loads. If a natural frequency of a structure matches the frequency of the synchronised movement, or an integer multiple of it, then resonance can occur that greatly amplifies the dynamic response.

Three alternative procedures are recommended for such design situations, as listed below:

(1) In addition to using Table 6.2 of EN 1991-1-1 to obtain the appropriate imposed loads the design should ensure that resonance of the structure is avoided by limiting its natural frequencies, so that the vertical frequency is greater than 8.4 Hz and the horizontal frequency is greater than 4.0 Hz. These frequencies should be evaluated for the appropriate mode of vibration of an empty structure.
(2) Structural elements subject to dancing and jumping should be designed to resist the anticipated dynamic loading. The deformation should not exceed limits appropriate to the structure type. Detailed design should be carried out accounting for dynamic response of the structure and a range of load frequencies and types, and with the help of specialist advice and specialist guidance documents, e.g. BRE Digest 426.[10] Alternatively, item (3) should be followed.

(3) Use specific guidance as required by the certifying authority for the type of structure under consideration. In certain cases, e.g. structures intended primarily for providing spectator facilities, the relevant certifying authority may refer to specific guidance documents as appropriate and sufficient for compliance with their requirements.

Example of 'dynamic loads from machinery'

Dynamic effects caused by the operation of machinery depend on the type of machinery and the structural form, and dynamic loads and potential resonant excitation of such structures should be investigated. For actions induced by cranes and machinery see EN 1991-3.[11]

Example of 'lightweight structures and long-span structures'

Where these structures are used as concourses and public spaces, they are likely to be subject to inadvertent or deliberate synchronised movement by people, causing dynamic excitation. The design provisions should take account of the nature and intended use of the structure, the potential number of people and their possible behaviour.

cl.2.2(4): 1991-1-1 No comment is necessary for ***Clause 2.2(4): 1991-1-1*** for the consideration of forklifts and helicopters in this chapter. However *Clause 5.1.3(3) of EN 1990* may be consulted.

An action which causes significant acceleration to a structure or its members is classified a *cl.2.2(5)P: 1991-1-1* dynamic action according to ***Clause 2.2(5)P: 1991-1-1***. To determine the effects of a dynamic action on the design a dynamic analysis should be used. See also Clause 5.3 of EN 1990.

CHAPTER 3

Design situations

This chapter is concerned with the general concepts of design situations relating to EN 1991-1-1: *Eurocode 1 – Actions on structures: Part 1.1: General Actions – Densities, self-weight, imposed loads for buildings*. The material described in this chapter is covered in the following clauses:

- General *Clause 3.1: 1991-1-1*
- Permanent loads *Clause 3.2: 1991-1-1*
- Imposed loads *Clause 3.3: 1991-1-1*
- Additional provisions for buildings *Clause 3.4: 1991-1-1*

3.1. General

EN 1990 Clause 3.2 identifies the following design situations for the verification of ultimate limit states: *cl.3.1(1)P: 1991-1-1*

- persistent design situations, which refer to the conditions of normal use
- transient design situations, which refer to temporary conditions applicable to the structure, e.g. during execution or repair
- accidental design situations, which refer to exceptional conditions applicable to the structure or to its exposure, e.g. explosions
- seismic design situations.

Each of these design situations is linked to a particular expression for the combination of action effects as follows:

- persistent and transient design situations, which refer to *expressions (6.10), or (6.10a) and (6.10b) in EN 1990*
- accidental design situations, which refers to *expression (6.11b) in EN 1990*
- seismic design situations, which refers to *expression (6.12b) in EN 1990.*

In addition, permanent and imposed loads need to be determined for the verification of serviceability limit states and the following expressions for the combination of action effects given in EN 1990:

- the characteristic combination which refers to *expression (6.14b) of EN 1990*
- the frequent combination which refers to *expression (6.15b) of EN 1990*
- the quasi-permanent combination which refers to *expression (6.16b) of EN 1990.*

The relevant permanent and imposed loads that need to be identified for each of the above design situations for ultimate and serviceability limit state verifications include the following:

(1) For self-weight, the characteristic value of a permanent action G. If the variability of G can be considered as small, one single value G_k may be used. If the variability of G cannot be considered as small, two values shall be used: an upper value $G_{k,sup}$ and a lower value $G_{k,inf}$ (see section 2.1 of this Part of this Designers' guide).

(2) For variable actions, the characteristic value (Q_k) which usually corresponds to an upper value with an intended probability of not being exceeded during a specific reference period (normally 50 years for buildings). This return period is directly linked to the service lifetime assumed in the design of the structure. In particular for a service lifetime greater than one year but lower than 50 years, the return period should be taken as 50 years; for a longer service lifetime, the return period should be taken as the service lifetime itself.

Depending upon the design situation being considered for the ultimate or serviceability limit states (described in *EN 1990 Clauses 6.4 and 6.5* and its Designers' Guide[1]) other representative values of a variable action need to be determined as follows:

- The combination value, represented as the product $\psi_0 Q_k$.
- The frequent value, represented as the product $\psi_1 Q_k$. For buildings, the frequent value is generally chosen so that the time it is exceeded is 0.01 of the reference period.
- The quasi-permanent value, represented as a product $\psi_2 Q_k$. Quasi-permanent values are also used for the calculation of long-term effects. For loads on building floors, the quasi-permanent value is usually chosen so that the proportion of the time it is exceeded is 0.50 of the reference period. The quasi-permanent value can alternatively be determined as the value averaged over a chosen period of time.

The notations ψ_0, ψ_1 and ψ_2 are factors for the combination, frequent and quasi-permanent value, respectively of a variable action. Values for ψ_0, ψ_1 and ψ_2 are given in *EN 1990 Table A1*.

Example 3.1

Consider the combination of actions in expression (6.10) for the persistent or transient design situation (EN 1990 Clause 6.4.3.2).

$$\sum_{j\geq 1} \gamma_{G,j} G_{k,j} \text{"+"} \gamma_P P \text{"+"} \gamma_{Q,1} Q_{k,1} \text{"+"} \sum_{i>1} \gamma_{Q,i} \psi_{0,i} Q_{k,i} \qquad (6.10)$$

Assume that the variability of self-weight is small; and three variable actions are being considered as follows: wind (leading) and imposed and snow accompanying.

The following permanent and imposed loads need to be determined for the design situation in this example:

- the characteristic value of a permanent action G
- for wind the characteristic value (Q_k)
- for imposed and snow loads the combination value, represented as a product $\psi_0 Q_k$. (For this example $0.7Q_k$ and $0.5Q_k$ in accordance with EN 1990: Table A.1.)

3.2. Permanent loads

The self-weight of structural members should be determined including non-structural members (e.g. claddings and finishes, unless the expected variability of the finishes is considered significant). This is taken into account in combinations of actions as a single action with a characteristic value G_k. Upper and lower design values (i.e. when considering the unfavourable (with $\gamma_G = 1.35$) and favourable effects (with $\gamma_G = 1.0$) for self-weight) are obtained by using appropriate partial factors in accordance with EN 1990.

cl.3.2(1): 1991-1-1

Figure 3.1 illustrates how the upper and lower design values are considered in the combination of action effects in accordance with *EN 1990 Table A1.2 (B) Note 3*, for the design of the central span of the beam. This case applies when a good standard of supervision

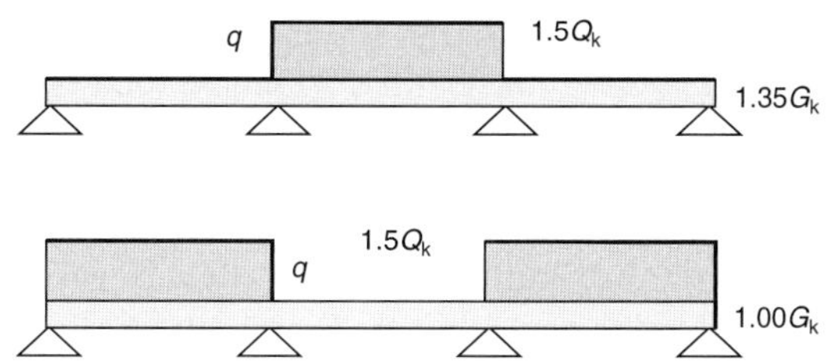

For mid-span effects of actions on central span with controlled conditions

Fig. 3.1. Applying upper and lower design values (e.g. the self-weight of continuous slabs, with good standard of supervision during execution)

during execution is envisaged and when significant deviations from a specified member thickness are not expected. Furthermore, permanent loads are supposed to originate from one source only.

With regard to *EN 1990 Table A1.2 (B) Note 3*, where the results of verification may be

- very sensitive to variations of the magnitude of a permanent action from G; and if
- the variability of G cannot be considered as small due to deviations from the design during the execution process,

allowance has to be made for the construction tolerances.

This may be significant, for example for the following two cases.

(1) For in-situ concrete where self-weight is dominant, and when the quality of site control may not be high and hence deviation from a specified member thickness may occur.
(2) For a thin reinforced concrete member (e.g. <150 mm thickness). For these cases, two values for G should be used in any verification: an upper value $G_{k,sup}$ (where the effects of the action are unfavourable) and a lower value $G_{k,inf}$ (where the effects of the action are favourable).

As explained in Clause 2.1 of this Designers' Guide and according to EN 1990, $G_{k,inf}$ is the 5% fractile and $G_{k,sup}$ is the 95% fractile of the statistical distribution for G, which may be assumed to be normal (Gaussian).

Figure 3.2 illustrates how the upper and lower design values are considered in the combination of action effects in accordance with *EN 1990 Table A1.2 (B) Note 3* where significant deviations from a specified member thickness may occur. Upper and lower design values (i.e. when considering the unfavourable (with $\gamma_G = 1.35$ applied to $G_{k,sup}$) and favourable effects (with $\gamma_G = 1.0$ applied to $G_{k,inf}$) for self-weight) are obtained by using appropriate partial factors in accordance with EN 1990.

As for the example in Fig. 3.1, permanent loads are assumed to originate from one single source.

The favourable or unfavourable effects of permanent loads in an area where it is intended to remove or add structural or non-structural elements (e.g. removable surfacing or

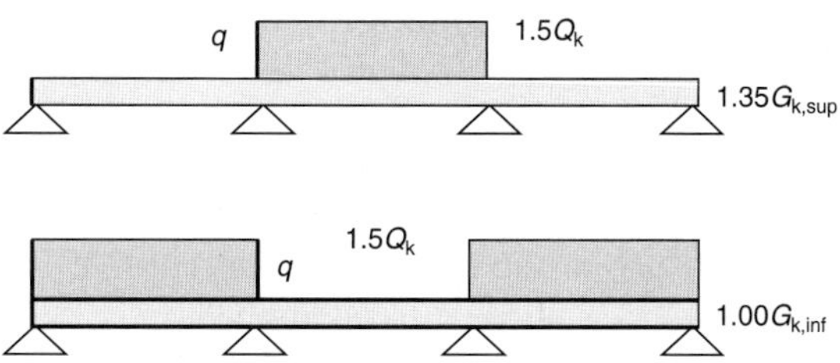

For mid-span effects of actions on central span with none or low controlled conditions

Fig. 3.2. Applying upper and lower design values (e.g. the self-weight of thin continuous slabs, or where a good standard of supervision during execution is not expected)

coverings within a specific room in a building) should be determined accordingly. Generally a load case considering unfavourable effects of a permanent load on a member should include the non-structural elements; but the load case determining favourable effects of a permanent action should not include the non-structural elements (***Clause 3.2(2): 1991-1-1***).

cl.3.2(2): 1991-1-1

The favourable or unfavourable effects of permanent loads in an area where it is intended to add new coatings and/or distribution conduits after execution should be determined accordingly. Generally a load case within a design situation determining unfavourable effects of a permanent load on a member, should include the new coatings and/or distribution conduits; but the load case determining favourable effects of a permanent action should not include them.

cl. 3.2(3): 1991-1-1

The favourable or unfavourable effects of permanent loads in an area where the effects of water level are critical (e.g. a roof slab supporting a swimming pool or a water tank) should be determined accordingly. Generally a load case within a design situation determining unfavourable effects of a permanent load on a member should include the highest possible water level. The load case determining favourable effects of a permanent action should consider the lowest possible water level, which may be an empty pool or tank.

cl.3.2(4)P: 1991-1-1

Note to cl.3.2(4)P: 1991-1-1

Advice on varying water levels in the ground when determining the effects of earth pressure are explained in EN 1997 and its Designers' Guide.[9]

The favourable or unfavourable effects of permanent loads for a building structure used for storage of bulk materials, where the source and moisture content of the material can effect its self-weight, should be determined accordingly. Generally a load case within a design situation determining unfavourable effects of a permanent load on a member should include the highest possible weight of the material. The load case determining favourable effects of a permanent action should consider the lowest possible weight of the material (***Clause 3.2(5): 1991-1-1***).

cl.3.2(5): 1991-1-1

Note to cl.3.2(5)P: 1991-1-1

No comment is necessary to the ***Note to Clause 3.2(5)P: 1991-1-1***.

3.3. Imposed loads

3.3.1. General

cl.3.3.1(1)P: 1991-1-1

The favourable or unfavourable effects of the different categories of imposed loads in an area should be determined accordingly. For example, for the design of beams and floor slabs the imposed loads q, from which Q_k is determined (see Figs 3.1 and 3.2) should be applied in various arrangements within one storey to give the most unfavourable forces and moments.

cl.3.3.1(2)P: 1991-1-1

If different categories of imposed loads in buildings (e.g. floors and balconies) are considered in a design situation where the imposed loads act simultaneously with other variable actions, then the total imposed loads can be considered in the load combination as a single action (see Chapter 1).

cl.3.3.1(3): 1991-1-1

No direct comment is necessary to ***Clause 3.3.1(3): 1991-1-1***. Further information is given in EN 1991-2 and its relevant Designers' Guide.[8]

cl.3.3.1(4): 1991-1-1

No direct comment is necessary to ***Clause 3.3.1(4): 1991-1-1***. The subject matter of this clause is treated in Section 5.1.3 of the Designers' Guide to EN 1990.[6]

3.3.2. Additional provisions for buildings

When determining the effects of actions on a roof, EN 1991-1-1 stipulates that imposed loads need not be considered in the same combination as wind actions and/or snow loads as they are not likely to occur simultaneously on a roof.

cl.3.3.2(1): 1991-1-1

cl.3.3.2(4)P: 1991-1-1

In a combination of the effects of actions when the imposed load is considered as an accompanying action, only one of the reduction coefficients (ψ from EN 1990, Table A1.1) or α_n from *Clause 6.3.1.2 (11)* is allowed to be applied.

cl.3.3.2(3): 1991-1-1

No comment necessary for ***Clause 3.3.2(3): 1991-1-1***.

The imposed loads to be considered for serviceability limit state verifications should be specified in accordance with the service conditions and the requirements concerning the performance of the structure, and the appropriate representative values of the actions

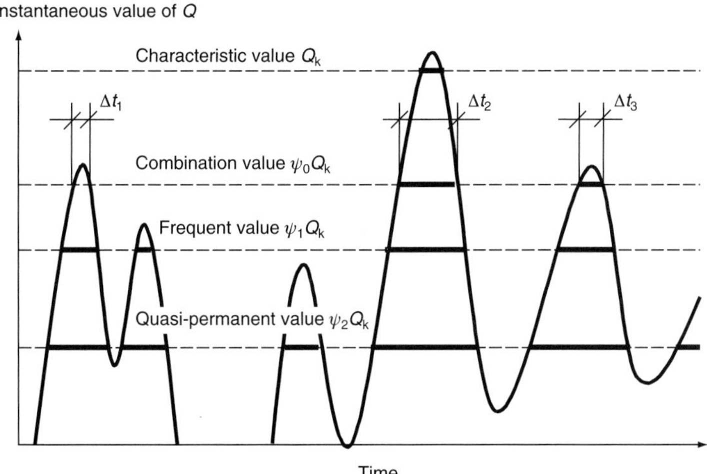

Fig. 3.3. Representative values of variable actions

$\psi_0 Q_k$, $\psi_1 Q_k$ and $\psi_2 Q_k$ should be used. See *Clause 4.1.3 of EN 1990* and the explanation in the Designers' Guide to EN 1990.[6] *cl.3.3.2(4): 1991-1-1*

The combination value $\psi_0 Q_k$ is associated with the combination of actions for irreversible serviceability limit states. Examples where this representative value should be considered are for the functioning of the structure or to its finishes or to non-structural parts (e.g. partition walls, claddings).

The frequent value $\psi_1 Q_k$ is primarily associated with the frequent combination in the serviceability limit states. Examples where this representative value should be considered are for the comfort of the user, or the functioning of machinery.

The main use of quasi-permanent values $\psi_2 Q_k$ is the assessment of long-term effects, for example cosmetic cracking and appearance of a building or creep effects in timber structures, and for verification of frequent and quasi-permanent combinations (long-term effects) of serviceability limit states. *cl.4.1.3(1)P: 1991-1-1*

See Fig. 3.3. for an explanation of the representative values of the actions $\psi_0 Q_k$, $\psi_1 Q_k$ and $\psi_2 Q_k$.

Two examples, regarding Clauses 3.1 and 3.2 of EN 1991-1-1 are given below showing the load arrangements for the design of the following: *cl.3.1: 1991-1-1* *cl.3.2: 1991-1-1*

- Example 3.2. A cantilever beam against overturning (reference Table A1.2(A) of EN 1990)
- Example 3.3. A three-span continuous floor slab (reference Table A1.2(B) of EN 1990).

N.B. Reference needs to be made to Tables A1.2(A) and A1.2(B) of Annex A of EN 1990.

Example 3.2. A cantilever beam against overturning (reference Table A1.2(A) of EN 1990)

A simply supported beam with a cantilever part.[6] Only two actions are considered, represented by uniformly distributed loads: the self-weight of the beam g and a free variable action q. When studying a possible loss of equilibrium, the variable action is applied only to the cantilever part (see Fig. 3.4).

In accordance with Clause 6.4.3.1(4)P of EN 1990 '*where the results of a verification may be very sensitive to variations of the magnitude of a permanent action from place to place in the structure, the unfavourable and the favourable parts of this action shall be considered as individual actions*'. This applies in particular to the verification of static equilibrium and analogous limit states. *cl.6.4.3.1(4)P: EN 1990*

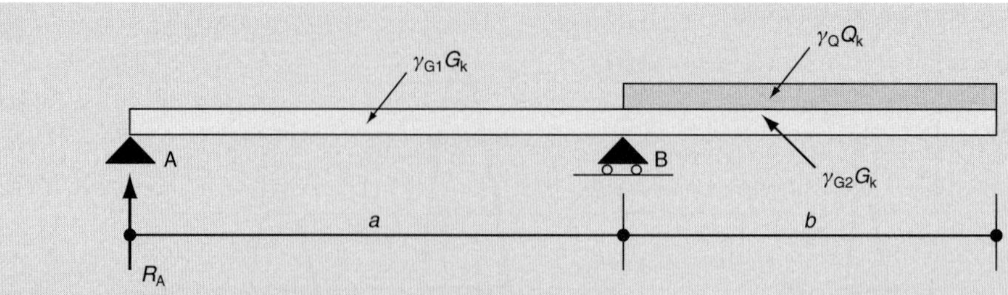

Fig. 3.4. Simply supported beam with cantilever part

In a general manner, the expression of the vertical reaction at support A is:

$$R_A = \frac{1}{2}aq_g(\gamma_{G1} - \beta^2\gamma_{G2} - \beta^2 x\gamma_Q) \quad \text{with} \quad \beta = \frac{b}{a} \quad \text{and} \quad x = \frac{q_q}{q_g}$$

and the static equilibrium is considered as verified if $R_A \geq 0$. In the following, it is assumed $\beta^2 = 0.5$ for the establishment of the numerical example. Where the static equilibrium is not ensured through the EQU verification (i.e. case A in Table 3.1), it is assumed that an anchor is placed in A. This anchor should be designed for an appropriate resistance (STR limit state) and an appropriate stability in the ground (GEO limit state) *cl.6.4.1(1)P: EN 1990* (for an explanation of EQU, STR and GEO see Section 6.4 of the Designers' Guide to EN 1990[6]). For STR limit states verifications for the anchor, EN 1990 recommends the factors to be used are $\gamma_{g,inf} = 1.00$, $\gamma_{g,sup} = 1.35$ and $\gamma_q = 1.50$, as given in Table A1.2(B) (i.e. case B in Table 3.1). Being a calculation related to equilibrium, the designer should not use the single-source principle. Alternatively, EN 1990 also recommends, through note 2 of Table A1.2(A), the factors to be used are $\gamma_{g,inf} = 1.15$, $\gamma_{g,sup} = 1.35$ and $\gamma_q = 1.50$ (i.e. case C of Table 3.1).

As can be seen from Table 3.1, if an anchor or other restraint is required, the alternative method (i.e. case C) provides a much lower level of safety then using case B. Therefore the authors recommend the use of case B only, if equilibrium in case A is not guaranteed. For case B the single-source principle should not be used.

Table 3.1. A cantilever beam against overturning

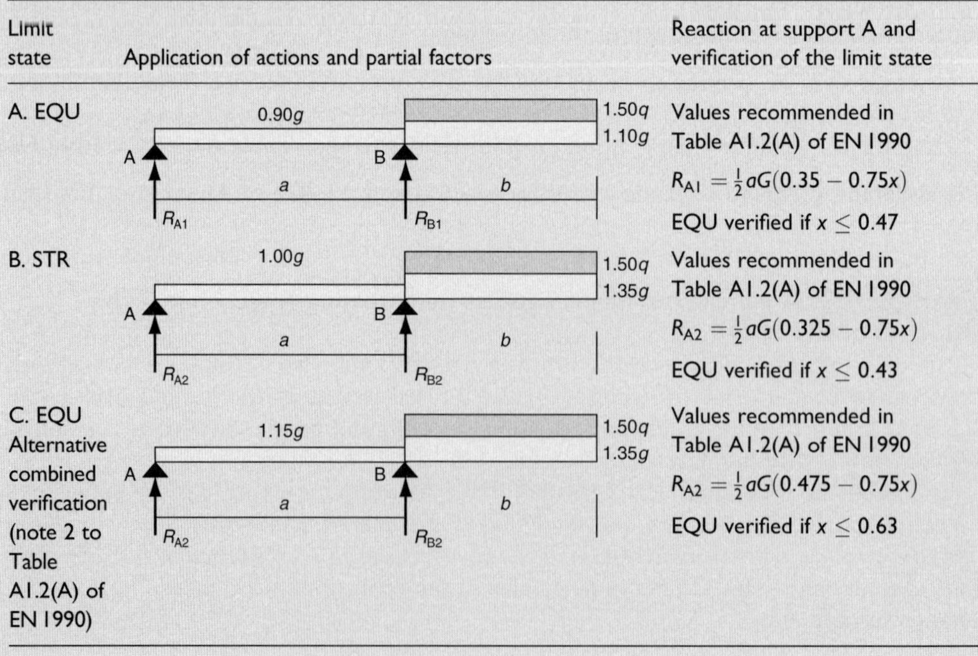

Limit state	Application of actions and partial factors	Reaction at support A and verification of the limit state
A. EQU	0.90g; 1.50q; 1.10g; A; B; a; b; R_{A1}; R_{B1}	Values recommended in Table A1.2(A) of EN 1990 $R_{A1} = \frac{1}{2}aG(0.35 - 0.75x)$ EQU verified if $x \leq 0.47$
B. STR	1.00g; 1.50q; 1.35g; A; B; a; b; R_{A2}; R_{B2}	Values recommended in Table A1.2(A) of EN 1990 $R_{A2} = \frac{1}{2}aG(0.325 - 0.75x)$ EQU verified if $x \leq 0.43$
C. EQU Alternative combined verification (note 2 to Table A1.2(A) of EN 1990)	1.15g; 1.50q; 1.35g; A; B; a; b; R_{A2}; R_{B2}	Values recommended in Table A1.2(A) of EN 1990 $R_{A2} = \frac{1}{2}aG(0.475 - 0.75x)$ EQU verified if $x \leq 0.63$

Example 3.3. A three-span continuous floor slab (reference Table A1.2(B) of EN 1990)

A steel beam and composite floor slab on three spans of equal length *a* (see Fig. 3.5). The analysis will consider Expressions 6.10 and 6.10.a/6.10.b of EN 1990, and the load arrangements and γ_G and γ_Q used are shown in Tables 3.2 to 3.5 below (***Clause 6.4.1(1)P: EN 1990***). *cl.6.4.1(1)P: EN 1990*

Assume low variability in self-weight, giving a single value of $G_k = 25$ kN/m. Office loading, for which $Q_k = 10$ kN/m. Prestress is not present and geotechnical actions are not present.

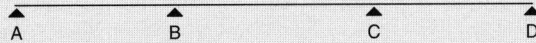

Fig. 3.5. Three-span continuous floor slab

Table 3.2. Using Expression 6.10 and Table A1.2(B) of EN 1990. Span BC

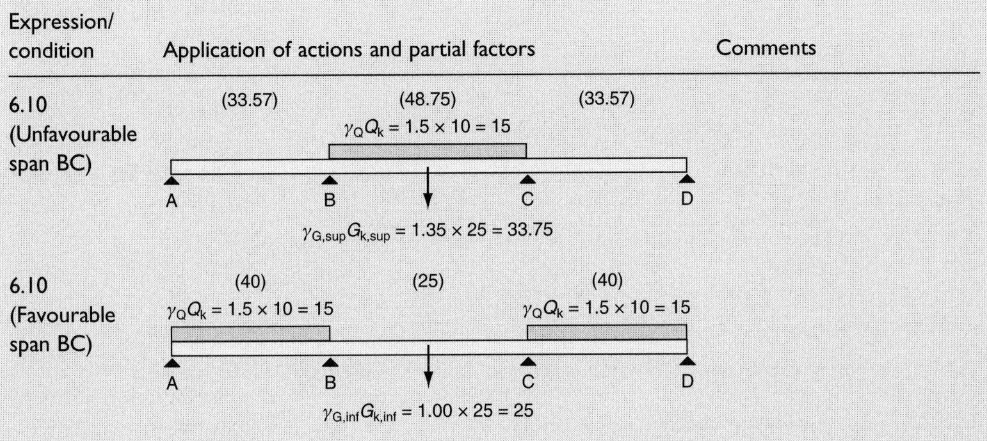

Expression/ condition	Application of actions and partial factors	Comments
6.10 (Unfavourable span BC)	(33.57) (48.75) (33.57); $\gamma_Q Q_k = 1.5 \times 10 = 15$ on span BC; A B C D; $\gamma_{G,sup} G_{k,sup} = 1.35 \times 25 = 33.75$	
6.10 (Favourable span BC)	(40) (25) (40); $\gamma_Q Q_k = 1.5 \times 10 = 15$ on span AB; $\gamma_Q Q_k = 1.5 \times 10 = 15$ on span CD; A B C D; $\gamma_{G,inf} G_{k,inf} = 1.00 \times 25 = 25$	

Table 3.3. Using Expression 6.10 and Table A1.2(B) of EN 1990. Span AB

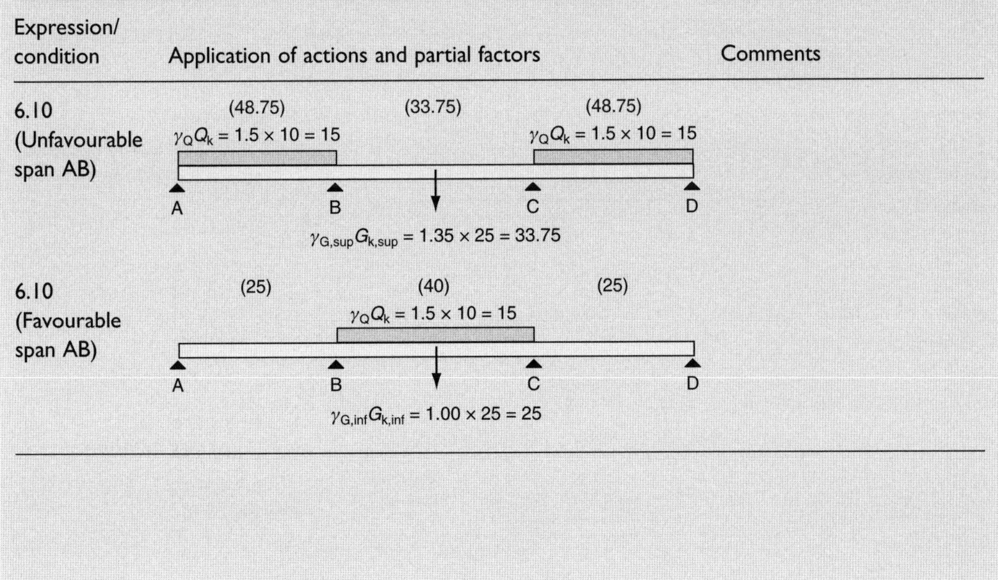

Expression/ condition	Application of actions and partial factors	Comments
6.10 (Unfavourable span AB)	(48.75) (33.75) (48.75); $\gamma_Q Q_k = 1.5 \times 10 = 15$ on span AB; $\gamma_Q Q_k = 1.5 \times 10 = 15$ on span CD; A B C D; $\gamma_{G,sup} G_{k,sup} = 1.35 \times 25 = 33.75$	
6.10 (Favourable span AB)	(25) (40) (25); $\gamma_Q Q_k = 1.5 \times 10 = 15$ on span BC; A B C D; $\gamma_{G,inf} G_{k,inf} = 1.00 \times 25 = 25$	

Table 3.4. Using Expressions 6.10a and 6.10b and Table A1.2(B) of EN 1990. Span BC

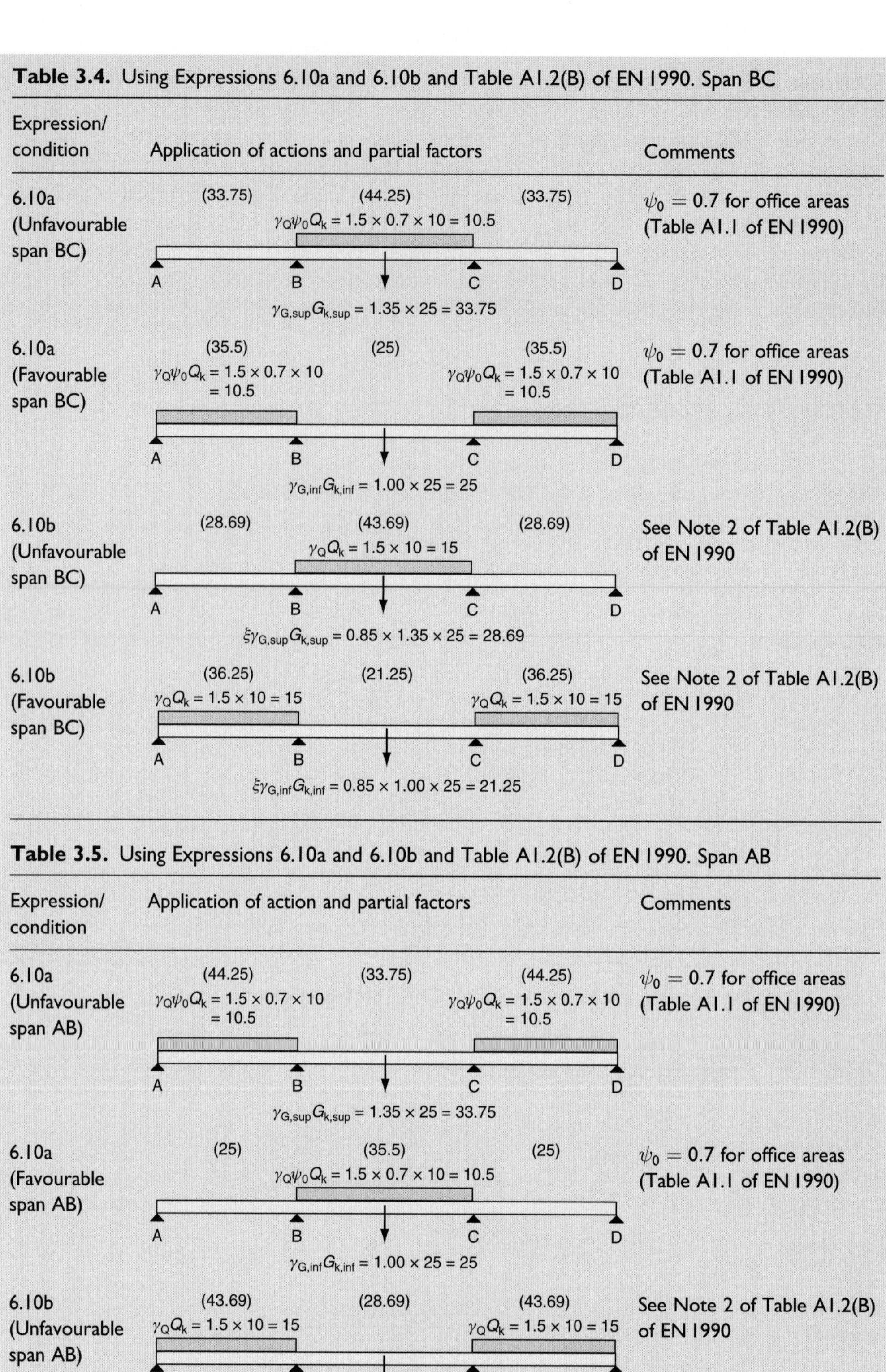

Expression/ condition	Application of actions and partial factors	Comments
6.10a (Unfavourable span BC)	(33.75) (44.25) (33.75); $\gamma_Q\psi_0Q_k$ = 1.5 × 0.7 × 10 = 10.5 (span BC); A B C D; $\gamma_{G,sup}G_{k,sup}$ = 1.35 × 25 = 33.75	$\psi_0 = 0.7$ for office areas (Table A1.1 of EN 1990)
6.10a (Favourable span BC)	(35.5) (25) (35.5); $\gamma_Q\psi_0Q_k$ = 1.5 × 0.7 × 10 = 10.5 (span AB); $\gamma_Q\psi_0Q_k$ = 1.5 × 0.7 × 10 = 10.5 (span CD); A B C D; $\gamma_{G,inf}G_{k,inf}$ = 1.00 × 25 = 25	$\psi_0 = 0.7$ for office areas (Table A1.1 of EN 1990)
6.10b (Unfavourable span BC)	(28.69) (43.69) (28.69); γ_QQ_k = 1.5 × 10 = 15 (span BC); A B C D; $\xi\gamma_{G,sup}G_{k,sup}$ = 0.85 × 1.35 × 25 = 28.69	See Note 2 of Table A1.2(B) of EN 1990
6.10b (Favourable span BC)	(36.25) (21.25) (36.25); γ_QQ_k = 1.5 × 10 = 15 (span AB); γ_QQ_k = 1.5 × 10 = 15 (span CD); A B C D; $\xi\gamma_{G,inf}G_{k,inf}$ = 0.85 × 1.00 × 25 = 21.25	See Note 2 of Table A1.2(B) of EN 1990

Table 3.5. Using Expressions 6.10a and 6.10b and Table A1.2(B) of EN 1990. Span AB

Expression/ condition	Application of action and partial factors	Comments
6.10a (Unfavourable span AB)	(44.25) (33.75) (44.25); $\gamma_Q\psi_0Q_k$ = 1.5 × 0.7 × 10 = 10.5 (span AB); $\gamma_Q\psi_0Q_k$ = 1.5 × 0.7 × 10 = 10.5 (span CD); A B C D; $\gamma_{G,sup}G_{k,sup}$ = 1.35 × 25 = 33.75	$\psi_0 = 0.7$ for office areas (Table A1.1 of EN 1990)
6.10a (Favourable span AB)	(25) (35.5) (25); $\gamma_Q\psi_0Q_k$ = 1.5 × 0.7 × 10 = 10.5 (span BC); A B C D; $\gamma_{G,inf}G_{k,inf}$ = 1.00 × 25 = 25	$\psi_0 = 0.7$ for office areas (Table A1.1 of EN 1990)
6.10b (Unfavourable span AB)	(43.69) (28.69) (43.69); γ_QQ_k = 1.5 × 10 = 15 (span AB); γ_QQ_k = 1.5 × 10 = 15 (span CD); A B C D; $\xi\gamma_{G,sup}G_{k,sup}$ = 0.85 × 1.35 × 25 = 28.69	See Note 2 of Table A1.2(B) of EN 1990
6.10b (Favourable span AB)	(21.25) (36.25) (21.25); γ_QQ_k = 1.5 × 10 = 15 (span BC); A B C D; $\xi\gamma_{G,inf}G_{k,inf}$ = 0.85 × 1.00 × 25 = 21.25	See Note 2 of Table A1.2(B) of EN 1990

Conclusion for Example 3.3

The different applications of the actions considered in this example suggest that the expression 6.10b, when used with unfavourable partial factors, gives the maximum bending moment in the spans BC and AB.

However, it is not clear, from the above values, which of expressions 6.10a and 6.10b would give the more onerous application of the actions for the bending moment at the supports. It has to be ascertained by calculating the actual bending moments for each application.

CHAPTER 4

Densities of construction and stored materials

This chapter is concerned with densities of construction and stored materials in EN 1991-1-1: *Eurocode 1 – Actions on structures: Part 1.1: General Actions – Densities, self-weight, imposed loads for buildings*. The material described in this chapter is covered in the following clauses:

- General *Clause 4.1: 1991-1-1*

4.1. General

Where available, the design should use characteristic values of densities of construction and stored materials. Where characteristic values are not available, mean values should be used as characteristic values. Annex A of EN 1991-1-1 (see Chapter 7 of the EN 1991-1-1 part of this Designers' Guide) gives mean values. It should be noted that when a range is given, EN 1991-1-1 recognises that the mean value will be highly dependent on the source of the material, moisture variation etc. These values should be selected and used considering each individual project and design situation (see Chapter 3 of the EN 1991-1-1 part of this Designers' Guide). *cl.4.1.(1): 1991-1-1*

For materials (e.g. new and innovative materials) which are not covered by the Tables in Annex A of EN 1991-1-1, the characteristic value of the density needs to be determined in accordance with EN 1990 Clause 4.1.2 and agreed for each individual project. Further information is given in EN 1990 and its relevant Designers' Guide.[6] *cl.4.1.(2): 1991-1-1*

Sections 2.1 and 3.2 of the EN 1991-1-1 part of this Designers' Guide explain how the *cl.4.1.(3): 1991-1-1*
requirements of (***Clause 4.1.(3): 1991-1-1***), and EN 1990 Clause 4.1.2 should be met. *cl.4.1.(3): 1991-1-1*

No comment necessary for ***Clause 4.1.(4): 1991-1-1***. *cl.4.1.(4): 1991-1-1*

CHAPTER 5

Self-weight of construction works

This chapter is concerned with the general concepts of self-weight of construction works relating to EN 1991-1-1: *Eurocode 1 – Actions on structures: Part 1.1: General Actions – Densities, self-weight, imposed loads for buildings*. The material described in this chapter is covered in the following clauses:

- Representation of actions — *Clause 5.1: 1991-1-1*
- Characteristic values of self-weight — *Clause 5.2: 1991-1-1*

5.1. Representation of actions

The self-weight of the construction works effects the structure throughout its design working life normally without substantial changes (see Fig. 5.1). For most cases, in particular for the design of buildings it may be represented by a single characteristic value and be calculated on the basis of the nominal dimensions and the characteristic values of the densities. *cl.5.1.(1): 1991-1-1*

The self-weight of the construction works includes:

- the structure (see examples for concrete, masonry, steel and timber in Section 5.2 below); and
- non-structural elements, see below, which may include fixed services. *cl.5.1.(2): 1991-1-1*

The effects of actions due to the self-weight of non-structural elements may change during the lifetime of the structure due to maintenance, repairs and refurbishment. Such changes may affect the load and its spatial distribution. At the time of design, it should be anticipated whether such changes will occur, but the changes in magnitude and spatial distribution will not generally be known. Removal of such components does not affect the general load-carrying capabilities of a member or structure but it may change the static equilibrium of the action–structure system.

Clause 5.1.(3) of EN 1991-1-1 lists examples of non-structural elements as below: *cl.5.1.(3): 1991-1-1*

- roofing
- surfacing and coverings
- partitions and linings
- handrails, safety barriers, parapets and kerbs
- wall cladding
- suspended ceilings
- thermal insulation
- bridge furniture
- fixed services.

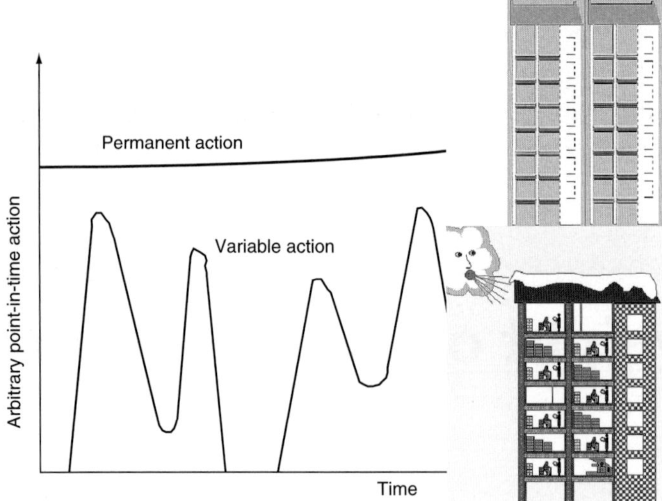

Fig. 5.1. Example of variation of permanent and variable actions during the design working life of a structure

Note to cl.5.1.(3): 1991-1-1 There will be occasions when a manufacturer will need to be consulted (e.g. for safes and other industrial equipment).

cl.5.1.(4): 1991-1-1 Clause 5.1.(4) of EN 1991-1-1 lists examples of fixed services as below:

- equipment for lifts and moving stairways
- heating, ventilating and air-conditioning equipment
- electrical equipment
- pipes without their contents
- cable trunking and conduits.

cl.5.1.(5): 1991-1-1 Loads due to some types of non-structural elements may be treated as imposed loads (e.g. movable partitions).

5.2. Characteristic values of self-weight

5.2.1. General

cl.5.2.1.(1)P: 1991-1-1 Characteristic values of self-weight and of the dimensions and densities are determined in accordance with EN 1990, 4.1.2.

The characteristic values of the self-weight for structures or members of different materials are described below.

Concrete structures

The factors which determine the characteristic values of the self-weight for concrete structures or members include the cross-section dimensions, the characteristic values of the densities for concrete, reinforcement, steel detail etc. Random deviations are mainly caused by the random variability of the dimensions and the density of concrete. The self-weight of a concrete member is a substantial part of the total load on a structure.

Masonry structures

The factors which determine the characteristic values of the self-weight for masonry structures or members, include:

- the dimensions of the bricks and blocks
- ancillary elements (e.g. wall ties)
- the bulk density of bricks or blocks (depending on the voids)
- the property of the mortar

- the pattern of masonry and thickness of the joints
- the extent of mortar penetration in the voids
- moisture contents.

Random deviations are mainly associated by the first four factors above. The self-weight of a masonry member is a substantial part of the total load on a structure.

Steel structures

The factors which determine the characteristic values of the self-weight for steel structures or members include the weight of the individual parts and components, and the weight of the connections. Random deviations are mainly caused by the variability of the dimensions (material thickness). In comparison with the random deviations of structures of other materials the random deviation of the self-weight of metal structures is small. The self-weight of a steel member is often a small part of the total load on a structure.

Timber structures

The factors which determine the characteristic values of the self-weight for timber structures or members, include:

- the weight density of the timber
- the dimensions of the individual parts
- the weight of connecting parts
- ancillary elements (e.g. the weight of metal parts)
- moisture contents.

Random deviations are mainly associated by the first two factors above. The self-weight of a timber member is important as timber structures are often designed to carry relatively low loads.

When determining the characteristic values for self-weight, any required nominal dimensions should be those as shown on the drawings. *cl.5.2.1.(2): 1991-1-1*

5.2.2. Additional provisions for buildings

Self-weights for manufactured elements (e.g. flooring systems, façades, lifts and other equipment for buildings) are not given in EN 1991-1-1 but the code advises to contact the appropriate manufacturer. *cl.5.2.2.(1): 1991-1-1*

For determining the effect of the self-weight due to movable partitions, an equivalent uniformly distributed load shall be used and added to the imposed load. See, in this Designers' Guide, Chapter 6 of the EN 1991-1-1 part. *cl.5.2.2.(2)P: 1991-1-1*

5.2.3. Additional provisions specific for bridges

The clauses in this section of EN 1991-1-1 are discussed in the Designers' Guide to EN 1991-1-1 Actions on Bridges[8]. *cl.5.2.3: 1991-1-1*

CHAPTER 6

Imposed loads on buildings

This chapter is concerned with the general concepts of imposed loads on buildings relating to EN 1991-1-1: *Eurocode 1 – Actions on structures: Part 1.1: General Actions – Densities, self-weight, imposed loads for buildings*. The material described in this chapter is covered in the following clauses.

- Representation of actions *Clause 6.1: 1991-1-1*
- Load arrangements *Clause 6.2: 1991-1-1*
- Characteristic values of Imposed Loads *Clause 6.3: 1991-1-1*

6.1. Representation of actions

Imposed loads on floors or roofs of buildings are be caused by the following:

(1) Furniture and movable objects (e.g. movable partitions, storage, equipment, the contents of containers); these loads at certain points in time are subjected to considerable instantaneous changes in their magnitudes, mainly due to change of occupancy, change of use etc. Between these instantaneous changes the load varies very slowly with time and the magnitudes of the variation are generally small (see Fig. 6.1(a)).
(2) Normal use by persons; these loads are often periodical and only present during a relatively small part of the time, e.g. for school rooms only for about $\frac{1}{4}$ of a 24-hour day (see Fig. 6.1(b)). In some cases the loads from persons may also cause dynamic effects, e.g. dancing halls and gyms used for aerobics.
(3) Vehicles including industrial vehicles, and helicopters in the case of some roofs.
(4) Anticipated rare events, such as concentrations of persons or of furniture, or the moving or stacking of objects which may occur during reorganisation or refurbishment, or the moving and stacking of storage items. These special situations occur during a short or moderate period of time but sufficiently during the design working life of a building to make it necessary to take them into account in the minimum values of imposed loads given in EN 1991-1-1 (see Fig. 6.1(c)). *cl.6.1.(1): 1991-1-1*

The imposed load on floors may, in special situations, be considerably increased for a short or moderate period of time. Examples of such situations could be:

- The gathering of people during special planned events, such as parties. During such events people tend to cluster into groups.
- The crowding of people under emergency-type situations, such as in an exit hall or a fire escape.

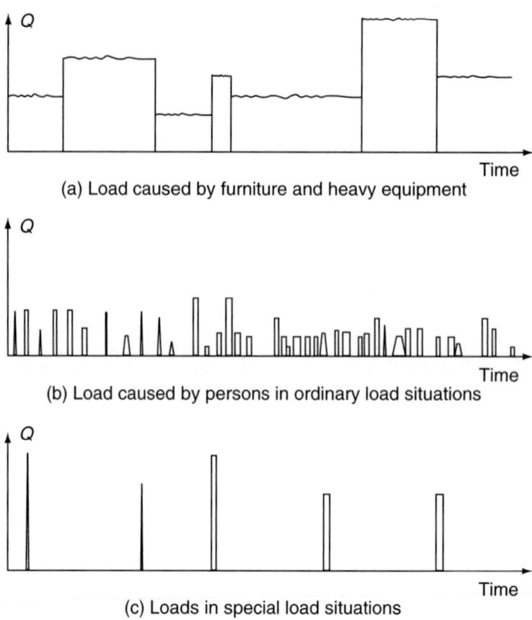

Fig. 6.1. Time variability of the magnitude Q of the load

The density and intensity of people on a floor, as shown in Fig. 6.2,[12] can have a significant influence on the imposed load on a floor.

Imposed loads may be regarded as consisting of two different parts. The first part is associated with *ordinary load situations* occurring many times a year, and caused by the loads described in Fig. 6.1(a) and (b) above. The second part is associated with *special load situations*, and caused by the loads described in Fig. 6.1(c) above where the load may be considerably increased. The *ordinary* and *special* load situations are also called *sustained and intermittent loads*.

cl.6.1.(22): 1991-1-1 The principal background to the modelling of imposed loads in EN 1991-1-1 is CIB Report 116.[13] The imposed loads specified EN 1991-1-1 are modelled by uniformly distributed loads (to determine general effects on a floor or a roof), line loads (normally for horizontal loads acting on parapets) or concentrated loads (for the determination of local effects) or combinations of these loads.

cl.6.1.(3): 1991-1-1 The values of imposed loads, on floor and roof areas in buildings are dependent on the use of the floor or roof, and are categorised in EN 1991-1-1 accordingly.

The categorisation is into ten main categories A, B, C, D, E, F, G, H, I, K as follows:

- Category A: Areas for domestic and residential activities
- Category B: Office areas
- Category C: Areas where people may congregate (with the exception of areas defined under category A, B and D)
- Category D: Shopping areas
- Category E1: Areas susceptible to accumulation of goods, including access areas
- Category E2: Industrial use
- Category F: Traffic and parking areas for light vehicles ($\leq$30 kN gross vehicle weight and $\leq$8 seats not including driver)
- Category G: Traffic and parking areas for medium vehicles (>30 kN, $\leq$160 kN gross vehicle weight, on two axles)
- Category H: Roofs not accessible except for normal maintenance and repair
- Category I: Roofs accessible with occupancy according to categories A to D
- Category K: Roofs accessible for special services, such as helicopter landing areas.

Values of loads for heavy equipment (e.g. in communal kitchens, radiology rooms, boiler rooms etc.) are not given in EN 1991-1-1, and the designer is asked to agree these with the

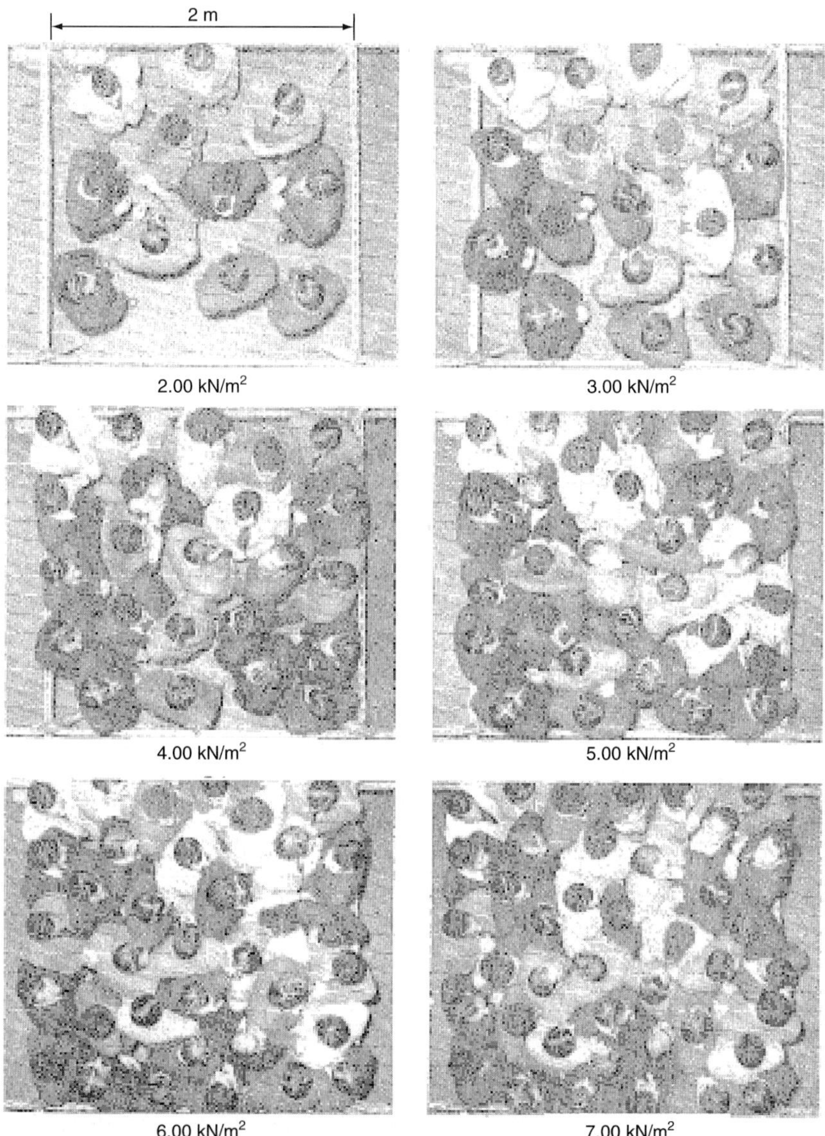

Fig. 6.2. Load values associated with various intensities of people loading

client and/or the relevant authority. Heavy equipment comes under Category E2 in EN 1991-1-1, and advice on value loads for heavy equipment is given in Section 6.3.2 and Table 6.6) of this Designers' Guide. *cl.6.1.(4): 1991-1-1*

6.2. Load arrangements

6.2.1. Floors, beams and roofs

For the design of a floor member in one storey or a roof of a building, the imposed load is taken into account as a free action applied at the most unfavourable part of the influence area of the action effects considered. See Fig. 6.3. *cl.6.2.1.(1)P: 1991-1-1*

Where the effects of the actions on the member from other storeys are considered relevant, they may be assumed as fixed actions, and be distributed uniformly (see Fig. 6.3(b)). A more rigorous 'chessboard' arrangement of the actions may be also used (see Fig. 6.3(a)). See also Examples 3.2 and 3.3 in this part of this Designers' Guide. *cl.6.2.1.(2): 1991-1-1*

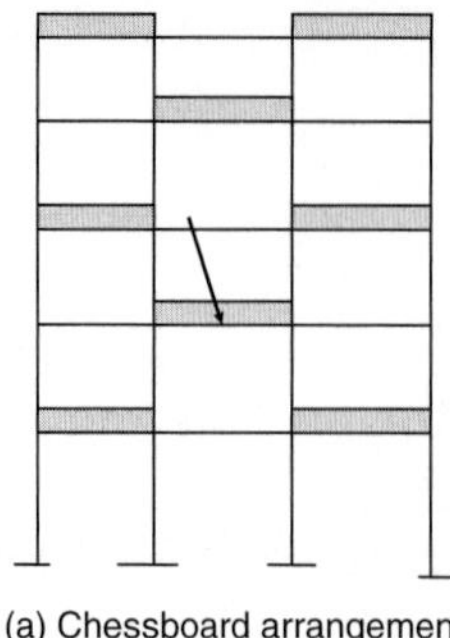

(a) Chessboard arrangement

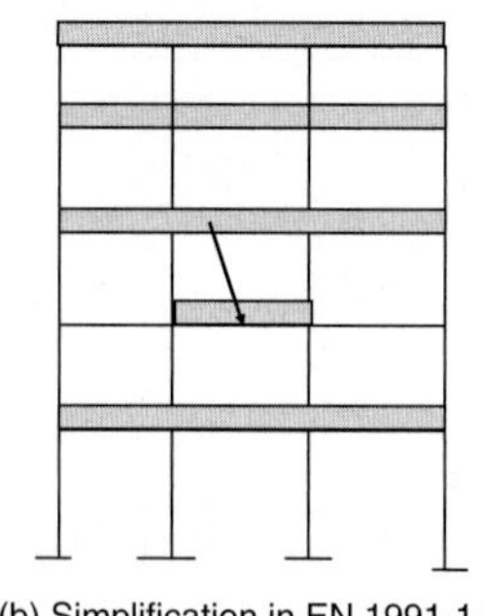

(b) Simplification in EN 1991-1-1

Fig. 6.3. Mid-span bending moment of a floor structure (arrow points to element under consideration)

cl.6.2.1.(3)P: 1991-1-1

In addition, to ensure a minimum local resistance of the floor member a separate verification needs to be performed with a concentrated load. Unless required by the design, a concentrated load need not normally be combined with the uniformly distributed loads or other variable actions.

cl.6.2.1.(4): 1991-1-1

The imposed loads on an appropriate member (e.g. a beam) may be reduced according to the magnitude of the area supported, by a reduction factor α_A. See Clause 6.3.1.2(10) for the value and conditions of use of α_A.

6.2.2. Columns and walls

cl.6.2.2.(1): 1991-1-1

For the design of vertical elements (e.g. columns or walls), loaded from several storeys, the total imposed loads on the floor of each storey, according to EN 1991-1-1, should be assumed to be distributed uniformly. The design however should consider the case when an alternate span is not loaded with a variable action, the effects of which may be to produce tension on one face of a column and/or additional compression on the other face.

cl.6.2.2.(2): 1991-1-1

The total imposed loads on a vertical element from several storeys may be reduced by a factor α_n. See Clause 6.3.1.2(11) for the values and conditions of use of α_n.

6.3. Characteristic values of imposed loads

6.3.1. Residential, social, commercial and administration areas

6.3.1.1. Categories

Areas in residential, social, commercial and administration buildings are divided into categories:

- A (areas for domestic and residential activities)
- B (office area)
- C (area where people may congregate)
- D (shopping area).

cl.6.3.1.1.(1)P: 1991-1-1

See Table 6.1 which gives an example of the style of presentation of Table 6.1 of EN 1991-1-1.

Table 6.1. Example of style of Table 6.1 of EN 1991-1-1

Category	Specific use	Example
A	–	–
B	–	–
C	–	–
D	Shopping areas	D1: Areas in general retail shops D2: Areas in department stores

Within Table 6.1 Categories C and D have been sub-categorised reflecting the different specific uses within a category as follows: *Table 6.1 and cl.6.3.1.1.(1)P: 1991-1-1*

(1) For category C:

- C1: areas with tables, e.g. restaurants
- C2: areas with fixed seats, e.g. churches, theatres and lecture rooms
- C3: areas without obstacles for moving people such as museums
- C4: for example gyms
- C5: areas susceptible to large crowds such as concert halls and stands.

(2) For Category D:

- D1: general retail shops
- D2: departmental stores.

Sub-categories C1 to C4, depending on their anticipated use, may be categorised as C5 (e.g. area susceptible to large crowds). This recognises that areas covered by C1 to C4 may themselves be susceptible to large crowds during the design working life of the structure. *Note 2 to Table 6.1 cl.6.3.1.1(1)P: 1991-1-1*

The National Annex may additionally sub-categorise Catagories A, B, C1 to C5, D1 and D2. Reference should therefore be made to the appropriate National Annex. In particular, users in the UK should note that the BSI National Annex offers a comprehensive heavy sub-categorisation for all categories A to D as described below. *Note 2 to Table 6.1 cl.6.3.1.1(1)P: 1991-1-1*

(1) Category A is sub-categorised into seven sub-categories:

- A1 (All usages within self-contained dwelling units (a unit occupied by a single family or a modular student accommodation unit with a secure door and comprising not more than six single bedrooms and an internal corridor.) Communal areas (including kitchens) in blocks of flats with limited use. For communal areas in other blocks of flats, see A5, A6 and C3)
- A2 (Bedrooms and dormitories except those in self-contained single family dwelling units and in hotels and motels)
- A3 (Bedrooms in hotels and motels; hospital wards; toilet areas)
- A4 (Billiard/snooker rooms)
- A5 (Balconies in single family dwelling units and communal areas in blocks of flats with limited use)
- A6 (Balconies in hostels, guest houses, residential clubs and communal areas in blocks of flats) and
- A7 (Balconies in hotels and motels)

(2) Category B is sub-categorised into two sub-categories:

- B1 (General office use)
- B2 (At or below ground-floor level)

(3) Category C1 (Areas with tables) is sub-categorised into three sub-categories:

- C11 (Public, institutional and communal dining rooms and lounges, cafes and restaurants)
- C12 (Reading rooms with no book storage)
- C13 (Classrooms)

(4) Category C2 (Areas with fixed seats) is sub-categorised into two sub-categories:

- C21 (Assembly areas with fixed seating)
- C22 (Places of worship)

(5) Category C3 (Areas without obstacles for moving people) is sub-categorised into nine sub-categories:

- C31 (Corridors, hallways, aisles in institutional-type buildings not subjected to crowds or wheeled vehicles, hostels, guest houses, residential clubs, and communal areas in blocks of flats)
- C32 (Stairs, landings in institutional-type buildings not subjected to crowds or wheeled vehicles, hostels, guest houses, residential clubs, and communal areas in blocks of flats)
- C33 (Corridors, hallways, aisles in all buildings not covered by C31 and C32, including hotels and motels and institutional buildings subjected to crowds)
- C34 (Corridors, hallways, aisles in all buildings not covered by C31 and C32, including hotels and motels and institutional buildings subjected to wheeled vehicles, including trolleys)
- C35 (Stairs, landings in all buildings not covered by C31 and C32, including hotels and motels and institutional buildings subjected to crowds)
- C36 (Walkways – light duty (access suitable for one person, walkway width approx 600 mm))
- C37 (Walkways – general duty (regular two-way pedestrian traffic))
- C38 (Walkways – heavy duty (high-density pedestrian traffic including escape routes))
- C39 (Museum floors and art galleries for exhibition purposes)

(6) Category C4 (Areas with possible physical activities) is sub-categorised into two sub-categories:

- C41 (Dance halls and studios, gymnasia, stages)
- C42 (Drill halls and drill rooms)

(7) Category C5 (Areas susceptible to large crowds) is sub-categorised into two sub-categories:

- C51 (Assembly areas without fixed seating, concert halls, bars and places of worship)
- C52 (Stages in public assembly areas)

(8) Category D in the UK NA to EN 1991-1-1 is similar to EN 1991-1-1.

UK NA to EN 1991-1-1: Table 6.1(BS)

The sub-categorisation is similar to BS 6399: Part 1[14]. The necessity of such a heavy sub-categorisation is debatable but given here to inform the designer.

cl.6.3.1.1(2)P: 1991-1-1

For guidance on situations where dynamic effects need to be considered where it is anticipated that the occupancy will cause significant dynamic effects, see Section 2.2 of this Designers' Guide.

6.3.1.2. Values of actions

cl.6.3.1.2(1)P: 1991-1-1

For each of the categories of loaded areas given in Table 6 of EN 1991-1-1 and described in Section 6.3.1.1 of this part of the Designers' Guide, Table 6.1 of EN 1991-1-1 specifies characteristic values q_k (uniformly distributed load) and Q_k (concentrated load), which, as either confirmed or modified by the National Annex must be used for the design of members of structures for the appropriate categories.

In Table 6.2 of EN 1991-1-1 a range of values for q_k and Q_k are given for each category. The actual value to be used for each appropriate country is set by the National Annex. In EN 1991-1-1 the recommended values within the range are underlined. Table 6.2 of this Designers' Guide gives an example to illustrate the style of Table 6.2 of EN 1991-1-1.

Notations q_k and Q_k are intended for separate application. Notation q_k is intended for determination of general effects and Q_k for local effects. The National Annex may define different conditions of use of this table. See in particular the BSI National Annex.[15]

The values given in Table 6.2 of EN 1991-1-1 should be regarded as minimum values and when considered necessary, q_k and Q_k should be increased in the design. This can apply for stairs, which are also used for emergencies where, depending on the occupancy of a building, there may be crowding of people under emergency-type situations (e.g. a fire alarm).

Table 6.2. Example to illustrate the style of Table 6.2 of EN 1991-1-1. Imposed loads on floors, balconies and stairs in buildings

Categories of loaded areas	q_k (kN/m^2)	Q_k (kN)
Category A		
Category B		
Category C		
Category D		
• D1	**4.0** to 5.0	3.5 to 7.0 **(4.0)**
• D2	4.0 to **5.0**	3.5 to **7.0**

Note: Bold type indicates recommended values

Balconies will also attract larger loads, for example at social gatherings, or watching street events, where clustering of people is common. *cl.6.3.1.2(2): 1991-1-1*

When verifying for local effects, the concentrated load Q_k from Table 6.2 of EN 1991-1-1, acting alone, should be used. *cl.6.3.1.2(3): 1991-1-1*

Concentrated loads should be assumed to act at positions on a member to give the greatest moment, shear or deflection. Concentrated loads should be applied to individual members and assumed to act on them unless there is evidence that adequate interaction exists to ensure that the load can be shared or spread (e.g. a concrete slab with adequate depth).

No comment necessary for ***Clause 6.3.1.2(4): 1991-1-1***. *cl.6.3.1.2(4): 1991-1-1*

When used for the calculation of local effects such as crushing and punching, the appropriate concentrated load Q_k should be assumed to act at a position, and over an area of application of a floor, balcony or stair, appropriate to its cause. Where this cannot be foreseen, both the ***Note to Clause 6.3.1.2(5): 1991-1-1*** and BS 6399: Part 1[14] recommends a square contact area with a 50 mm side. *Note to cl.6.3.1.2(5): 1991-1-1* *cl.6.3.1.2(5)P: 1991-1-1*

For specific applications, for example for traffic loads and helicopters, EN 1991-1-1 gives specific recommendations (see Sections 6.3.2.4, 6.3.3 and 6.3.4 of this part of this Designers' Guide).

No comment is necessary for ***Clause 6.3.1.2(6)P: 1991-1-1***. *cl.6.3.1.2(6)P: 1991-1-1*

In accordance with EN 1991-1-1, floors that may be subjected to multiple use have to be designed for the most unfavourable category of loading which produces the highest effects of actions (e.g. bending moment, shear force or deflection) for the member under consideration. *cl.6.3.1.2(7)P: 1991-1-1*

In EN 1991-1-1, movable partitions are considered as variable actions. Provided that a floor allows a lateral distribution of loads (e.g. a concrete floor), the self-weight of movable partitions is considered as a uniformly distributed load q_k which is added to the imposed loads of floors obtained from Table 6.2 of EN 1991-1-1. The value of q_k is dependent on the self-weight of the partitions. EN 1991-1-1 gives values for q_k for movable partitions with a self-weight: *cl.6.3.1.2(8): 1991-1-1*

- $\leq$1.0 kN/m
- $\leq$2.0 kN/m
- $\leq$3.0 kN/m wall length.

As an example:

- for movable partitions with a self-weight $\leq$2.0 kN/m wall length: $q_k = 0.8$ kN/m^2.

Values for q_k for other self-weighs of movable partitions are given in ***Clause 6.3.1.2(8): 1991-1-1***. *cl.6.3.1.2(8): 1991-1-1*

When selecting a value for the self-weight of the partition it would be prudent for the design to anticipate any alterations that may occur during the design working life if feasible.

For heavier partitions than those described above (i.e. >3.0 kN/m wall length), the design should take account of:

Table 6.3. Comparison between EN 1991-1-1 and the UK National Annex for α_A

A (m^2)	α_A (EN 1991-1-1 with $\psi_0 = 0.7$)	α_A (EN 1991-1-1 with $\psi_0 = 1.0$)	α_A (BSI NA for EN 1991-1-1)
40	0.75	0.96	0.96
80	0.63	0.84	0.92
120	0.59	0.80	0.88
160	0.56	0.78	0.84
240	0.54	0.76	0.76

- the locations and directions of the partitions
- the structural form of the floors.

cl.6.3.1.2(9): 1991-1-1

As described in Clause 6.2.1 of this Designers' Guide, the imposed loads on an appropriate member (e.g. a beam, floors, and accessible roofs) may be reduced according to the magnitude of the area supported, by a reduction factor α_A.

This reduction factor α_A may only be applied to the q_k values for imposed loads in the following Tables in EN 1991-1-1:

- Tables 6.2 and 6.10 for floors (Categories A to D)
- Table 6.9 for accessible roofs (Category I).

The following expression (expression 6.1 of EN 1991-1-1) is recommended in the Note 1 to (*Clause 6.3.1.2(10): 1991-1-1*), for the determination of the value for the reduction factor α_A for categories A to D:

cl.6.3.1.2(10): 1991-1-1

$$\alpha_A = \frac{5}{7}\psi_0 + \frac{A_0}{A} \leq 1.0 \quad \text{(Expression 6.1 of EN 1991-1-1)}$$

with the restriction for categories C and D: $\alpha_A \geq 0.6$

where:

ψ_0 is the factor according to EN 1990 Annex A1 Table A1.1
$A_0 = 10.0\ m^2$
A is the loaded area.

Note 2 to (*Clause 6.3.1.2(10): 1991-1-1*) allows the National Annex to give an alternative method.

cl.NA 2.5 UK NA to 1991-1-1

The National Annex for use in the UK substitutes expression 6.1 of EN 1991-1-1 by the expression below (expression NA.1 from UK National Annex to EN 1991-1-1) for the determination of the reduction factor α_A.

$$\alpha_A = 1.0 - A/1000 \geq 0.75 \quad \text{(Expression NA.1 from UK NA to EN 1991-1-1)}$$

where A is the area (m^2) supported.

It is emphasised that *Clause NA 2.5 UK NA to EN 1991-1-1* states that '*Loads that have been specifically determined from knowledge of the proposed use of the structure do not qualify for reduction.*'

Table 6.3 shows a comparison between the EN 1991-1-1 and the UK National Annex values for α_A.

Figure 6.4 shows the comparison for α_A graphically for EN 1991-1-1 (CEN), the UK National Annex (UK) and current practices (i.e. before implementation of the Eurocodes) taken from the codes of practices for France (FR), Germany (DE), Czech Republic (CR) and Finland (FI).

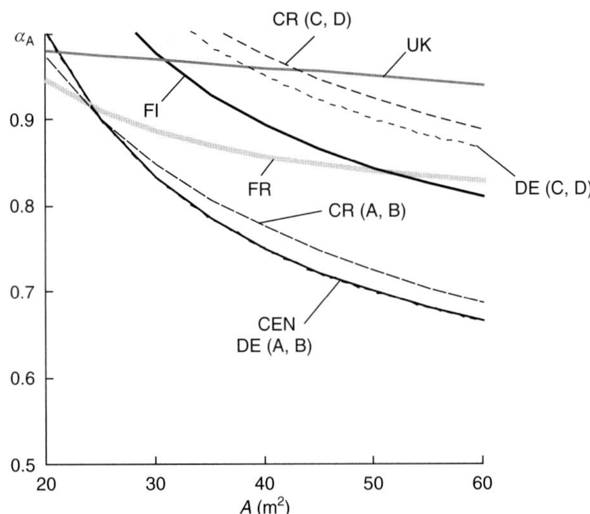

Fig. 6.4. Comparison of α_A values

As described in Clause 6.2.2 of this Designers' Guide, the imposed loads on columns and walls can be reduced by a reduction factor α_n. This reduction factor α_n may only be applied to the q_k values for imposed loads in the following Tables in EN 1991-1-1:

- Tables 6.2, for floors (Categories A to D).

The following expression (expression 6.2 of EN 1991-1-1) is recommended in Note 1 to (*Clause 6.3.1.2(11): 1991-1-1*), for the determination of the value for the reduction factor α_n, where the total imposed loads from several storeys may be multiplied by the reduction factor α_n. *Note to cl.6.3.1.2(11): 1991-1-1*

$$\alpha_n = \frac{2 + (n-2)\psi_0}{n} \quad \text{(Expression 6.2 of EN 1991-1-1)}$$

where:

n is the number of storeys (>2) above the loaded structural elements from the same category
ψ_0 is in accordance with EN 1990, Annex A1, Table A1.1.

Note 2 to (*Clause 6.3.1.2(11): 1991-1-1*) allows the National Annex to give an alternative method.

The National Annex for use in the UK substitutes expression 6.2 of EN 1991-1-1 by the expression below (expression NA.2 from UK National Annex to EN 1991-1-1) for the determination of the reduction factor α_n. *cl.NA 2.6 UK NA to 1991-1-1*

$$\begin{aligned} \alpha_n &= 1.1 - n/10 && \text{for } 1 \le n \le 5 \\ \alpha_n &= 0.6 && \text{for } 5 < n \le 10 \quad \text{(Expression NA.2 from UK NA)} \\ \alpha_n &= 0.5 && \text{for } n > 10 \end{aligned}$$

where n is the number of storeys with loads qualifying for reduction.

Clause NA 2.6 UK NA to EN 1991-1-1 states that '*Loads that have been specifically determined from knowledge of the proposed use of the structure do not qualify for reduction*'. *cl.NA 2.6 UK NA to 1991-1-1*

Table 6.4. Comparison between EN 1991-1-1 and the UK National Annex for α_n

n (number of storeys of building)	α_n (EN 1991-1-1); $\psi = 0.7$	α_n (BSI NA for EN 1991-1-1)
1	1	1
2	1	0.9
3	0.9	0.8
4	0.85	0.7
5	0.82	0.6
6	0.8	0.6
7	0.79	0.6
8	0.78	0.6
9	0.77	0.6
10	0.76	0.6

Table 6.4 shows a comparison between the EN 1991-1-1 and the UK National Annex values for α_n.

Figure 6.5 shows the comparison for α_n graphically for EN 1991-1-1 (CEN), the UK National Annex (UK) and current practices taken from the codes of practice (i.e. before implementation of the Eurocodes) for France (FR), Germany (DE), and the Czech Republic (CR).

No advice is given in EN 1991-1-1 on whether the reduction factors α_A and α_n can both be used simultaneously. The UK National Annex gives the following guidance:

- Load reductions based on area may be applied if $\alpha_A < \alpha_n$.
- However, the reductions given by α_A cannot be used in combination with those determined for α_n.

The following advice is recommended by the authors.

Use of α_A for the reduction of values of q_k for the design of beams, floors, roofs etc. for appropriate floor area.

The total imposed loads from several storeys, can be reduced by α_n. However, the value of the imposed loads from floors etc. should be the value of q_k before any reduction by α_A.

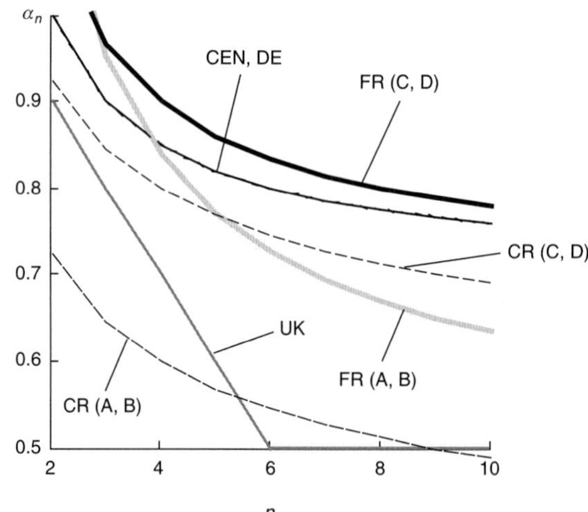

Fig. 6.5. Comparison of α_n values

6.3.2. Areas for storage and industrial activities

6.3.2.1. Categories

Areas for storage and industrial activities may be separated into the two categories in EN 1991-1-1 as follows:

- Category E1 for areas susceptible to accumulation of goods, including access areas (e.g. areas for storage use including storage of books and other documents).
- Category E2 for Industrial use.

cl.6.3.2.1(1)P: 1991-1-1

Table 6.5. Categories of storage and industrial use from the UK National Annex to EN 1991-1

Category	Specific use	Examples (sub-categories)
E1	Areas susceptible to accumulation of goods, including access areas	E1: General areas for static equipment not specified elsewhere (institutional and public buildings) E11: Reading rooms with book storage, e.g. libraries E12: General storage other than those specified E13: File rooms, filing and storage space (offices) E14: Stack rooms (books) E15: Paper storage for printing plants and stationery stores E16: Dense mobile stacking (books) on mobile trolleys, in public and institutional buildings E17: Dense mobile stacking (books) on mobile trucks, in warehouses E18: Cold storage
E2	Industrial use	

6.3.2.2. Values for actions

For the loaded areas, as categorised E1 and E2, EN 1991-1-1 gives characteristic values q_k (uniformly distributed load) and Q_k (concentrated load). q_k is intended for determination of general effects and Q_k for local effects.

For Category E1, EN 1991-1-1 provides only one set of values, namely $q_k = 7.5\,\text{kN/m}^2$ and $Q_k = 7.0\,\text{kN}$. No further advice is provided in EN 1991-1-1. *cl.6.3.2.2(1)P: 1991-1-1*

However, EN 1991-1-1 states that the recommended values for q_k and Q_k given above may be changed if necessary according to the usage (e.g. for storage it refers to Annex A of EN 1991-1-1) for the particular project or by the National Annex. It is also stated the National Annex may define different conditions of use. *Note to cl.6.3.2.2(1)P: 1991-1-1*

The UK National Annex expands the limited information given in EN 1991-1-1 for Categories E1 and E2 with the information in Table 6.5 (reproduced from ***Table 6.3: National Annex to BS EN 1991-1-1***). *Table 6.3: to 1991-1-1*

Table 6.6a shows a representation of imposed floor load values due to storage from the UK National Annex. The other values are in the UK National Annex (Table 6.4: National Annex to BS EN 1991-1-1). *Table 6.4: to 1991-1-1*

BS EN 1991-1-1: 2002, 6.3.2.2 does not specify values for imposed loads on floors for areas of industrial use (i.e. Category E2 from BS EN 1991-1-1: 2002, Table 6.3). *cl.6.3.2.2(1): 1991-1-1*

However, BS EN 1991-1-1: 2002, 6.1(4) does state that loads for heavy equipment (e.g. in communal kitchens, radiology rooms and boiler rooms) should be agreed between the client and/or the relevant authority. *cl.6.1(4): 1991-1-1*

Table 6.6a. Imposed floor loads due to storage from the UK National Annex to EN 1991-1-1

Category of loaded areas	q_k (kN/m^2)	Q_k (kN)
Category E1	2.0	1.8
• E11	–	–
• E12	2.4 for each metre of storage height	7.0
• E13	5.0	4.5
• E14	–	–
• E15	–	–
• E16	–	–
• E17	–	–
• E18	5.0 for each metre of storage height but with a minimum of 15.0	9.0

Table 6.6b. Imposed loads on floors for areas of industrial use

Category of loaded area	Examples of specific use	q_k (kN/m^2)	Q_k (kN)
E2 (industrial use)	Communal kitchens except those covered by occupancy class A in Table NA.3	3.0	4.5
	Operating theatres, X-ray rooms, utility rooms	2.0	4.5
	Work rooms (light industrial) without storage	2.5	1.8
	Kitchens, laundries, laboratories	3.0	4.5
	Rooms with mainframe computers or similar equipment	3.5	4.5
	Machinery halls, circulation spaces therein	4.0	4.5
	Cinematographic projection rooms	5.0	To be determined for specific use
	Factories, workshops and similar buildings (general industrial)	5.0	4.5
	Foundries	20.0	To be determined for specific use
	Catwalks	–	1.0 at 1 m centres
	Fly galleries (i.e. access structures used in theatres to hang scenery, curtains, etc.)	4.5 kN/m run distributed uniformly over width	–
	Ladders	–	1.5 rung load

In the absence of such agreement, the minimum imposed loads given in Table 6.6b above, reproduced from BS PD 6688-1[16] may be used for Category E2 loaded areas.

cl.6.3.2.(2)P: 1991-1-1

For industrial activities the characteristic value of the imposed load that is taken in the design should be the maximum value taking account of the dynamic effects when appropriate (e.g. for forklifts, see 6.3.2.3 and for other transport vehicles see 6.3.2.4). The loading arrangement needs to produce the most unfavourable conditions allowed in use.

Note to cl.6.3.2.(2)P: 1991-1-1

No further advice required for the Note to Clause 6.3.2.(2)P: 1991-1-1.

cl.6.3.2.(3): 1991-1-1

In the absence of the information in EN 1991-1-1 or its appropriate National Annex, the characteristic values of vertical loads in storage areas should be derived by taking into account the density of the storage item which can generally be obtained from Annex A of EN 1991-1-1 and the upper design values for stacking heights. When stored material exerts horizontal forces on walls etc. (e.g. in silos), the horizontal force can be determined in accordance with EN 1991-4.

cl.6.3.2.(4): 1991-1-1

The design needs to take into account the stabilising and de-stabilising effects of the effects of filling and emptying.

cl.6.3.2.(5): 1991-1-1

As for example specified for Category E17 in the BSI National Annex to EN 1991-1-1 (see Table 6.6 of this Designers' Guide), loads for storage areas for books and other documents should be determined from the loaded area and the height of the book cases using the appropriate values for density from Annex A of EN 1991-1-1 (***Clause 6.3.2.(5): 1991-1-1***).

cl.6.3.2.(6): 1991-1-1

No further guidance required for ***Clause 6.3.2.(6): 1991-1-1***.

cl.6.3.2.(7): 1991-1-1

Actions due to industrial vehicles (e.g. forklifts and transport vehicles) should be considered as concentrated loads as described below. Where these industrial vehicle loads occur simultaneously with the categories of loads which are given in Tables 6.2, 6.4. and 6.8 of EN 1991-1-1 (see Table 6.2, Clause 6.3.2 and Table 6.10 of this Designers' Guide), the appropriate imposed distributed loads given in Tables 6.2, 6.4 and 6.8 need only be used with the concentrated loads from the forklifts (***Clause 6.3.2.(7): 1991-1-1***).

Table 6.7. Dimensions of forklift according to classes FL

Class of forklift	Net weight (kN)	Hoisting load (kN)	Width of axle *a* (m)	Overall width *b* (m)	Overall length *l* (m)
FL 1	21				
FL 2	31				
FL 3	44				
FL 4	60	40	1.20	1.40	4.00
FL 5	90	60	1.50	1.90	4.60
FL 6	110				

6.3.2.3. *Actions induced by forklifts*

EN 1991-1-1 describes six forklifts classes, namely FL 1 to FL 6, depending on their net weight, dimensions and hoisting loads. Table 6.7 provides a limited form of Table 6.5 of EN 1991-1-1. *cl.6.3.2.3(1): 1991-1-1*

The static vertical axle load Q_k of a forklift is dependent upon the class of the forklifts FL 1 to FL 6 and is obtained from Table 6.8 below which reproduces Table 6.6 of EN 1991-1-1. *cl.6.3.2.3(2): 1991-1-1*

To determine the dynamic characteristic value of the action, the static vertical axle load Q_k, obtained from Table 6.8 (Table 6.6 of EN 1991-1-1) needs to be increased by the dynamic factor φ using expression (6.3) of EN 1991-1-1. *cl.6.3.2.3(3): 1991-1-1*

$$Q_{k,dyn} = \varphi Q_k \quad \text{(Expression 6.3 of EN 1991-1-1)}$$

where:

$Q_{k,dyn}$ is the dynamic characteristic value of the action
φ is the dynamic magnification factor
Q_k is the static characteristic value of the action.

Notation φ for forklifts takes into account the inertial effects caused by acceleration and deceleration of the hoisting load and is be taken as in EN 1991-1-1 as follows:

$\varphi = 1.40$ for pneumatic tyres *cl.6.3.2.3(4): 1991-1-1*
$\varphi = 2.00$ for solid tyres.

When the type of tyre is not known at the design stage, $\varphi = 2.00$ should be used.

For forklifts having a net weight greater than 110 kN, EN 1991-1-1 recommends that the dynamic loads should be defined by a more accurate analysis. A good source would be the manufacturers of forklifts. *cl.6.3.2.3(5): 1991-1-1*

It is recommended that details of specific forklifts be checked with the clients or the manufacturers in case they differ from those described in Table 6.7.

The load arrangements for the vertical axle load Q_k and $Q_{k,dyn}$ of a forklift should be in accordance with Fig. 6.6 (reproduced from Fig. 6.1 of EN 1991-1-1). All measurements in the diagram are in metres. *cl.6.3.2.3(6): 1991-1-1*

For situations where the design needs to consider the horizontal loads due to acceleration or deceleration of forklifts, EN 1991-1-1 recommends that these are taken as 30% of the vertical axle loads Q_k given in Table 6.8 below which reproduces Table 6.6 of EN 1991-1-1. Dynamic magnification factors are not needed to be applied to horizontal loads for forklifts (Note Clause 6.3.2.3(7): 1991-1-1). *cl.6.3.2.3(7): 1991-1-1*

6.3.2.4. *Actions induced by transport vehicles*

EN 1991-1-1 provides only general information on actions induced by transport vehicles, and these are outlined below.

Table 6.8. Axle loads of forklifts

Class of forklifts	Axle load Q_k (kN)
FL 1	26
FL 2	40
FL 3	63
FL 4	90
FL 5	140
FL 6	170

cl.6.3.2.4(1): 1991-1-1 The actions from transport vehicles that move on floors freely or guided by rails need to be determined considering the pattern of wheel loads.

cl.6.3.2.4(2): 1991-1-1 The static values of the vertical wheel loads are defined in terms of permanent weights and payloads. Their spectra should be used to define combination factors and fatigue loads.

cl.6.3.2.4(3): 1991-1-1 The vertical and horizontal wheel loads need to be determined for each specific case.

cl.6.3.2.4(1): 1991-1-1 The load arrangements including the dimensions relevant for the design are determined for the specific case.

Note to cl.6.3.2.4(1): 1991-1-1 Appropriate load models from EN 1991-2 may be used where relevant. In the absence of any information it is recommended that contact areas for loading be obtained from vehicle manufacturers or hirers.

6.3.2.5. Actions induced by special devices for maintenance

EN 1991-1-1 provides only general information on actions induced by special devices for maintenance.

cl.6.3.2.5(1): 1991-1-1 These special devices for maintenance can be modelled as loads from transportation vehicles, as explained above in Section 6.3.2.4.

cl.6.3.2.5(2): 1991-1-1 As for loads from transportation vehicles, the load arrangements including the dimensions relevant for the design should be determined for the specific case.

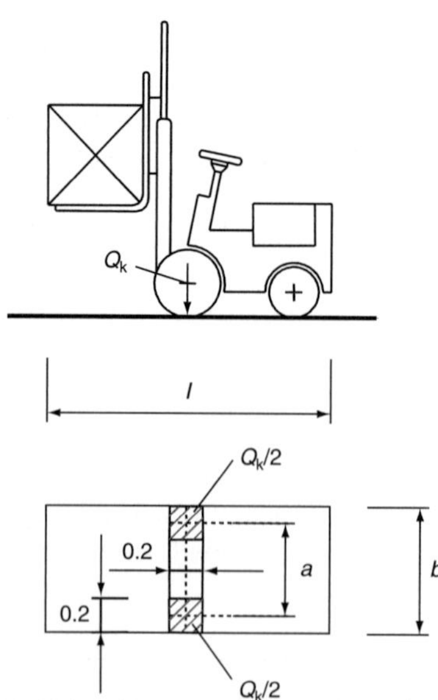

Fig. 6.6. Dimensions of forklifts

Table 6.9. Traffic and parking areas in buildings

Categories of traffic areas	Specific use	Examples
F	Traffic and parking areas for light vehicles ($\leq$30 kN gross vehicle weight and $\leq$8 seats not including driver)	Garages, parking areas, parking halls
G	Traffic and parking areas for medium vehicles (>30 kN, $\leq$160 kN gross vehicle weight, on 2 axles)	Access routes, delivery zones, zones accessible to fire engines ($\leq$160 kN gross vehicle weight)

6.3.3. Garages and vehicle traffic areas (excluding bridges)

6.3.3.1. Categories

Traffic and parking areas in buildings are required to be divided into two categories according to their specified accessibility for vehicles as shown in Table 6.9 (reproduced from Table 6.7 of EN 1991-1-1. *cl.6.3.3.1(1): 1991-1-1*

Access to areas designed to Category F should be limited by physical means built into the structure (Note 1 of Table 6.7 EN 1991-1-1).

Consideration should be given to posting appropriate warning signs for areas designed to Categories F or G (Note 2 of Table 6.7 EN 1991-1-1).

6.3.3.2. Values of actions

The load model which has to be used is:

- a single axle with a load with a total concentrated load Q_k with dimensions according to Fig. 6.7 (***Figure 6.2 of EN 1991-1-1***); and *Figure 6.2 1991-1-1*
- a uniformly distributed load q_k.

The recommended characteristic values for q_k and Q_k are given in Table 6.10 (***Table 6.8 of EN 1991-1-1***), where q_k and Q_k are Nationally Determined Parameters. *Table 6.8 and cl.6.3.3.2(1): 1991-1-1*

In accordance with EN 1991-1-1, q_k is intended for determination of general effects and Q_k for local effects. *Note to cl.6.3.2.5(1): 1991-1-1*

EN 1991-1-1 does not give guidance as to whether q_k and Q_k should be considered in simultaneity or not. However, EN 1991-1-1 does state that the National Annex may define different conditions of use of this table. *Note to cl.6.3.2.5(1): 1991-1-1*

The axle load should be applied as shown in Fig. 6.7 in the positions which will produce the most adverse effects of the action. *cl.6.3.3.2(2): 1991-1-1*

The BSI National Annex for EN 1991-1-1 has selected the recommended value for q_k and but different values for Q_k than those recommended (***Clause NA2.9: BS EN 1991-1-1***) (***Table NA6: BS EN 1991-1-1-1***). *Table 3.4 and cl.6.3.3.2(2): 1991-1-1*

The BSI National Annex for EN 1991-1-1 requires that q_k and Q_k are not applied simultaneously. *cl.NA2.9: and Table NA6 1991-1-1*

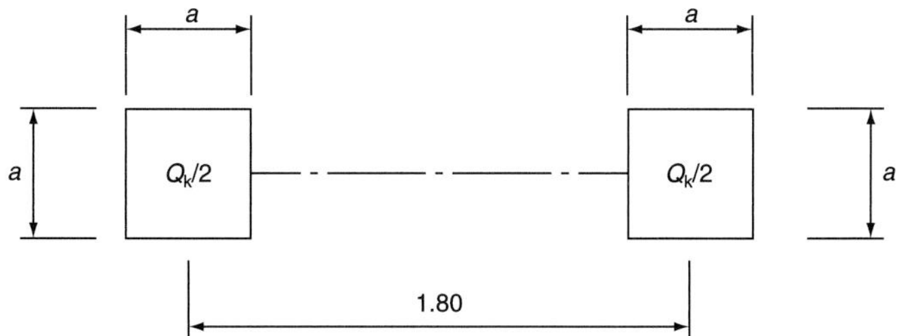

Fig. 6.7. Dimensions of axle load (for Category F, a = 100 mm and for Category G a + 200 mm (see Table 6.10 below) (***Note of Table 6.7 EN 1991-1-1***))

Table 6.10. Imposed loads on garages and vehicle traffic areas

Categories of traffic areas	q_k (kN/m^2)	Q_k (kN)
Category F Gross vehicle weight: ≤30 kN	q_k	Q_k
Category G 30 kN < gross vehicle weight ≤160 kN	5.0	Q_k

Note 1: For Category F, q_k may be selected within the range 1.5 to **2.5** kN/m^2 and Q_k may be selected within the range 10 to **20** kN.
Note 2: For Category G, Q_k may be selected within the range 40 to **90** kN.
Note 3: Where a range of values is given in Notes 1 and 2, the value may be set by the National Annex. The recommended values are in bold type.

Regarding the information given in Table 6.10 for Category F, $Q_k = 10$ to 20 kN represents the side of a vehicle being jacked to change a wheel. For Category G a 160 kN vehicle represents a fire engine.

6.3.4. Roofs

6.3.4.1. Categories

cl.6.3.4.1(1)P: 1991-1-1
cl.6.3.4.1(2): and cl.6.3.4.1(3): 1991-1-1

Roofs need to be categorised according to their accessibility into three categories as shown in Table 6.11 (based on Table 6.9 of EN 1991-1-1).

No further information required for Clause 6.3.4.1(2).

No further information required for Clause 6.3.4.1(3).

6.3.4.2. Values of actions

Category H roofs

Table 6.10 1991
cl.6.3.4.2(1): 1991-1-1

For roofs of Category H (i.e. roofs not accessible except for normal maintenance and repair) the minimum characteristic values q_k and Q_k that have to be used are given in Table 6.12. These load values are related to the projected area of the roof under consideration (***Clause 6.3.4.2(1): 1991-1-1***).

In Table 6.12 q_k and Q_k are Nationally Determined Parameters.

It is recommended by EN 1991-1-1 for Category H roofs that q_k may be selected within the range 0.00 kN/m^2 to 1.0 kN/m^2 and Q_k may be selected within the range 0.9 kN to 1.5 kN in accordance with Note 1 of Table 6.10 of EN 1991-1-1.

The recommended values for these Nationally Determined Parameters are:

$$q_k = 0.4\,\text{kN/m}^2, \quad Q_k = 1.0\,\text{kN}$$

EN 1991-1-1 allows q_k to be varied by the National Annex dependent upon the roof slope in accordance with Note 2 of Table 6.10 of EN 1991-1-1. It should be noted that this reduction applies to q_k only. For reductions for snow loads on pitched roofs see the Part of this Designers' Guide relating to snow loads.

q_k may be assumed to act on an area A which may be set by the National Annex. The recommended value for A is 10 m^2, within the range of zero to the whole area of the roof in accordance with Note 3 of Table 6.10 of EN 1991-1-1.

Table 6.11. Categorisation of roofs

Categories of loaded area	Specific use
H	Roofs not accessible except for normal maintenance and repair (see Table 6.12)
I	Roofs accessible with occupancy according to Categories A to G (see Tables 6.2, 6.6 and 6.10)
K	Roofs accessible for special services, such as helicopter landing areas (see Table 6.13)

Table 6.12. Imposed loads on roofs of Category H

Roof	q_k (kN/m^2)	Q_k (kN)
Category H	q_k	Q_k

As an example of National Choices the BSI National Annex specifies values of q_k and Q_k different from the EN 1991-1-1 recommended values given above. Furthermore, for more pitched roofs the uniformly distributed load q_k which is specified at 0.6 kN/m^2 for pitches of up to 30° reduces to $q_k = 0$ for roofs of pitch greater that 60° in the UK National Annex. A value of $Q_k = 0.9$ kN is selected. It is debatable whether a reduction for increasing roof slope can be justified for imposed loads considering that they are already made for snow loads.

cl.NA.2.10 of 1991-1-1

Regarding Note 3 to Table 6.10, it is recommended by the BSI National Annex that the value of A should be the whole area of the roof (***Clause NA.2.10 of BS EN 1991-1-1: 2002***).

cl.NA.2.10 of 1991-1-1

The minimum values given in Table 6.12 (Table 6.10 of EN 1991-1-1) or those specified in a National Annex do not take into account uncontrolled accumulations of construction materials that may occur during maintenance (***Clause 6.3.4.1(2): 1991-1-1***). Further information on these loads may be obtained from EN 1991-1-6: *Actions during execution* and described in the Part relating to EN 1991-1-6 of this Designers' Guide (***Note to Clause 6.3.4.1(2): 1991-1-1***).

cl.6.3.4.1(2): and Note to cl.6.3.4.1(2): 1991-1-1

Roofs – all categories

To ensure that the worst case for the effects of actions are taken into account in the design of roofs, separate verifications need to be performed for the concentrated load Q_k and the uniformly distributed load q_k, acting independently.

cl.6.3.4.1(3)P: 1991-1-1

Roofs, other than those with roof sheeting, need to be designed to resist 1.5 kN on an area based on a 50 mm sided square. Roof elements with a profiled or discontinuously laid surface, should be designed so that the concentrated load Q_k acts over the effective area provided by load-spreading arrangements.

cl.6.3.4.1(4): 1991-1-1

Category K roofs

For Category K roofs, the actions from helicopters on landing areas are specified in EN 1991-1-1 Table 6.11 reproduced in this Guide as Table 6.13. Dynamic amplification factors (see below) have to be applied to the characteristic load values given for the two classes of helicopter in Table 6.13.

cl.6.3.4.1(5): 1991-1-1

A dynamic factor φ has to be applied to the characteristic take-off load Q_k to take account of impact effects. EN 1991-1-1 specifies $\varphi = 1.40$.

cl.6.3.4.1(6): 1991-1-1

Roofs – general

Access ladders and walkways are assumed to be loaded according to Table 6.12 (***Table 6.10 of EN 1991-1-1***) or the appropriate clause in a National Annex for a roof slope $<20°$. Therefore considering the recommended values for Table 6.12 of EN 1991-1-1 these values are:

Table 6.10 1991-1-1

$$q_k = 0.4\,\text{kN/m}^2, \quad Q_k = 1.0\,\text{kN}$$

Table 6.13. Imposed loads on roofs of Category K for helicopters

Class of helicopter	Take-off load Q of helicopter	Take-off load Q_k	Dimension of the loaded area (m × m)
HC1	$Q \leq 20$ kN	$Q_k = 20$ kN	0.2 × 0.2
HC2	$20\text{ kN} < Q \leq 60$ kN	$Q_k = 60$ kN	0.3 × 0.3

For walkways which are part of a designated escape route, q_k should be according to Table 6.2 of EN 1991-1-2 and this Designers' Guide. Therefore for a roof acting as a roof restaurant considering the recommended values from Table 6.2 of EN 1991-1-1, $q_k = 3.0\,kN/m^2$. EN 1991-1-1 does not recommend a value to use for Q_k for walkways designated as escape routes for Category I roofs. It is considered reasonable to use the recommended values in Table 6.2.

cl.6.3.4.1(7): 1991-1-1

However, for walkways for service (e.g. for maintenance access) a *minimum* characteristic value for Q_k of 1.5 kN is specified by EN 1991-1-1.

For the design of frames and coverings of access hatches (other than glazing), the supports of ceilings and similar structures, EN 1991-1-1 specifies the following:

cl.6.3.4.1(8): 1991-1-1

- Without access: no imposed load.
- With access: $0.25\,kN/m^2$ distributed over the whole area or the area supported, and the concentrated load of 0.9 kN so placed so as to produce maximum stresses in the affected member.

Horizontal loads on parapets and partition walls acting as barriers

Table 6.12 1991-1-1 cl.6.4.(1)P: 1991-1-1

The characteristic values of the line load q_k acting at the height of the partition wall or parapets but not higher than 1.20 m are given in Table 6.14 reproduced from ***Table 6.12 of EN 1991-1-1***. It will be apparent from Table 6.14 that every line load value is a Nationally Determined Parameter and needs to be selected by the appropriate National Annex.

Table 6.12 1991-1-1 cl.6.4.(2): 1991-1-1

As guidance to Table 6.14 (***Table 6.12 of EN 1991-1-1***) EN 1991-1-1 reminds the user that for areas susceptible to significant overcrowding associated with public events, e.g. for sports stadia, stands, stages, assembly halls or conference rooms, the line load should be taken according to Category C5.

Table 6.12 1991-1-1 Table NA8 from 1991-1-1 cl.NA.2.11 1991-1-1: 2002

The British National Annex sub-categorises Categories A; B and C1; C2 to C4 and D; and C5 from Table 6.14 (***Table 6.12 of EN 1991-1-1***). See Table 6.15 (***Table NA8 from BS EN 1991-1-1***). The table only gives partial information on the loads, which can be obtained from BS EN 1991-1-1.

Table 6.14. Horizontal loads on partition walls and parapets

Loaded areas	q_k (kN/m)
Category A	q_k
Category B and C1	q_k
Categories C2 to C4 and D	q_k
Category C5	q_k
Category E	q_k
Category F	See Annex B
Category G	See Annex B

Note 1: For Categories A, B and C1, q_k may be selected within the range 0.2 to 1.0 (**0.5**).

Note 2: For Categories C2 to C4 and D, q_k may be selected within the range 0.8 kN/m to **1.0** kN/m.

Note 3: For Category C5, q_k may be selected within the range **3.0** kN/m to 5.0 kN/m.

Note 4: For Category E, q_k may be selected within the range 0.8 kN/m to **2.0** kN/m. For areas of Category E, the horizontal loads depend on the occupancy. Therefore the value of q_k is defined as a minimum value and should be checked for the specific occupancy.

Note 5: Where a range of values is given in Notes 1, 2, 3 and 4, the value may be set by the National Annex. The recommended value is shown in bold face.

Note 6: The National Annex may prescribe additional point loads Q_k and/or hard or soft body impact specifications for analytical or experimental verification.

Table 6.15. Horizontal loads on partition walls and parapets (from BS EN 1991-1-1)

Category of loaded area	Sub-category	Examples	q_k (kN/m)
A (including sub-categories in Table NA.2)	(i)	All areas within or serving exclusively one dwelling including stairs, landings etc. but excluding external balconies and edges of roofs [see (vii)]	
	(ii)	Residential areas not covered by (i)	
B and C1 (including sub-categories in Table NA.2)	(iii)	Areas not susceptible to overcrowding in office and institutional buildings, reading rooms and classrooms including stairs	
	(iv)	Restaurants and cafés	
C2, C3, C4 and D (including sub-categories in Table NA.2)*	(v)	Areas having fixed seating within 530 mm of the barrier, balustrade or parapet	1.5
	(vi)	Stairs, landings, balustrades, corridors and ramps	0.74
	(vii)	External balconies and edges of roofs Footways within building curtilage and adjacent to basement/sunken areas	0.74
	(viii)	All retail areas	1.5
C5 (including sub-categories in Table NA.2)	(ix)	Footways or pavements less than 3 m wide adjacent to sunken areas Footways or pavements greater than 3 m wide adjacent to sunken areas	
	(x)	Theatres, cinemas, discotheques, bars, auditoria, shopping malls, assembly areas, studios	
	(xi)	Grandstands and stadia	See requirements of the appropriate certifying authority
E (including sub-categories in Table NA.4)	(xii)	Industrial and storage buildings except as given by (xiii) and (xiv)	
	(xiii)	Light pedestrian traffic routes in industrial and storage buildings except designated escape routes	
	(xiv)	Light access stairs and gangways not more than 600 mm wide	
F and G	(xv)	Pedestrian areas in car parks including stairs, landings, ramps, edges or internal floors, footways, edges of roofs	
	(xvi)	Horizontal loads imposed by vehicles	See BS EN 1991-1-1: 2002, Annex B

* For areas where large crowds might occur, see C5.

Table 6.15. Horizontal loads on partition walls and parapets (from EN 1991-1-1)

CHAPTER 7

Annex A (informative) Tables for nominal density of construction materials, and nominal density and angles of repose for stored materials

Annex A of EN 1991-1-1 is an informative annex which gives the following tables for nominal density of construction materials, and nominal density and angles of repose for stored materials. Annex A will mean that BS 648[17] will be withdrawn by the BSI in the UK. *Annex A 1991-1-1: 2002*

(1) Table A.1 – Construction materials – concrete and mortar
(2) Table A.2 – Construction materials – masonry
(3) Table A.3 – Construction materials – wood
(4) Table A.4 – Construction materials – metals
(5) Table A.5 – Construction materials – other materials
(6) Table A.6 – Bridge materials
(7) Table A.7 – Stored materials – building and construction
(8) Table A.8 – Stored products – agricultural
(9) Table A.9 – Stored products – foodstuffs
(10) Table A.10 – Stored products – liquids
(11) Table A.11 – Stored products – solid fuels
(12) Table A.12 – Stored products – industrial and general

For those materials where their density is dependent upon the moisture content, unless otherwise stated it may be assumed that the tabulated value includes an allowance for this for the situation in which the material is normally used.

For certain materials, means are given for converting dry densities to densities in situ, taking account of the potential for absorption of moisture.

For some materials the density has significant variability according to its source; for these items a range of representative values is given in the tables.

A representative value for densities of the more commonly used construction materials (from Tables A.1 to A.4) is given in Table 7.1 of this Designers' Guide.

Table 7.1. Examples of nominal density of some construction materials (from Tables A.1, A.3 and A.4 of EN 1991-1-1)

	Materials	Density γ (kN/m^3)
Table A.1 1991-1-1	Concrete (see EN 206)	
	Lightweight	
	Density class LC 1.0	9.0 to 10.0 (see (1) and (2))
	Density class LC 2.0	18.0 to 20.0 (see (1) and (2))
	Normal weight	24.0 (see (1) and (2))
	(1) Increase by 1 kN/m^3 for normal percentage of reinforcing and pre-stressing steel	
	(2) Increase by 1 kN/m^3 for unhardened concrete	
	Mortar	
	Cement mortar	19.0 to 23.0
	Materials	Density γ (kN/m^3)
Table A.3 1991-1-1	Wood (see EN 338 for timber strength classes)	
	Timber strength class C14	3.5
	Timber strength class C30	4.6
	Timber strength class D50	7.8
	Timber strength class D70	10.8
	Glued laminated timber (glulam) (see EN 1194 for timber strength classes)	
	Homogeneous glulam GL24h	3.7
	Homogeneous glulam GL36h	4.4
	Combined glulam GL24c	3.5
	Combined glulam GL36c	4.2
	Materials	Density γ (kN/m^3)
Table A.4 1991-1-1	Metals	
	Aluminium	27.0
	Iron, cast	71.0 to 72.5
	Iron, wrought	76.0
	Steel	77.0 to 78.5

CHAPTER 8

Annex B of EN 1991-1-1: Vehicle barriers and parapets for car parks

Annex B of E 1991-1-1: *Vehicle barriers and parapets for car parks* is an informative annex. Its use is required by the UK National Annex. *cl.NA.3.2 1991-1-1:*

The horizontal loads that barriers and parapets in car parking areas should be designed to resist are explained in this Annex B of EN 1991-1-1 and are explained below. *cl.B(1) 1991-1-1*

The horizontal characteristic force F (in kN), normal to and uniformly distributed over any length of 1.5 m of a barrier for a car park, required to withstand the impact of a vehicle is given by the expression below. *cl.B(2) 1991-1-1: 2002*

$$F = 0.5mv^2/(\delta_c + \delta_b)$$

where:

m is the gross mass of the vehicle (in kg)
v is the velocity of the vehicle (in m/s) normal to the barrier
δ_c is the deformations of the vehicle (in mm)
δ_b is the deformations of the barrier (in mm).

For car parks designed on the basis that the gross mass of the vehicles using the car park will not exceed 2500 kg, the following values are used to determine the force F:

$m = 1500$ kg
$v = 4.5$ m/s
$\delta_c = 100$ mm unless better evidence is available.

For a rigid barrier, for which δ_b may be given as zero, the characteristic force F appropriate to vehicles up to 2500 kg gross mass is taken as 150 kN. *cl.B(3) 1991-1-1*

For car parks designed on the basis that the gross mass of the vehicles using the car park can exceed 2500 kg, the following values are used to determine the force F: *cl.B(4) 1991-1-1*

$m =$ the actual mass of the vehicle for which the car park is designed (in kg)
$v = 4.5$ m/s
$\delta_c = 100$ mm unless better evidence is available.

The force F determined in either B(3) or B(4) of EN 1991-1-1 may be considered to act at bumper height of the vehicle. For vehicles whose gross mass does not exceed 2500 kg this height may be taken as 375 mm above the floor level. *cl.B(5) 1991-1-1*

Example

The graph below (Fig. 8.1) shows the variation of F for various deformations for the vehicle and the barrier for car of masses of 1500 and 2500 kg, assuming:

Thus for $m = 1500\,\text{kg}$, $\delta_b = 0$ and $\delta_c = 100$, $F = 150\,\text{kN}$
Thus for $m = 2500\,\text{kg}$, $\delta_b = 0$ and $\delta_c = 100$, $F = 250\,\text{kN}$

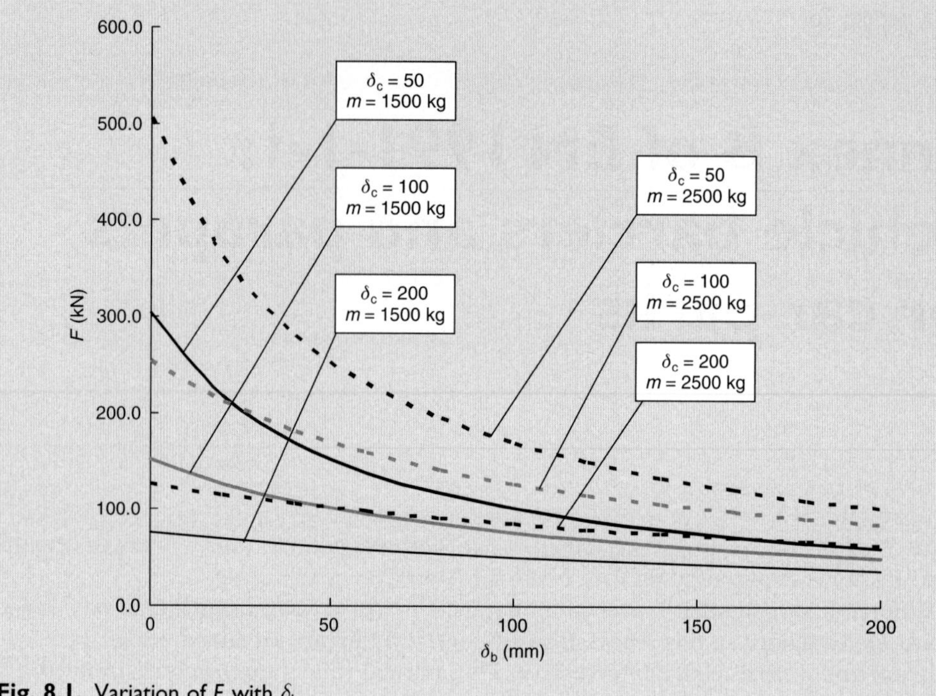

Fig. 8.1. Variation of F with δ_b

EN 1991-1-1 does not give guidance for bumper height for vehicles greater that 2500 kg. For car parks designed for these heavier vehicles the information may be obtained from vehicle manufacturers or EN 1991-1-7.

cl.B(6) 1991-1-1 For the design of barriers to access ramps of car parks, these have to withstand one half of the force determined in B(3) or B(4) of EN 1991-1-1, acting at a height of 610 mm above the ramp.

cl.B(7) 1991-1-1 Opposite the ends of straight ramps intended for downward travel which exceed 20 m in length, the barrier needs to withstand twice the force determined in B(3) acting at a height of 610 mm above the ramp (***Clause B(7) of BS EN 1991-1-1: 2002***).

References to Part I

1. EN 1991-1-1. *Eurocode 1. Actions on Structures – Part 1-1: General Actions – Densities, self-weight, imposed loads for buildings.* European Committee for Standardisation, Brussels, 2002.
2. *Guidance paper L – Application and Use of Eurocodes.* European Commission Enterprise Directorate-General ENTR/G5, Brussels, 27 November 2003.
3. ISO 2394: 1998. *General Principles on Reliability for Structures.* International Organisation for Standardisation, Geneva, 1998.
4. ISO 3898: 1997. *Bases for Design of Structures – Notations – General symbols.* International Organisation for Standardisation, Geneva, 1997.
5. ISO 8930: 1987. *General Principles on Reliability for Structures – List of equivalent terms.* International Organisation for Standardisation, Geneva, 1987.
6. Gulvanessian, H., Calgaro, J.-A. and Holický, M. *Designers' Guide to EN 1990. Eurocode: Basis of Structural Design.* Thomas Telford, London, 2002, ISBN 0 7277 3011 8.
7. ISO 9194: 1987. *Bases for Design of Structures – Actions Due to the Self-weight of Structures, Non-structural Elements and Stored Materials – Density.* International Organisation for Standardisation, Geneva, 1987.
8. Calgaro, J.-A., Tschumi, M. and Gulvanessian, H. *Designers' Guide to EN 1991 for Bridges.* Actions on Bridges. Thomas Telford, London, 2009.
9. Frank, R., Bauduin, C., Driscoll, R., Kavvadas, M., Krebs Ovesen, N., Orr, T. and Schuppener, B. *Designers' Guide to EN 1997-1. Eurocode 7: Geotechnical Design – General rules.* Thomas Telford, London, 2004, ISBN 9 780 7277 3154 8.
10. Building Research Establishment. *The Response of Structures to Dynamic Crowd Loads.* BRE Digest 246, BRE, Watford, 2004.
11. BS EN 1991-3: 2006. *Eurocode 1. Actions on Structures – Part 3: Actions induced by cranes and machinery.* European Committee for Standardisation, Brussels, 2006.
12. Ferry Norges, J. and Castanheta, M. *Structural Safety.* Laboratorio, Nacional de Engenharia Civil, Lisbon, 1971.
13. International Council for Research and Innovation in Building and Construction. Committee W81. *Actions on Structures. Live Loads in Buildings.* CIB report 116. CIB, Rotterdam, 1989.
14. BS 6399-1: 1996. *Loading for Buildings. Code of Practice for Dead and Imposed Loads.* British Standards Institution, London, 2006.
15. UK National Annex to EN 1991-1-1. *Eurocode 1. Actions on Structures – Part 1-1: General Actions – Densities, self-weight, imposed loads for buildings.* British Standards Institution, London, 2005.
16. PD 6688. *Background Paper to the UK National Annexes to BS EN 1991-1-1.* British Standards Institution, London, 2006.
17. BS 648: 1964. *Schedule of Weights of Building Materials.* British Standards Institution, London, 1964.

PART 2: EN 1991-1-2: Eurocode 1: Part 1.2: Actions on structures exposed to fire

CHAPTER 1

Eurocode 1 – Actions on structures: Part 1.2: General Actions – Actions on structures exposed to fire

EN 1991-1-1: *Eurocode 1 – Actions on structures: Part 1.2: General Actions – Actions on structures exposed to fire*[1] is one of the ten Parts of EN 1991. It describes the thermal and mechanical actions for the structural design of buildings exposed to fire, including the following aspects:

- safety requirements
- design procedures
- design aids.

This part of Eurocode 1 is outside the scope of this guide as it is comprehensively described in the *Designers' Guide to the Eurocode Fire Parts (EN 1991-1-2; EN 1992-1-2; EN 1991-1-3 and EN 1991-1-4)* by Colin Bailey *et al.*[2]

References

1. EN 1991-1-2: 2002. *Eurocode 1: Actions on Structures – Part 1-2: General Actions – Actions on structures exposed to fire*. European Committee for Standardisation, Brussels, 2002.
2. Moore, D., Bailey, C., Lennon, T. and Wang, Y. *Designers' Guide to the Eurocode Fire Parts (EN 1991-1-2; EN 1992-1-2; EN 1991-1-3 and EN 1991-1-4)*. Thomas Telford, London, 2007, ISBN 9 7807 2773 1579.

PART 3: EN 1991-1-3:
Eurocode 1:
Part 1.3: Snow loads

CHAPTER 1

General

This chapter is concerned with the general aspects of EN 1991-1-3: *Eurocode 1 – Actions on structures: Part 1.3: General Actions – Snow loads.*[1] The material described in this chapter is covered in the following clauses:

- Scope — *Clause 1.1: 1991-1-3*
- Normative references — *Clause 1.2: 1991-1-3*
- Assumptions — *Clause 1.3: 1991-1-3*
- Distinction between Principles and Application Rules — *Clause 1.4: 1991-1-3*
- Design assisted by testing — *Clause 1.5: 1991-1-3*
- Terms and definitions — *Clause 1.6: 1991-1-3*
- Symbols — *Clause 1.7: 1991-1-3*

1.1. Scope

EN 1991-1-3: *Eurocode 1 – Actions on structures: Part 1.3: General Actions – Snow loads* is one of the ten Parts of EN 1991. It gives design guidance to determine the values of loads due to snow for the structural design of buildings and other civil engineering works. *cl.1.1(1): 1991-1-3*

EN 1991-1-3 only applies for sites at altitudes below 1500 m. Comparing with national codes, in the UK, BS 6399 Part 3 is applicable for sites up to 500 m. For Italy the limit is the same as EN 1991-1-3. *cl.1.1(2): 1991-1-3*

In the case of altitudes above 1500 m, advice may be found in the appropriate National Annex. *Note to cl.1.1(2): 1991-1-3*

Differing climatic situations will give rise to different conditions. In EN 1991-1-3, the following have been identified:

- non-exceptional falls and drifts
- exceptional falls and drifts.

Depending on the climatic conditions at the site, four possible situations and load arrangements are identified in Annex A and explained below. *cl.1.1(3): 1991-1-3*

(1) Case A: non-exceptional fall and non exceptional drifts
(2) Case B1: exceptional fall and non-exceptional drifts
(3) Case B2: non-exceptional fall and exceptional drifts
(4) Case B3: exceptional fall and exceptional drifts.

The national competent authority may choose for inclusion in the National Annex, the case applicable to particular locations for their own territory. For example: *Note to cl.1.1(3): 1991-1-3*

Table 1.1. Recommended return periods for the determination of the characteristic values of climatic actions

Design working life	Return period (years) to be used in design
≤3 days	2
≤3 months (but >3 days)	5
≤1 year (but >3 months)	10
>1 year (but less than 50 years)	50
>50 years	Actual agreed design working life

- the UK National Annex specifies the use of case B2
- the Italian National Annex specifies the use of case A for the whole Italian territory
- the Spanish National Annex gives the responsibility for the choice of which case to use to the designer, depending on the location and type of structure.

cl.1.1(4): 1991-1-3 Annex B is intended to be used primarily by those countries where exceptional snow drifts occur (e.g. UK and Ireland) and where cases B2 and B3 apply.

Note to cl.1.1(4): 1991-1-3 The use of Annex B is only allowed through the National Annex.

In Annex C the maps of the characteristic snow load values on the ground are given, and these are based on the results of a research work carried out between 1995 and 1998, under a contract specific to this Eurocode, to DGIII/D3 of the European Commission.[2,3]

cl.1.1(5): 1991-1-3 The maps are intended to give information to National technical Competent Authorities for those countries that were members of the European Union (EU) and European Free Trade Association (EFTA) in 1998, in order to help them redraft and update their national maps, based on the established harmonised procedures identified in the research, for inclusion in their National Annexes. It should be noted that Annex C of EN 1991-1-3 gives the maps of Czech Republic, Iceland and Poland, but these are not based on the findings of the research project.

cl.1.1(6): 1991-1-3 EN 1991-1-3, through its Annex D, gives guidance for determining ground snow loads for return periods different than 50 years. This annex is useful for determining the ground snow loads for structures with a design working life other than 50 years. The user must be warned that a structure with, for example, nine months' design life, must not be designed for ground snow loads determined for a nine-month return period. Table 1.1 from EN 1991-1-6 gives guidance for determination of return periods to be used for different design and is reproduced above. For further details see Chapter 3 of Part 6 of this Designers' Guide.

cl.1.1(7): 1991-1-3 The guidance in Annex E on bulk weight densities of snow may be used when an actual depth of snow needs to be converted into the relative action (***Clause 1.1(7): 1991-1-3***).

No comment is necessary on *Clause 1.1(8): 1991-1-3*, but see Section 1.2 below of this Designers' Guide.

1.2. Normative references

cls.1.1(8) and 1.2: 1991-1-3 No comment is necessary on the quoted references, with the exception of ISO 4355[4] which provides a comprehensive background to the information given in this code, and in particular gives guidance on dynamic forces on a snow fence from sliding snow, which is not covered in EN 1991-1-3.

1.3. Assumptions

cl.1.3: 1991-1-3 The statements and assumptions given in EN 1990 *Clause 1.3* apply to all the Eurocode parts and designers' guidance for these is given in the Designers' Guide to EN 1990.

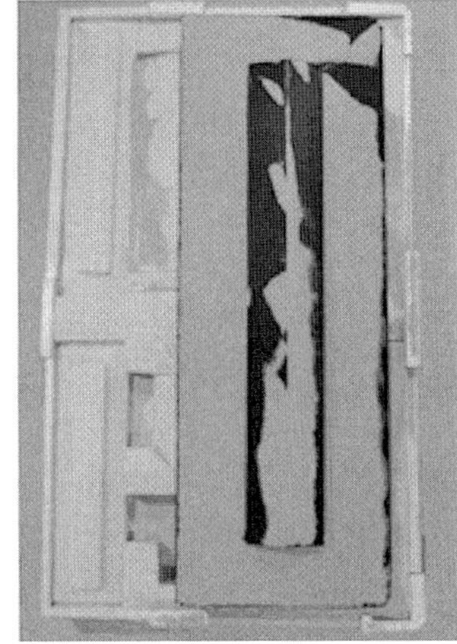

Fig. 1.1. Transversal cross-section of the central market hall in Livorno, Italy and wind tunnel tests on the roof structure

1.4. Distinction between Principles and Application Rules

The statements and assumptions given in EN 1990 *Clause 1.4* apply to all the Eurocode parts and designers' guidance, for these are given in Chapter 1 *Clause 1.3* of this Designers' Guide. *cl.1.4: 1991-1-3*

1.5. Design assisted by testing

For roofs for which the information given in EN 1991-1-3 cannot be directly used (e.g. for stadia roofs or large-span roofs or roofs with shapes not covered by EN 1991-1-3, for which there are no validated numerical methods) then design assisted by testing may be used. *cl.1.5: 1991-1-3*

With regard to roofs where the deposition patterns of snow is very sensitive to wind, then wind tunnel tests may be considered.

As an example, Fig. 1.1 shows the plan view of the wind tunnel model used for the definition of snow deposition patterns over an historical structure in Livorno, Italy. The roof structure of the central marketplace, built in 1896, is made by latticed steel beams, with a pitched shape, completed on the top by a duopitched structure. The assessment of the roof snow loads in combination with wind effects has been possible through wind tunnel tests.

Further examples of the use of wind tunnels to determine snow deposition patterns are given by Irwin *et al.*[5]

No further guidance is required on the note to *Clause 1.5: 1991-1-3*.

1.6. Terms and definitions

Most of the definitions given in EN 1991-1-3 derive from ISO 2394, ISO 3898, and ISO 8930 and ISO 4355. In addition reference should be made to EN 1990 which provides a basic list of terms and definitions which are applicable to EN 1990 to EN 1999, thus ensuring a common basis for the Eurocode suite. *cl.1.6: 1991-1-3*

For the structural Eurocode suite, attention is drawn to the following key definitions, which may be different from current national practices:

- '*Action*' means a load, or an imposed deformation (e.g. temperature effects or settlement).
- '*Effects of Actions*' or '*Action effects*' are internal moments and forces, bending moments, shear forces and deformations caused by actions.

From the many definitions provided in EN 1990, those that apply for use with EN 1991 are described in Chapter 1 *Clause 1.4(a), (b), (c) and (d): 1991-1-1.*

The following comments are made to help the understanding of particular definitions in EN 1991-1-3.

Note to cl. 1.6.4: 1991-1-3

The Note to Clause 1.6.4 'characteristic value of snow load on the roof' explains that the coefficients chosen for converting snow load on the ground into snow load on the roof have been selected taking that the snow load on the roof has an annual probability of exceedence equal to or less then the annual probability of exceedence of the snow load on the ground, which the code states in *Clause 1.6.1: 1991-1-3* to be 0.02.

cl. 1.6.1: 1991-1-3

1.7. Symbols

cl. 1.7(1): 1991-1-3

The notation in *Clause 1.7: EN 1991-1-3* is based on ISO 3898.

EN 1990 *Clause 1.6* provides a comprehensive list of symbols, some of which may be appropriate for use with EN 1991-1-3. The symbols given in *Clause 1.7(2): 1991-1-3* are additional notations specific to this part of EN 1991-1-3.

CHAPTER 2

Classification of actions

This chapter is concerned with the classification of actions of EN 1991-1-3: *Eurocode 1 – Actions on structures: Part 1.3: General Actions – Snow loads.*

EN 1991-1-3 classifies snow loads as variable, fixed actions in most cases. See EN 1990: 2002, 4.1.1(1)P and 4.1.1(4).

A fixed action has a fixed distribution in space such that its magnitude and direction are determined unambiguously for the whole structure when determined at one point on it. Deposits of snow either undrifted or drifted are considered fixed actions. *cl.2.(1)P: 1991-1-3*

EN 1991-1-3 classifies snow loads as variable static actions in most cases. See EN 1990: 2002, 4.1.1(4).

A static action is an action that does not cause significant acceleration of the structure or its elements. *cl.2.(2): 1991-1-3*

As stated in EN 1990 Clause 4.1.1(2), snow loads may be considered in exceptional cases as accidental actions dependent on the site locations.

In EN 1991-1-3 exceptional snow falls and exceptional snow drifts are treated as accidental actions and explained in Chapter 3 of this Part of the Designers' Guide. *cls.2.(3) and 2(4): 1991-1-3*

EN 1990 defines an accidental action as an action usually of short duration but of significant magnitude that is unlikely to occur on a given structure during the design working life. For these cases the accidental design situation, described in expression 6.11b of EN 1990, should be used in the design. See also Chapter 3 of this Part of the Designers' Guide.

No further guidance is required to the *notes to Clauses 2.(3) and 2(4): 1991-1-3.*

CHAPTER 3

Design situations

This chapter is concerned with the general concepts of design situations relating to EN 1991-1-3: *Eurocode 1 – Actions on structures: Part 1.3: General Actions – Snow loads*. The material described in this chapter is covered in the following clauses:

- General *Clause 3.1: 1991-1-3*
- Normal conditions *Clause 3.2: 1991-1-3*
- Exceptional conditions *Clause 3.3: 1991-1-3*

3.1. General

EN 1990 Clause 3.2 identifies the following design situations for the verification of ultimate limit states: *cl.3.1(1)P: 1991-1-3*

- persistent design situations, which refer to the conditions of normal use
- transient design situations, which refer to temporary conditions applicable to the structure, e.g. during execution or repair
- accidental design situations, which refer to exceptional conditions applicable to the structure or to its exposure, e.g. exceptional snow falls and drifts
- seismic design situations.

Each of these design situations is linked to a particular expression for the combination of action effects as follows:

- persistent and transient design situations, which refer to *expressions (6.10), or (6.10a) and (6.10b) in EN 1990*
- accidental design situations, which refers to *expression (6.11b) in EN 1990*
- seismic design situations, which refers to *expression (6.12b) in EN 1990.*

In addition snow loads need to be determined for the verification of serviceability limit states using the following expressions for the combination of action effects given in EN 1990:

- the characteristic combination which refers to *expression (6.14b) of EN 1990*
- the frequent combination which refers to *expression (6.15b) of EN 1990*
- the quasi-permanent combination which refers to *expression (6.16b) of EN 1990.*

In all the ULS and SLS verifications for situations including snow loads, the characteristic value (Q_k in EN 1991-1-3 is described as s_k) which usually corresponds to an upper value with an intended probability of not being exceeded during a specific reference period (normally 50 years for buildings) is used. Depending upon the design situation being considered for the ultimate or serviceability limit states (described in *EN 1990 Clauses 6.4 and 6.5*) and its Designers' Guide, other representative values of a variable action need to be determined. See Chapter 4 of this Part of this Designers' Guide.

cl.3.1(2): 1991-1-3 When considering the local effects (e.g. drifting at projections and obstructions or snow overhanging the edge of a roof) EN 1991-1-3 specifies that the persistent/transient design situation should be used for normal snow falls.

In locations where exceptional drifting occurs, the accidental design situation is used according to Annex B of EN 1991-1-3.

3.2. Normal conditions

Clause 1.1 of this Part of this Designers' Guide explains four possible situations and load arrangements for treating snow loads dependent on the location being considered and its climatic features. The National Annex will specify the use of case A or cases B1, B2 and B3 as appropriate. Case A refers to the normal situation and cases B1 to B3 refer to exceptional situations.

Clause 3.2(1): 1991-1-3 covers locations where exceptional snow falls and drifts are not expected. This condition is considered the normal condition where the transient/persistent design situation should be used, in accordance with EN 1990, for both the undrifted and drifted snow arrangements. See also Chapter 5 of this Part of this Designers' Guide.

No further explanation is needed for the note to *Clause 3.2(1): 1991-1-3*.

3.3. Exceptional conditions

(a) Case B.1 Exceptional snow falls but no exceptional drifts

Clause 3.3(1): 1991-1-3 covers locations where exceptional snow falls are likely to be expected but not exceptional drifts. This condition is considered exceptional and designated Case B1 in EN 1991-1-3. In this case the designer should follow the procedure summarised below.

The transient/persistent design situation should be used for both the undrifted and the drifted snow load arrangements, making reference to the characteristic ground snow load value given in the appropriate National Annex to EN 1991-1-3. In this case the roof snow load is obtained according to the specifications given in Clause 5.2(3)P(a) and Section 5.3 of EN 1991-1-3.

In addition to the above verification the designer should use the accidental design situation for both the undrifted and the drifted snow load arrangements, making reference to the exceptional ground snow load, determined according to Section 4.3 of EN 1991-1-3. Again the roof snow load is determined according to the rules given in Clause 5.2(3)P(a) and Section 5.3 of EN 1991-1-3.

Expressions of the roof snow loads to be used are given below; see also Chapters 4 and 5 of this Part of this Designers' Guide.

Persistent/transient design situation

(1) Undrifted $\mu_i C_e C_t s_k$.
(2) Drifted $\mu_i C_e C_t s_k$.

Accidental design situation (where snow is the accidental action)

(3) Undrifted $\mu_i C_e C_t C_{esl} s_k$.
(4) Drifted $\mu_i C_e C_t C_{esl} s_k$.

No further guidance is required to the *note 1 to Clause 3.3(1): 1991-1-3*.

Note 2 to cl.3.3(1): 1991-1-3 The above applies for the overall verification of the roof's structure and of the building. For local effects described in Section 6 of EN 1991-1-3, such as snow drifting at projections and obstructions or snow loads on snowguards, the National Annex may specify either the persistent/transient or the accidental design situation may apply.

(b) Case B.2 No exceptional snow falls but exceptional drifts

Clause 3.3(2): 1991-1-3 covers locations where exceptional snow falls are not expected and exceptional drifts are likely to be expected. This condition is considered exceptional and designated Case B2 in EN 1991-1-3. In this case the designer should follow the procedure summarised below.

The transient/persistent design situation should be used for the undrifted snow load arrangements; the same design situation should be used for the drifted snow load arrangements, except for roof shapes indicated in Annex B of EN 1991-1-3. The characteristic ground snow load value given in the appropriate National Annex to EN 1991-1-3 is the reference ground load value. In this case the roof snow load is obtained according to the specifications given in Clause 5.2(3)P(a) and Section 5.3 of EN 1991-1-3.

In addition to the above verification the designer should use the accidental design situation for the drifted snow load arrangement for roof shapes indicated in Annex B to EN 1991-1-3 only. In this case the exceptional snow drift is the accidental action and its value is to be calculated according to specifications given in Annex B of EN 1991-1-3.

Expressions of the roof snow loads to be used are given below; see also Chapters 4 and 5 and Annex B of this Part of this Designers' Guide.

Persistent/transient design situation

(1) Undrifted $\mu_i C_e C_t s_k$.

(2) Drifted $\mu_i C_e C_t s_k$ (except for roof shapes in Annex B of EN 1991-1-3).

Accidental design situation (where snow is the accidental action)

(3) Drifted $\mu_i s_k$ (for roof shapes in Annex B of EN 1991-1-3).

No further guidance is required to the *note to Clause 3.3(1): 1991-1-3.*

(c) Case B.3 Exceptional snow falls and exceptional drifts

Clause 3.3(3): 1991-1-3 covers locations where both exceptional snow falls and exceptional snow drifts are likely to be expected. This condition is considered exceptional and designated Case B3 in EN 1991-1-3. In this case the designer should follow the procedure summarised below.

The transient/persistent design situation should be used similarly to the above Case B2.

In addition to the above verification the designer should use the accidental design situation for the undrifted snow load arrangements, making reference to the exceptional ground snow load, determined according to Section 4.3 of EN 1991-1-3. The roof snow load is obtained according to the specifications given in Clause 5.2(3)P(a) and Section 5.3 of EN 1991-1-3.

Furthermore, the designer should use the accidental design situation for the drifted snow load arrangement for roof shapes indicated in Annex B to EN 1991-1-3 only. In this case the exceptional snow drift is the accidental action and its value is to be calculated according to specifications given in Annex B of EN 1991-1-3.

Expressions of the roof snow loads to be used are given below; see also Chapters 4 and 5 and Annex B of this Designers' Guide.

Persistent/transient design situation

(1) Undrifted $\mu_i C_e C_t s_k$.

(2) Drifted $\mu_i C_e C_t s_k$ (except for roof shapes in Annex B of EN 1991-1-3).

Accidental design situation (where snow is the accidental action)

(3) Undrifted $\mu_i C_e C_t C_{esl} s_k$.

(4) Drifted $\mu_i s_k$ (for roof shapes in Annex B of EN 1991-1-3).

Note 1 to cl.3.3(3): 1991-1-3

No further guidance is required to the *note 1 to Clause 3.3(3): 1991-1-3.*

Note 2 to cl.3.3(3): 1991-1-3

As in Case B1, for local effects described in Section 6 of EN 1991-1-3, the National Annex may specify either the persistent/transient or the accidental design situation may apply.

CHAPTER 4

Snow load on the ground

This chapter is concerned with the snow load on the ground in EN 1991-1-3: *Eurocode 1 – Actions on Structures: Part 1.3: General Actions – Snow loads*. The material described in this chapter is covered in the following clauses:

- Characteristic values *Clause 4.1: 1991-1-3*
- Other representative values *Clause 4.2: 1991-1-3*
- Treatment of exceptional snow loads on the ground *Clause 4.3: 1991-1-3*

4.1. Characteristic values

The characteristic value of the snow load on the ground (s_k) should be determined according to the main definitions given in EN 1990: 2002, Section 4.1.2 (7)P and Clause 1.6.1 of EN 1991-1-3. In these definitions s_k is intended as the upper value of a random variable, for which a given statistical distribution function applies, with the annual probability of exceedence set to 0.02. (i.e. a probability of not being exceeded on the unfavourable side during a 'reference period' of 50 years). *cl.4.1(1): 1991-1-3*

The designer can generally obtain the characteristic value of snow load on the ground (s_k) from a national snow map provided in the appropriate National Annex. To cover unusual local conditions however, the designer is permitted to determine different snow load values from that specified in the maps and this may have to be agreed with the client and the relevant authority for an individual project. See also the guidance in this Designers' Guide for Clause 4.1(2) of EN 1991-1-3. *Note 1 to cl.4.1(1): 1991-1-3*

The characteristic values provided by the maps in the national annexes must be taken as the minimum value for a particular location.

An example of an unusual local condition could be that of a construction site located in a valley where, due to the local climate and wind effects, snow accumulates during the winter. As normal for the elaboration of snow maps, the local climatic influences of such specific locations may not necessarily have been taken into account. Therefore based on experience and judgement and in agreement with the client and the relevant authority, the value for s_k given in the map for this particular location should be increased.

Annex C of EN 1991-1-3 provides snow maps for the European Union (EU) and the European Free Trade Association (EFTA) countries of 1998. It also provides maps for the Czech Republic, Iceland and Poland. All the maps with the exception of those for the Czech Republic, Iceland and Poland have been developed with a common basis in order to eliminate, or reduce, inconsistencies occurring at borders between countries. Chapter 9 of this Part of this Designer's Guide gives an explanation of the background of maps and provides appropriate references.

For countries who have based their maps on the principles of the maps given in Annex C of EN 1991-1-3, such as France, Germany, Italy and UK, the method, for which the principles of calculation are given below, may be used to determine different values for a particular location.

Before applying the method the designer is advised to check the background of a particular national map with the relevant standard authority.

Note 2 to cl.4.1(1): 1991-1-3 It is the intention that all the National Annexes eventually base their maps on those provided in Annex C. The maps in the National Annex should be used to determine s_k. The maps in Annex C may only be used directly with the agreement of the National Competent Authority.

cl.4.1(2): 1991-1-3 and note 2 to cl.4.1(2): 1991-1-3 For some locations it may be necessary to refine the data for the determination of s_k. For example for a location where the value of s_k in the map is determined by interpolation from surrounding weather station measurements, where snow records are available for at least 20 years, characteristic ground snow load may be determined for that particular location (***Clause 4.1(2): 1991-1-3 and note 2 to Clause 4.1(2): 1991-1-3***).

For the maps in Annex C of EN 1991-1-3 a common mean return interval, equal to 50 years, was agreed upon, and a common statistical model adopted for processing yearly maxima. The Gumbel Type I cumulative distribution function has been used together with the least squares method to get the best fitting regression curve. In the analysis both the zero and the non-zero values have been taken into account, through the so-called mixed distribution approach. In the research work[2,3] it is shown that the above method leads to the best correlation coefficients of the fitting curves for the majority of the 2600 weather stations analysed throughout Europe. See Fig. 4.1.

Note to cl.4.1(2): 1991-1-3 No more guidance required to *note 1 to Clause 4.1(2): 1991-1-3.*

At some locations isolated very heavy snow falls have been recorded in the past and these have resulted in abnormally large snow loads. Such snow falls disturb the statistical processing of the more regular snow data significantly and clearly do not fit in with the statistical distribution calculated for the remaining values. These snow loads are referred

cl.4.1(3): 1991-1-3 to as ‘exceptional snow loads’.

Exceptional values have been encountered in many areas of Europe, mainly in maritime and coastal areas and in mild climates where snow falls are intermittent and generally short-lived. Exceptional snow loads may be identified in the appropriate National

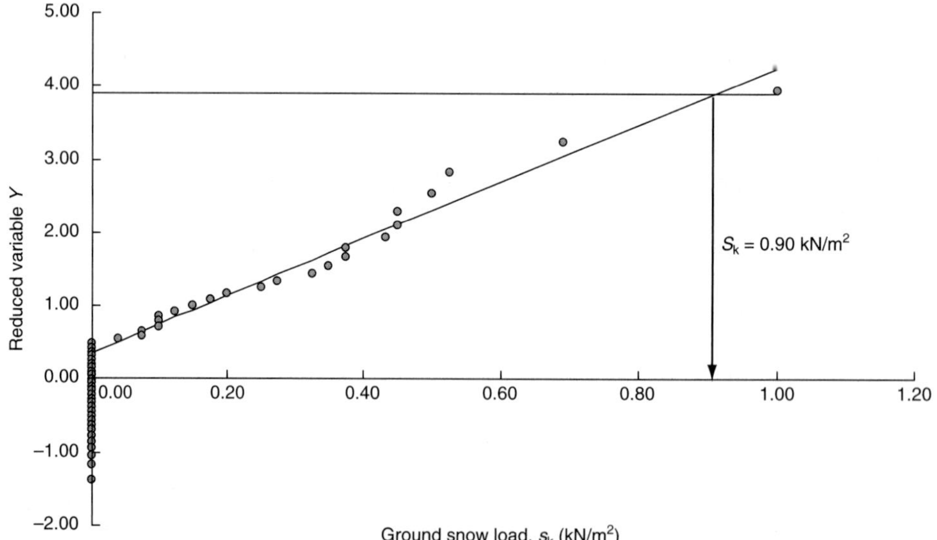

Fig. 4.1. Probability paper (Gumbel Type I) for the weather station at Arezzo, Italy. On the *x* axis ground snow loads s_k kN/m^2; on *y* axis the probability reduced variable $Y = -\ln\{-\ln[1 - m/(n+1)]\}$, where *m* is the plotting position of the yearly maximum and *n* the total number of years in the database. The intersection of the best fitting regression line of the non-zero values with the line at $Y = -\ln[-\ln(0.98)] = 3.90$ gives the characteristic value $s_k = 0.90$ kN/m^2

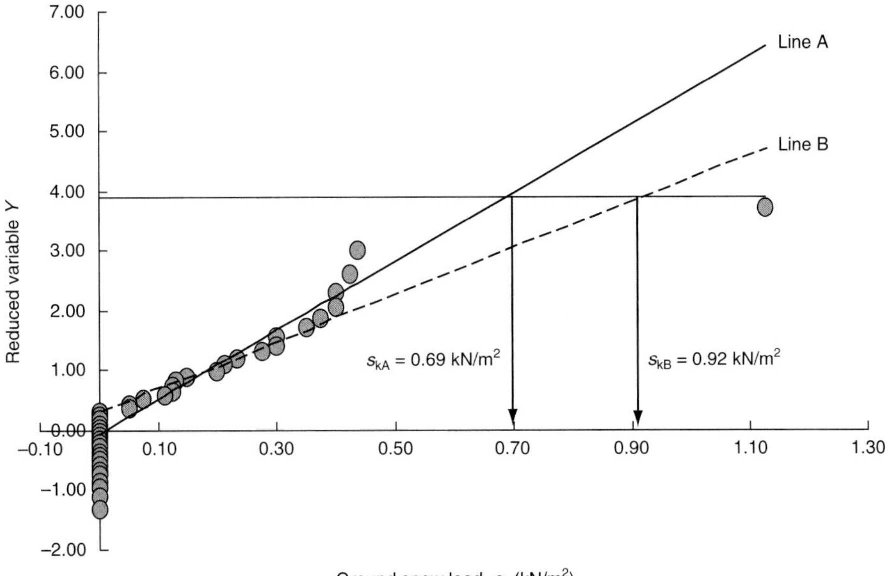

Fig. 4.2. Probability paper (Gumbel Type I) for the weather station at Vercelli, Italy (135 m a.s.l.), showing an exceptional value $s_k = 0.69\,kN/m^2$, $s_{max} = 1.13\,kN/m^2$, $s_{max}/s_k = 1.64$

Annexes, and below is given the background to their derivation, as this may be of interest to the user for understanding their concept. See also Cases B1 and B3 in Chapter 3 of this part of this Designers' Guide.

Figure 4.2 shows an example for exceptional snow loads according to the following definition established within the programme of research:

'If the ratio of the largest load value to the characteristic load value (determined without inclusion of that maximum value) is greater than 1.5, then the largest load value shall be treated as an exceptional snow load.'

In Fig. 4.2 line A is the curve obtained using a regression analysis considering all the values with the exception of the maximum value. In this case the characteristic value for s_{kA} is $0.69\,kN/m^2$.

Line B is the curve obtained using a regression analysis considering all the values including the maximum value. In this case the characteristic value for s_{kB} is $0.92\,kN/m^2$.

The maximum recorded value is $1.13\,kN/m^2$, which leads to a ratio of $s_{max}/s_{kA} = 1.64$. Therefore in accordance with the definition provided above for exceptional snow loads, the maximum value, being over 50% greater than s_{kA}, can be treated as an exceptional snow load for the particular location. The characteristic snow load for that location is $0.69\,kN/m^2$.

The opportunity to choose s_{kA} together with the exceptional snow load or s_{kA} alone or s_{kB} is a decision for the National Annex. See also Section 5.2 of this part of this Designers' Guide. For example in the UK no exceptional snow load locations were identified, and hence s_{kA} was used.

In Italy, although over approximately 130 weather stations were used for the data, only in eight cases exceptional snow loads were identified. Although these values are exceptional according to the above definition, they are not considered relevant to the Italian snow map, since the interpolation of the results for s_{kA} leads to values covering the maximum recorded values. In addition, although defined as exceptional, their absolute values are relatively low compared to values obtained at other locations in Italy. Therefore the Italian National Annex does not cover exceptional snow loads.

4.2. Other representative values

EN 1990, in its Clause 1.5.3.20, defines a representative value of an action (F_{rep}) as a value used for the verification of the limit state. A representative value may be the characteristic value (F_k) or an accompanying value ψF_k. Dependent upon the ultimate limit state or serviceability limit state verification, as described in EN 1990 Clauses 6.4 and 6.5, three values of ψ (ψ_0, ψ_1, ψ_2) need to be considered as appropriate. A brief description is given below of the use of the representative values.

The characteristic value described in Section 4.1 above is used as a leading variable action for the ultimate limit state verifications for persistent and transient design situations described in EN 1990 (expressions 6.10, 6.10a and 6.10b) and the characteristic combination for the serviceability limit state verification (expression 6.14b of EN 1990). For these two verifications all the accompanying actions are defined as the combination value (i.e. $\psi_0 Q_k$).

The other two representative values ($\psi_1 Q_k$ and $\psi_2 Q_k$) are used as appropriate for the verification of:

- accidental design situations for ULS (6.11b of EN 1990) when snow is not considered as an accidental action; the case of snow considered as an accidental action is described in Section 4.3 below
- seismic design situations for ULS (6.12b of EN 1990)
- frequent design situations for SLS (6.15b of EN 1990)
- quasi-permanent design situations for SLS (6.16b of EN 1990).

EN 1990, Clause 4.1.3 gives values to be used for snow loads as reproduced in Table 4.1 below.

cl.4.2(1): 1991-1-3 As seen in Table 4.1, the recommended values of the coefficients ψ_0, ψ_1 and ψ_2 for buildings are dependent upon the location of the site being considered. The values of ψ may be set by the National Annex.

These values have been obtained by recent research activities sponsored by the European Commission. Further details may be obtained from References 2 and 3.

In Example 4.1 below, snow is considered a normal variable load and therefore the persistent design situation and hence expression 6.10 of EN 1990 is used.

For the cases B.1, B.2 and B.3 (see Section 3.3 above) both persistent and accidental design situations need to be used as appropriate.

4.3. Treatment of exceptional snow loads on the ground

For specific locations where the National Annex specifies that exceptional snow falls on the ground can occur, Clause 4.3(1) of EN 1991-1-3 gives the following expression to calculate the design value s_{Ad} of exceptional snow load on the ground as follows:

$$s_{Ad} = C_{esl} s_k \tag{4.1}$$

Table 4.1. Recommended values of coefficients ψ_0, ψ_1 and ψ_2 for different locations for buildings

Regions	ψ_0	ψ_1	ψ_2
Finland Iceland Norway Sweden	0.70	0.50	0.20
Reminder of other CEN member states, for sites located at altitude $H > 1000$ m above sea level	0.70	0.50	0.20
Reminder of other CEN member states, for sites located at altitude $H \leq 1000$ m above sea level	0.50	0.20	0.00

Example 4.1

Consider the combination of actions given in expression (6.10) for the persistent or transient design situation (***EN 1990 Clause 6.4.3.2***). *1990 cl.6.4.3.2*

$$\sum_{j\geq 1} \gamma_{G,j} G_{k,j} \text{"+"} \gamma_P P \text{"+"} \gamma_{Q,1} Q_{k,1} \text{"+"} \sum_{i>1} \gamma_{Q,i} \psi_{0,i} Q_{k,i} \qquad (6.10)$$

Assume that the variability of self-weight is small, and three variable actions are being considered as follows: wind (*leading*) and imposed and snow *accompanying*.

The following permanent and imposed loads need to be determined for the design situation in this example:

- the characteristic value of a permanent action G_k
- for wind (the leading variable action) the characteristic value (Q_k)
- for imposed and snow loads (the accompanying actions) the combination value is represented by the product $\psi_0 Q_k$ (for this example $0.7Q_k$ for imposed loads and $0.5Q_k$ for snow loads, in accordance with EN 1990: Table A.1 for sites where $h \leq 1000$ m e.g. in Continental Europe).

where:

C_{esl} is the coefficient for exceptional snow loads which may be set by the National Annex. EN 1991-1-3 recommends a value for C_{esl} is 2.0 *cl.4.3(1): 1991-1-3*

s_k is the characteristic value of snow load on the ground for a given location.

In Example 4.2 below the snow load on the ground is considered an accidental action and therefore the accidental design situation and hence expression 6.11b of EN 1990 is used.

Example 4.2

Consider the combination of actions in expression (6.11b) for the accidental situation. *1990 cl.6.4.3.3*

$$\sum_{j\geq 1} G_{k,j} \text{"+"} P \text{"+"} A_d \text{"+"} (\psi_{1,1} \text{ or } \psi_{2,1}) Q_{k,1} \text{"+"} \sum_{i>1} \psi_{2,i} Q_{k,i} \qquad (6.11b)$$

Assume that the variability of self-weight is small; and three variable actions are being considered as follows: snow as an accidental action, wind (as the *leading variable action*) and imposed as the *accompanying variable action*.

The following permanent accidental and imposed loads need to be determined for the design situation in this example:

- the characteristic value of a permanent action G_k
- the value for A_d is the accidental snow load (i.e. s_{Ad})
- for wind the leading value (taken as $\psi_1 Q_k$) where $\psi_1 = 0.2$ as given in EN 1990 Table A1.1 *cl.4.3(1): EN 1991-1-3*
- for imposed load the accompanying value is represented as a product ($\psi_2 Q_k$) where $\psi_2 = 0.3$ as given in EN 1990 Table A1.1.

CHAPTER 5

Snow load on roofs

In this chapter the determination of snow loads on the roof, to be used for the verification of the whole structural resisting frame is described. The material described in this chapter is described in the following clauses:

- nature of the load — *Clause 5.1: EN 1991-1-3*
- load arrangements — *Clause 5.2: EN 1991-1-3*
- roof shape coefficients — *Clause 5.3: EN 1991-1-3*

5.1. Nature of the load

This section is based on the premise that the snow layers on a roof can have many different shapes depending on the following characteristics of the roof: *cl.5.1(1): 1991-1-3* *cl.5.1(2): 1991-1-3*

- its shape
- its thermal properties
- the roughness of its surface
- the amount of heat generated under the roof
- the proximity of nearby buildings
- the surrounding terrain
- the local meteorological climate, in particular its windiness, temperature variations, and likelihood of precipitation (either as rain or as snow).

5.2. Load arrangements

A snow layer on a roof is the result of subsequent events such as snow falls, wind interaction, partial snow melting due to warming of the upper layers, sliding off of part of snow from the roof, etc.

In the absence of wind, snow is deposited on the roof in a balanced way and generally a uniform cover is formed (initial deposition).

Falling snow is deposited on roofs in uniform layers only when the wind speed is low. It is known that with wind speeds in the range 4–5 m/s, much of the snow is deposited in areas of '*aerodynamic shade*' (see Fig. 5.1).

For situations where the wind velocity increases above this range, snow particles can be picked up from the snow cover and redeposited on the lee sides, or on lower roofs in the lee side, or behind obstructions on the roof (see Fig. 5.2).

To cover the above situations EN 1991-1-3 considers two idealised load arrangements due to snow as follows: *cl.5.2(1)P: 1991-1-3*

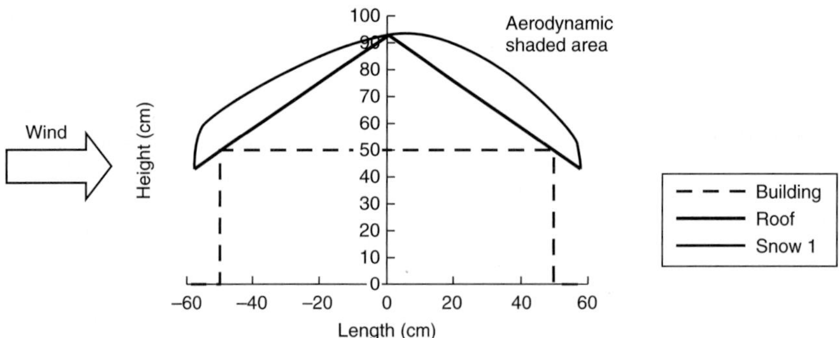

Fig. 5.1. Model in wind tunnel, wind from the left side with velocity of 4 m/s

cl.1.6.5: 1991-1-3 (1) Undrifted snow load on roofs, which describes the uniformly distributed snow load on the roof, affected only by the shape of the roof, before any redistribution of snow due to other climatic actions (see Fig. 5.3, for a case where the load arrangement is an undrifted state).

cl.1.6.6: 1991-1-3 (2) Drifted snow load which describes the snow load distribution resulting from snow having been moved from one location to another location on a roof, e.g. by the action of the wind (see Fig. 5.4 for a case where wind has caused drifting from one pitch to the other).

EN 1991-1-3 assumes snow depositions and drifting for the different climatic regions of Europe.

(a) Continental
Where the snow that falls is persistent and where snow falling in calm conditions may be followed by further snow, carried by another weather system driven by wind and there may be several repetitions of these events before there is any significant thawing (this case is covered in Section 5 of EN 1991-1-3).

(b) Maritime
Where all the snow usually melts and clears between the individual weather systems and where moderate to high wind speeds occur during the individual weather system. For example for the UK and Ireland it is assumed that all the layer of snow on the windward

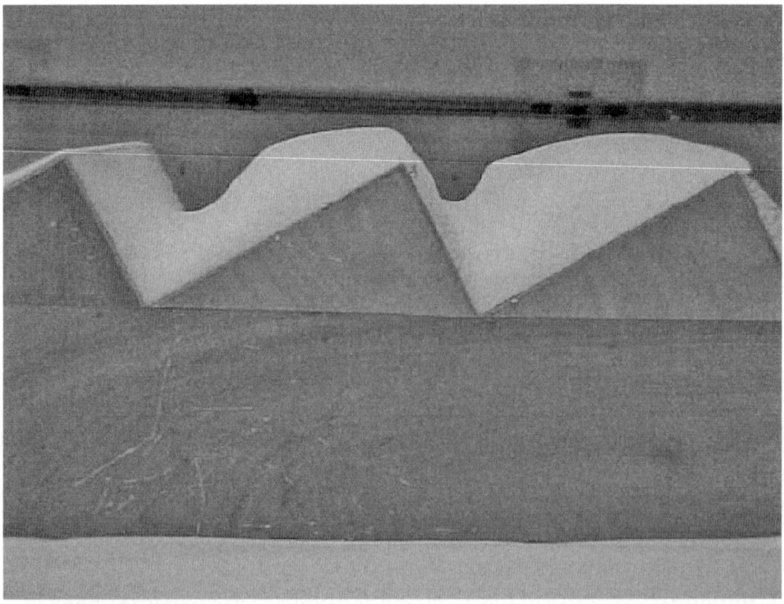

Fig. 5.2. Model in wind tunnel for multi-pitched roof, wind from the left side with 4 m/sec velocity

Fig. 5.3. Undrifted snow load

Fig. 5.4. Drifted snow load

side is available for drifting. The amount of the drifted load is considered to be of a high magnitude, compared to the ground snow load, and for this reason in these two countries the drifted snow is considered an exceptional load and treated as an accidental load using the accidental design situation (this case is covered by Annex B of EN 1991-1-3).

The National Annex may specify the use of alternative drift patterns dependent on the climatic variations described above for particular roof shapes in specific locations. *cl.5.2(2): 1991-1-3*

In EN 1991-1-3 snow load on the roof is determined as follows:

Step 1. By determining the characteristic snow load on the ground, which is generally obtained from national maps.
Step 2. Converting this load into an undrifted or drifted roof load for both persistent and, where required by the National Annex, accidental design situations by the use of:

- an appropriate shape coefficient which depends on the shape of the roof, and
- considering the influence of thermal effects from inside the building and the terrain around the building.

For the persistent/transient design situations, i.e. no exceptional snow falls or drifts:

$$s = \mu_i C_e C_t s_k \qquad (5.1 \text{ EN 1991-1-3})$$

For the accidental design situations, where exceptional snow load is the accidental action:

$$s = \mu_i C_e C_t s_{Ad} \qquad (5.2 \text{ EN 1991-1-3})$$

For the accidental design situations where exceptional snow drift is the accidental action and where Annex B applies:

$$s = \mu_i s_k \qquad (5.3 \text{ EN 1991-1-3})$$

where:

μ_i is the snow load shape coefficient (see Section 5.3 below and Annex B of EN 1991-1-3)
s_k is the characteristic value of snow load on the ground
s_{Ad} is the design value of exceptional snow load on the ground for a given location (see Section 4.3 above)
C_e is the exposure coefficient (see 'Exposure characteristics' below)
C_t is the thermal coefficient (see 'Thermal characteristics' below).

Dependent upon the specific locations, they can be four different cases that may need to be considered as shown in Table 5.1, as reproduced from Table A.1 of EN 1991-1-3. Decision for which case to use is found in the National Annex. For example in Italy *cl.5.2(3)P: 1991-1-3*

Table 5.1. Design situations and load arrangements for different locations

Normal	Exceptional conditions		
Case A	Case B1	Case B2	Case B3
No exceptional falls No exceptional drift	Exceptional falls No exceptional drift	No exceptional falls Exceptional drift	Exceptional falls Exceptional drift
3.2(1)	3.3(1)	3.3(2)	3.3(3)
Persistent/transient design situation: (1) Undrifted $\mu_i C_e C_t s_k$ (2) Drifted $\mu_i C_e C_t s_k$	Persistent/transient design situation: (1) Undrifted $\mu_i C_e C_t s_k$ (2) Drifted $\mu_i C_e C_t s_k$ Accidental design situation (where snow is the accidental action): (3) Undrifted $\mu_i C_e C_t C_{esl} s_k$ (4) Drifted $\mu_i C_e C_t C_{esl} s_k$	Persistent/transient design situation: (1) Undrifted $\mu_i C_e C_t s_k$ (2) Drifted $\mu_i C_e C_t s_k$ (except for roof shapes in Annex B) Accidental design situation (where snow is the accidental action): (3) Drifted $\mu_i s_k$ (for roof shapes in Annex B)	Persistent/transient design situation: (1) Undrifted $\mu_i C_e C_t s_k$ (2) Drifted $\mu_i C_e C_t s_k$ (except for roof shapes in Annex B) Accidental design situation (where snow is the accidental action): (3) Undrifted $\mu_i C_e C_t C_{esl} s_k$ (4) Drifted $\mu_i s_k$ (for roof shapes in Annex B)

Note 1: Exceptional conditions are defined according to the National Annex.
Note 2: For Cases B1 and B3 the National Annex may define design situations which apply for the particular local effects described in Section 6.

cl.A.1: 1991-1-3 Case A is used; in the UK Case B2 is used; and in France for certain locations Case B1 is used.

cl.5.2(4): 1991-1-3 Once the snow load value on the roof has been determined, this should be assumed to act vertically and referred to a horizontal projection of the roof area.

Load cases corresponding to severe imbalances on a roof resulting from snow removal, redistribution of snow on the roof, sliding, melting etc. should be considered. The results from these effects will cause load arrangement patterns which will include zero snow loads on specific parts of the roof. These considerations are very important for structures that are sensitive to the form of the load distribution, i.e. curved roofs, arches, domes, long-span flat roofs etc.

Figure 5.5 shows, as an example, the collapse of the roof of the Baltimore & Ohio Railroad Museum (Maryland, US), which occurred in February 2003, under a severe unbalanced load distribution on the roof surface. The roof structure was formed by simply supported latticed beams, disposed along radial lines. The beams were tied together by horizontal steel rings, unable to resist bending moments. The unbalanced load arrangement caused by wind (a trace of which is visible on the upper roof in Fig. 5.5), induced the collapse.

Another example of unbalanced load distribution is shown in Fig. 5.6.

cl.5.2(5): 1991-1-3 If artificial removal or redistribution is anticipated, EN 1991-1-3 allows the National Annex to produce suitable load arrangements if considered necessary.

cl.5.2(6): 1991-1-3 Another cause of possible increase in snow loads on roofs may occur in regions with possible rainfalls on the snow and consecutive melting and freezing. Where such a scenario is possible, the design snow loads on roofs should be increased, especially in cases where snow and ice can block the drainage system of the roof. There have allegedly been roof collapses causing fatalities due to this phenomenon.

In the US it is estimated that approximately 2000 roof collapses occur every year, a relevant majority of which are supposed to be caused by ponding effects on large flat roofs, due to the presence of heavy snow and rain. Figure 5.7 shows a partial roof collapse, which occurred in 2004 in Portland, Oregon.

EN 1991-1-3 asks the National Annexes to provide appropriate advice.

Fig. 5.5. Baltimore & Ohio Railroad Museum (Maryland, US) roof collapse

Exposure characteristics

Exposure of the structure depends on the wind speed in the area and the features of the surrounding areas. If the structure is surrounded by high trees or other higher structures, a so-called sheltered situation results. If on the other hand there is no obstruction in the vicinity of the roof and high wind speeds occur in the region, windswept conditions result. In addition EN 1991-1-3 asks for the consideration of the future development around the site, for example in areas where planning permissions for future developments have been granted.

EN 1991-1-3 defines the exposure coefficient C_e as a coefficient defining the reduction or increase of load on a roof of an unheated building, as a fraction of the characteristic snow load on the ground. *cl.1.6.9: 1991-1-3*

Fig. 5.6. Unbalanced snow deposition on spherical domes, covering oil tanks

Fig. 5.7. Partial flat roof collapse due to presence of snow and rain

cl.5.2(7): 1991-1-3 The exposure coefficient C_e is used in expressions 5.1 and 5.2 for determining the snow load on the roof, in cases where no exceptional snow drift occurs.

The choice for C_e is in the National Annex, the recommended value being 1.0 unless otherwise specified for different topographies.

Table 5.1 of EN 1991-1-3 is partially reproduced below as Table 5.2.

Thermal characteristics

cl.1.6.8: 1991-1-3 The rate at which snow on roofs melts after snowfall is influenced by the roof thermal transmittance. EN 1991-1-3 defines the thermal coefficient C_t as a coefficient defining the reduction of snow load on roofs as a function of the heat flux through the roof, causing snow melting.

cl.5.2(8): 1991-1-3 The thermal coefficient C_t is used in expressions 5.1 and 5.2 for determining the snow load on the roof, in cases where no exceptional snow drift occurs. The choice for C_t is in the National Annex, the recommended value being 1.0, unless otherwise specified.

Table 5.2. C_e values for different topographies

Topography	C_e
Windswept[a]	0.8
Normal[b]	1.0
Sheltered[c]	1.2

[a] A windswept topography is for flat unobstructed areas exposed on all sides without, or with little, shelter afforded by terrain, higher construction works or trees. This is illustrated in Fig. 5.8(a).

[b] A normal topography is for areas where there is no significant removal of snow by wind on construction work, because of terrain, other construction works or trees. This is illustrated in Fig. 5.8(b).

[c] A sheltered topography is for areas in which the construction work being considered is considerably lower than the surrounding terrain or surrounded by high trees and/or surrounded by higher construction works. This is illustrated in Fig. 5.8(c).

(a)

(b)

(c)

Fig. 5.8. Examples of different exposure conditions: (a) windswept condition; (b) normal condition; (c) sheltered condition

Regarding roofs with high thermal transmittance (>1 W/m^{2}K), such as glass-covered roofs, the effect of snow load reduction is important, especially in cases where there will be melting between the upper roof surface and the layer of snow, producing a thin layer of water, which reduces the friction between the snow layer and the top of the roof, causing sliding of snow from the roof.

A check that melting water can be drained from the roof surface should be carried out. Further guidance may be found in ISO 4355.[4]

5.3. Roof shape coefficients

5.3.1. General

Section 5.3 of EN 1991-1-3 gives roof shape coefficient μ, which is defined as the ratio of the snow load on the roof to the undrifted snow load on the ground, without the influence of exposure and thermal effects. *cl.1.6.7: 1991-1-3*

Section 5.3 gives the roof shape coefficients for undrifted and drifted snow load arrangements for the following roof shapes:

- monopitch in ***Clause 5.3.2*** *cl.5.3.2: 1991-1-3*
- pitched in ***Clause 5.3.3*** *cl.5.3.3: 1991-1-3*
- multi-span in ***Clause 5.3.4*** *cl.5.3.4: 1991-1-3*
- cylindrical in ***Clause 5.3.5*** *cl.5.3.5: 1991-1-3*
- roofs abutting and close to taller construction works in ***Clause 5.3.6***. *cl.5.3.6: 1991-1-3*

The drifted load arrangements given in this section refer to described roofs with the exception of the consideration of exceptional snow drifts, defined in Annex B of EN 1991-1-3.

The use of Annex B is allowed through the National Annex. *cl.5.3.1(1): 1991-1-3*

The advice given in this Designers' Guide on shape coefficients μ does not replicate the actual values given in the clauses for different roof shapes in Section 5.3 of EN 1991-1-3. The Designers' Guide will however give additional information as appropriate for each roof shape and also show the comparison for the normal and exceptional snow drifted cases. Additionally this section of this Part of this Designers' Guide will give a few examples.

For other roof shapes than those covered by this section, further guidance may be found in ISO 4355.[4] In special cases, such as cantilever roofs in stadia, or for other very large roofs or where severe load arrangements are anticipated, the shape coefficients can be determined by testing, using boundary layer wind tunnels.

Roofs that have an external geometry which may lead to increases in snow load, and which may be significant in comparison with a roof of linear profile, should be given special consideration regarding the determination of the snow load shape coefficients.

Fig. 5.9. Example of a roof having an external geometry that may lead to increases in snow load

This case applies, for example, for large flat roofs where the roof elements (e.g. precast prestressed RC beams) are placed in such a way as to be treated as multi-span roofs (see Section 5.3.4 below), and the clear length under the two adjacent pitches is greater than 3.5 m and the pitches angles are above 30°.

cl.5.3.1(2): 1991-1-3

An example of such a roof is shown in Fig. 5.9.

Shape coefficients μ for the following roof shapes:

- *cl.5.3.2: 1991-1-3* monopitch
- *cl.5.3.3: 1991-1-3* pitched
- *cl.5.3.4: 1991-1-3* multi-span

cl.5.3.1(3): 1991-1-3

are given in Fig. 5.10, which is reproduced from Fig. 5.1 of EN 1991-1-3.

In Fig. 5.10 α is the pitch angle.

cl.5.3.2(1): 1991-1-3

5.3.2. Monopitch roofs

No further guidance is needed for Clause 5.3.2(1).

cl.5.3.2(2): 1991-1-3

Independently of pitch angle, if on the roof there are snow fences or other obstructions, or if the lower edge of the roof is terminated with a parapet, the μ coefficient should not be reduced below 0.8, as sliding will not take place.

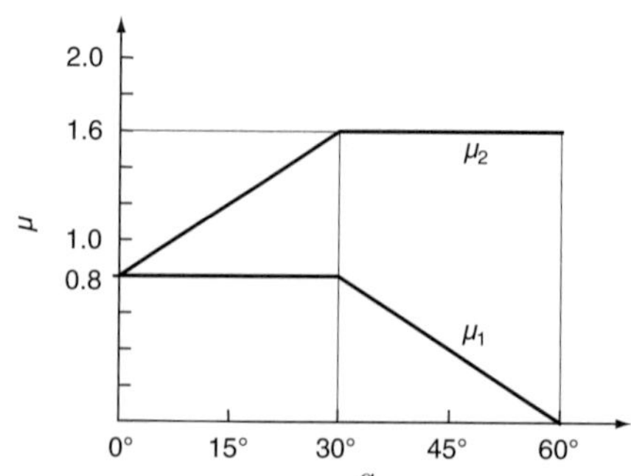

Fig. 5.10. Snow load shape coefficients

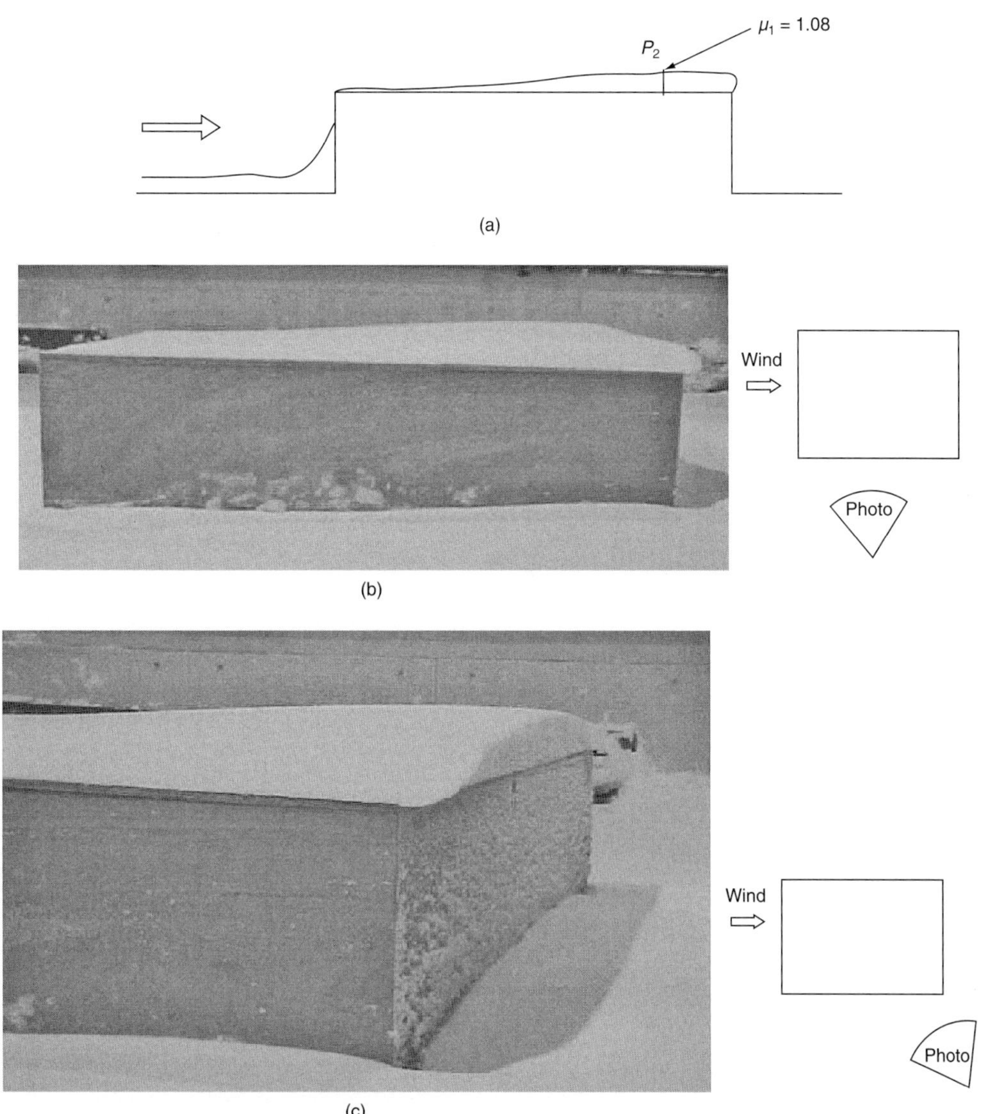

Fig. 5.11. Wind tunnel tests on a large flat roof: (a) diagrammatic representation of the snow deposition; (b) lateral view of the model; (c) angled front view of the model

Load arrangements given in Fig. 5.2 of EN 1991-1-3 that should be used for both the undrifted and drifted cases are calibrated for roof sizes which usually are encountered for general everyday buildings.

The designer should consider large flat roofs which are treated in EN 1991-1-3 as a monopitch roof with $\alpha = 0°$.

There is research evidence that for larger roofs (e.g. square or almost square roofs with length about 40 m) the snow layer may be non-uniform and the maximum value of the ratio between the roof and the ground snow loads reaches unity. Figure 5.11 illustrates an example of such a case which was obtained in a climatic wind tunnel.[3] In this case, as is shown in Fig. 5.11(a), the μ shape coefficient locally reaches 1.08. The average value for μ in this case is 0.77.

5.3.3. Pitched roofs

No further guidance is needed for Clause 5.3.3(1). *cl.5.3.3(1): 1991-1-3*

As for the case of monopitch roofs, independently of pitch angles, if on the roof there are snow fences or other obstructions, or if the lower edge of the roof is terminated *cl.5.3.3(2): 1991-1-3* with a parapet, the μ coefficient should not be reduced below 0.8, as sliding will not take place.

cls.5.3.3(3) and 5.3.3(4): 1991-1-3 No further guidance is needed for Clauses 5.3.3(3) and 5.3.3(4).

Note to cl.5.3.3(4): 1991-1-3 EN 1991-1-3 allows, through the National Annex, an alternative drifting load arrangement based on local conditions. An example of the use of an alternative drifting load arrangement is a roof located in the UK, where the shape coefficients are obtained for maritime climates.

Two examples are given below: the first uses the guidance in Clause 5.3.3 and the Italian National Annex; the second uses the guidance given in the UK National Annex to EN 1991-1-3.

Example 5.1

Consider a steel portal frame located in Milan, Italy, with dimensions as shown in the diagram below.

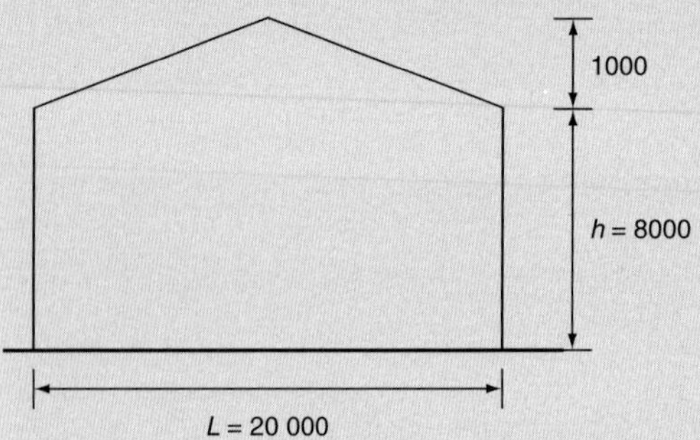

Determination of $s_k \mu_1 C_e$ and C_t.

Characteristic ground snow load:
From Italian Map: Zone I Mediterranean
$s_k = 1.60\,\text{kN/m}^2$

Shape coefficient:
Slope: 10%
$\alpha = 6°\ 10'$
$\mu_1(\alpha) = 0.8$

Exposure coefficient:
Normal topography $C_e = 1.0$

Thermal coefficient:
Insulated roof $C_t = 1.0$

The load arrangements for the undrifted case and the drifted case are determined using the guidance given in Clause 5.3.3 of EN 1991-1-3 and the following shape profiles for snow are obtained.

(a) Undrifted (case i)

(b) Drifted (case ii and iii)

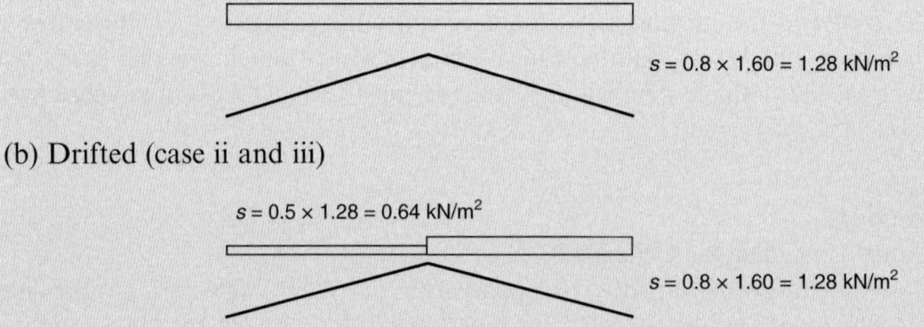

Example 5.2

Consider the same steel portal frame as in Example 5.1, located in the Scottish Highlands UK, see UK National Annex where $s_k = 1.60\,\text{kN/m}^2$, after allowing for altitude corrections.

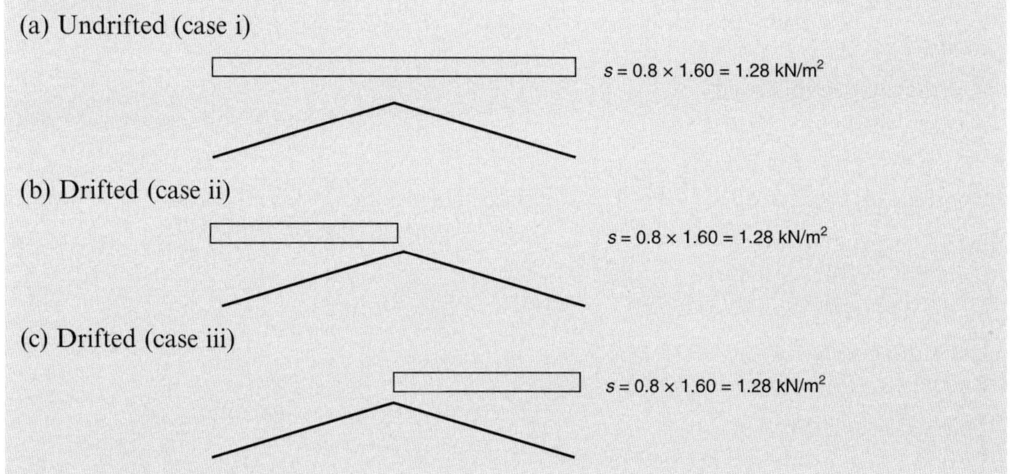

5.3.4. Multi-span roofs

cls.5.3.4(1) and 5.3.4(2): 1991-1-3

No further guidance needed for Clauses 5.3.4(1) and 5.3.4(2).

Note to cl.5.3.4(3): 1991-1-3

EN 1991-1-3 allows through the National Annex an alternative drifting load arrangement based on local conditions.

cl.B2(1): 1991-1-3

Figure 5.12(a) and (b) illustrate the different patterns for drifted load arrangements for normal and accidental conditions (for definition of symbols see EN 1991-1-3).

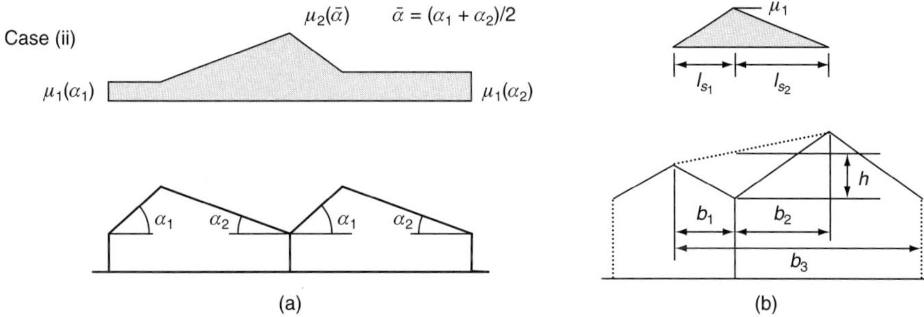

Fig. 5.12. Patterns for load arrangements: (a) normal drift; (b) accidental drift

An example of the use of normal load arrangements, for both undrifted and drifted, is given in Example 5.3 below with reference to a building located in Sweden.

Example 5.3

Consider a factory building with the following properties.
Location: Sweden – Snow load zone 2 – 300 m a.s.l.

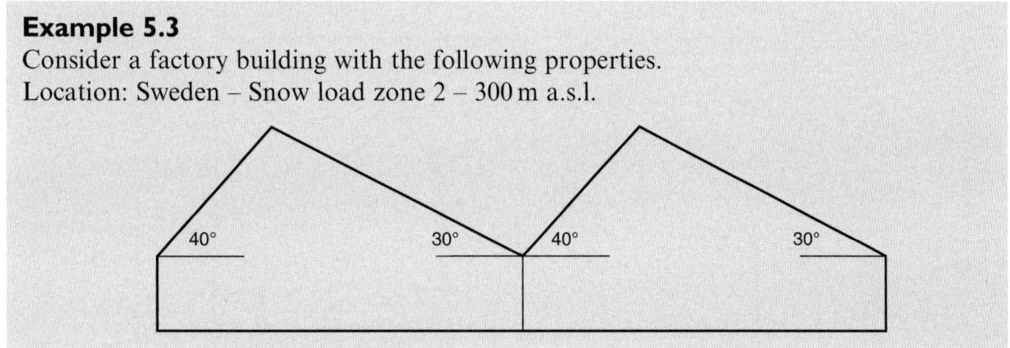

Determination of $s_k \mu_1 C_e$ and C_t.

Characteristic ground snow load:
From Swedish Map: Zone 2

$s_k = 0.790Z + 0.375A/336$

where:

Z is the zone number = 2
A is the altitude = 300 m a.s.l.
Giving $s_k = 2.85\,\text{kN/m}^2$

Considering Fig. 5.1 of EN 1991-1-3, the shape coefficients μ_1 and μ_2 are:

$\mu_1(\alpha_1) = \mu_1(40°) = 0.53$
$\mu_1(\alpha_2) = \mu_1(30°) = 0.80$
$\mu_2[(\alpha_1 + \alpha_2)/2] = \mu_2(35°) = 1.60$

Exposure coefficient:
Building surroundings normal $C_e = 1.0$

Thermal coefficient:
Effective heat insulation applied to roof $C_t = 1.0$

The load arrangements for the undrifted case and the drifted case are determined using the guidance given in Clause 5.3.4 of EN 1991-1-3 and the following shape profiles for snow are obtained.

(a) Undrifted (case i)

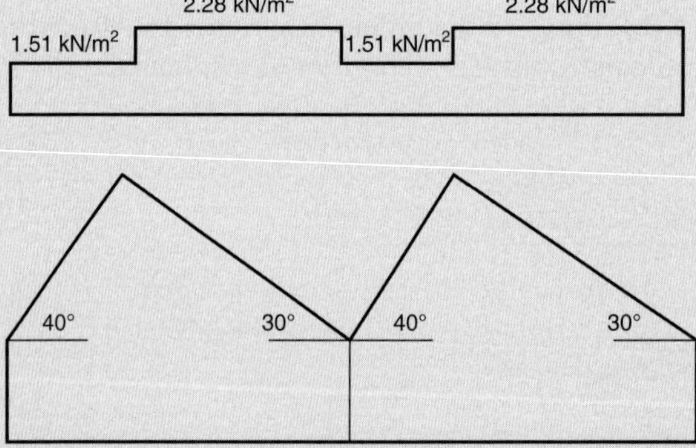

(b) Drifted (case ii)

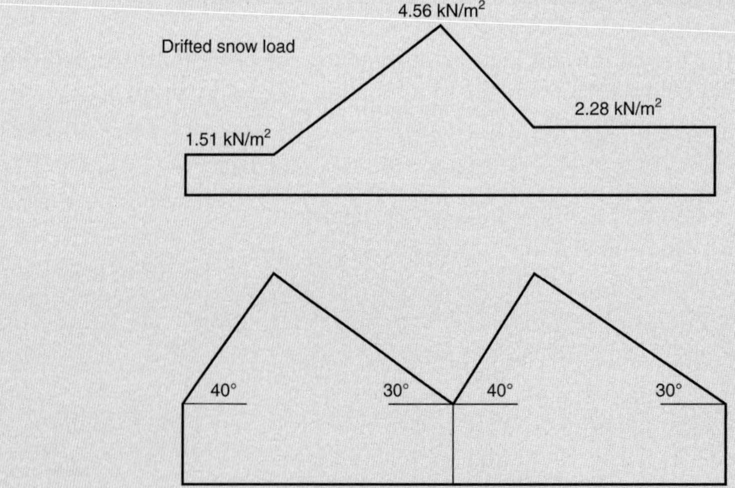

Fig. 5.13. Wind tunnel test on a cylindrical roof

The profiles given in Clause 5.3.4 are not applicable when the slope α is greater than 60° as in this case all the deposited snow on the roof is available for drifting.

This case should normally be covered by the individual National Annexes and is covered by ISO 4355.[4]

5.3.5. Cylindrical roofs

cl.5.3.5(1): 1991-1-3

No further guidance is needed for Clause 5.3.5(1).

The shape coefficient values given in this clause are calibrated also on the basis of wind tunnel tests, performed on different types of cylindrical roofs. Figure 5.13 shows an example of such wind tunnel tests.[3]

Note 2 to clause 5.3.5(1): 1991-1-3

All the shape coefficients given are intended for use in case of absence of snow fences on the roof surface. Specific rules for considering such effects may be given in the National Annex to EN 1991-1-3.

cls.5.3.5(2) and 5.3.5(3): 1991-1-3

No further guidance is needed for Clauses 5.3.5(2) and 5.3.5(3).

5.3.6. Roof abutting and close to taller construction works

cl.5.3.6(1): 1991-1-3

For normal conditions the shape coefficients to be considered for undrifted and drifted snow load arrangements are shown in Fig. 5.14.

Snow deposition patterns for the drifted load arrangement are influenced by two phenomena: the sliding of snow from the upper roof and the accumulation induced by the redistribution of snow due to wind.

The first phenomenon is possible when the slope of the upper roof is greater than a given value α, which in EN 1991-1-3 is taken equal to 15°. This is a general indication. It is important however to consider the thermal properties of the upper roof; in the case of a glass covering with high thermal transmittance the sliding effects remain possible even in the presence of lower pitch angles.

The shape coefficient μ_w, taking account of the snow accumulation due to wind, has been calibrated also on the basis of wind tunnel tests, an example of which is shown in Fig. 5.15.

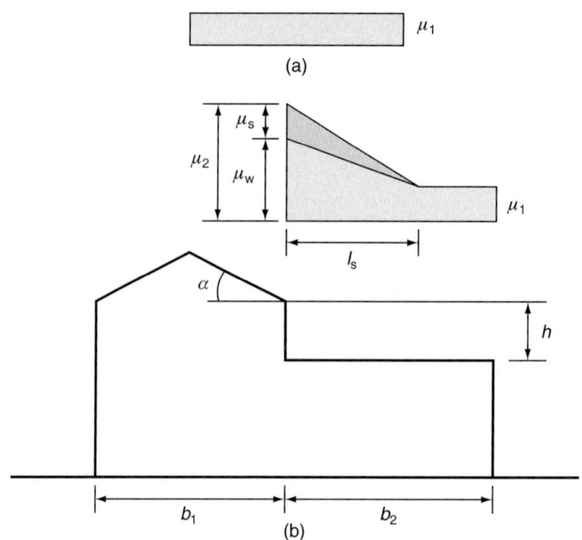

Fig. 5.14. Shape coefficients for (a) undrifted and (b) drifted snow load arrangements

The expression giving μ_w is:

$$\mu_w = (b_1 + b_2)/2h \leq \gamma h/s_k \qquad \text{(5.1 EN 1991-1-3)}$$

where γ is the weight density of snow, which for this calculation may be taken as $2\,kN/m^3$.

Note 1 to cl.5.3.6(1): 1991-1-3 The National Annex may specify a range for the variation of both μ_w and l_s. The recommended range for μ_w is $0.8 \leq \mu_w \leq 4$. The recommended range for the drift length l_s, to be taken equal to $2h$, is $5 \leq l_s \leq 15\,m$.

Note 2 to cl.5.3.6(1): 1991-1-3 Examples of the use of normal load arrangements for the drifted case are given in Examples 5.4 and 5.5.

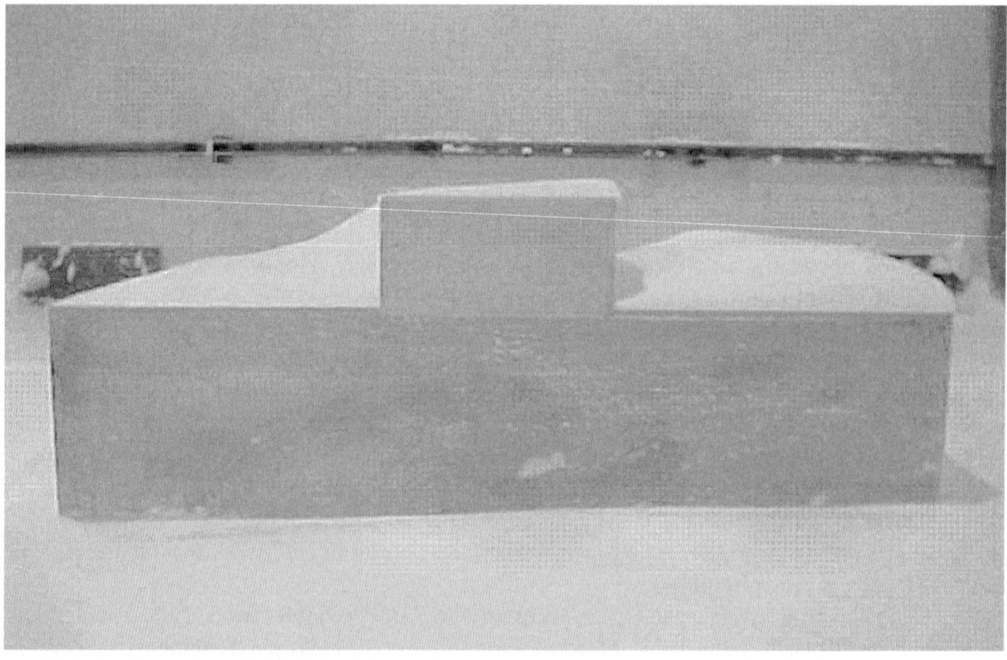

Fig. 5.15. Wind tunnel test on a multi-level flat roof (wind acting from the left side)

Example 5.4
Consider the building in Fig. 5.14, with the following dimensions:

$b_1 = 8.0\,m$
$b_2 = 10.0\,m$
$s_k = 0.8\,kN/m^2$
$\alpha = 10°$

From the above, as $\alpha = 10°$, $\mu_s = 0$.

For the calculation of the contribution of wind accumulation μ_w let us consider the variability of the distance h from the upper roof to the lower one within the range 0.5–7.0 m.

From the formula (5.1 *EN 1991-1-3*) we get the following variation of the shape coefficient μ_w.

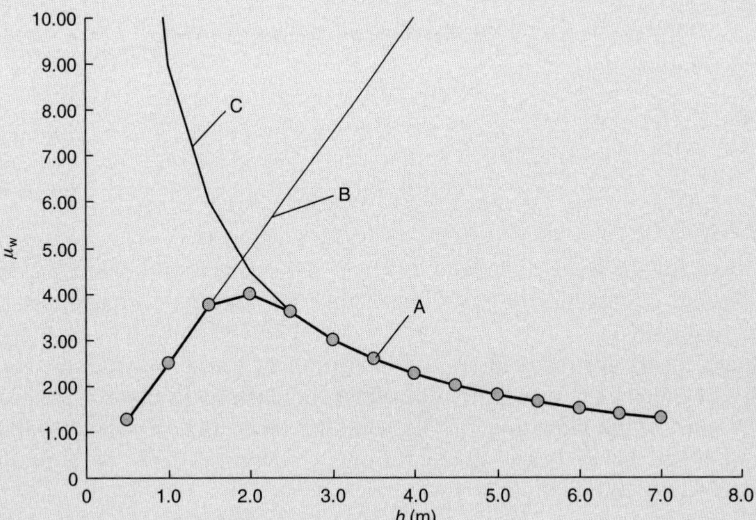

In the graph line A indicates the variation of the coefficient μ_w, which is limited by line B ($\gamma h/s_k$) and by the hyperbolic line C $[(b_1 + b_2)/2h]$.

For h equal to 5 m the wind accumulation effect results in a shape coefficient μ_w equal to 1.80 and the corresponding peak value of the snow load at the conjunction line between the two buildings is $s = \mu_w, s_k = 1.80 \times 0.80 = 1.44\,kN/m^2$.

For greater values of h the shape coefficient μ_w decreases due to the action wind vortexes, which are increasingly altered by the obstacle (see Fig. 5.15).

Example 5.5
Consider the building in the previous Example 5.4, with the following dimensions:

$b_1 = 30.0\,\text{m}$
$b_2 = 15.0\,\text{m}$

The variation of the shape coefficient μ_w is illustrated in the figure below.

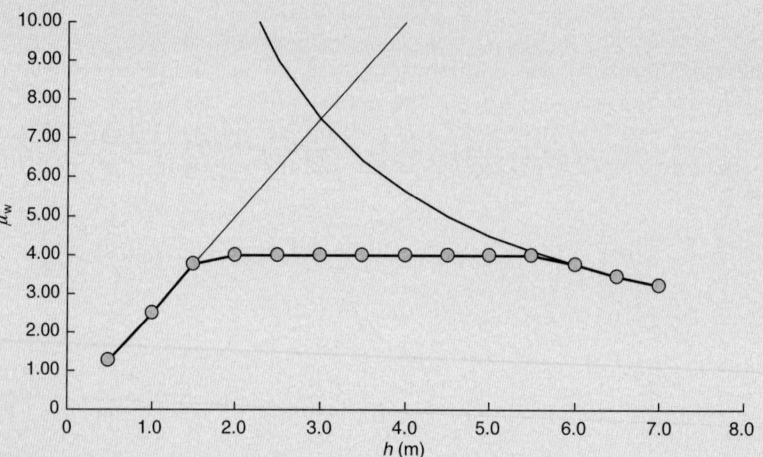

In this case, for the same difference in height $h = 5.0\,\text{m}$ examined in Example 5.4, the shape coefficient μ_w rises to the maximum value 4.0, and the corresponding peak snow load value to $3.2\,\text{kN/m}^2$.

The above effects are justified by the alteration of the aerodynamic shaded area in the vicinity of the change in height. The wider dimensions in plan of the two buildings result in a sensible modification of the wind vortexes and of the available snow for redistribution (see figure below taken during wind tunnel tests on a multi-level roof with a 'large' lower roof).

EN 1991-1-3 allows through the National Annex an alternative drifting load arrangement based on local conditions. *Note to cl.5.3.6(3): 1991-1-3*

Figure 5.16 illustrates the pattern for drifted load arrangements for accidental conditions (for definition of symbols see EN 1991-1-3). *cl.B3(1): 1991-1-3*

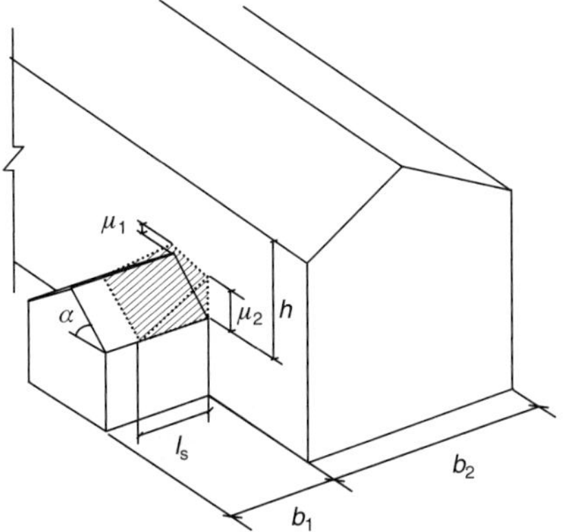

Fig. 5.16. Pattern for drifted load arrangements for accidental conditions

CHAPTER 6

Local effects

6.1. General

This chapter treats snow load arrangements specific for the verification of local effects, which may involve structural members or structural parts rather than the whole structural resisting frame, as for load cases treated in Chapter 5.

The cases covered allow the verification of:

- drifting at projections and obstructions *Clause 6.2: EN 1991-1-3*
- snow overhanging the edge of the roof *Clause 6.3: EN 1991-1-3*
- snow fences *Clause 6.4: EN 1991-1-3*

In all cases the design situations to be considered are persistent/transient. *cl.6.1(2): 1991-1-3*

It has to be recognised that there is an inconsistency in the above clause, which leads to a contradiction when the case of drifting at projections and obstructions is considered with reference to those locations where, through the National Annex, the use of Annex B is allowed. This will give alternative load arrangements which actually cover exceptional snow drifts.

To better clarify this point it has to be borne in mind that considering the load arrangement given in Annex B for drifting at projections and obstructions for local verifications, it is assumed that there is no snow elsewhere on the roof, notwithstanding Clause 6.1(2). It seems therefore reasonable to adopt the load arrangement indicated in Annex B only in accidental design situations.

6.2. Drifting at projections and obstructions

cl.6.2(1) and cl.6.2(2): 1991-1-3

No further guidance is needed for Clauses 6.2(1) and 6.2(2).

Note to cl.6.2.(1): 1991-1-3

EN 1991-1-3 allows, through the National Annex, an alternative drifting load arrangement based on local conditions.

cl.B4(1): 1991-1-3

Figures 6.1 and 6.2 illustrate the patterns for drifted load arrangements for accidental conditions (for definition of symbols see EN 1991-1-3).

cls.B4(2) and B4(3): 1991-1-3

A distinction is made for drifting patterns occurring at projections and obstructions other than parapets and for drifting occurring at parapets.

Example 6.1

Consider the obstruction on a flat roof illustrated in the figure below, located in a region where the characteristic ground snow load is equal to $s_k = 1.0\,\text{kN/m}^2$. Let us now consider the variation of the height h of the obstruction in a range between 0.0 and 1.5 m.

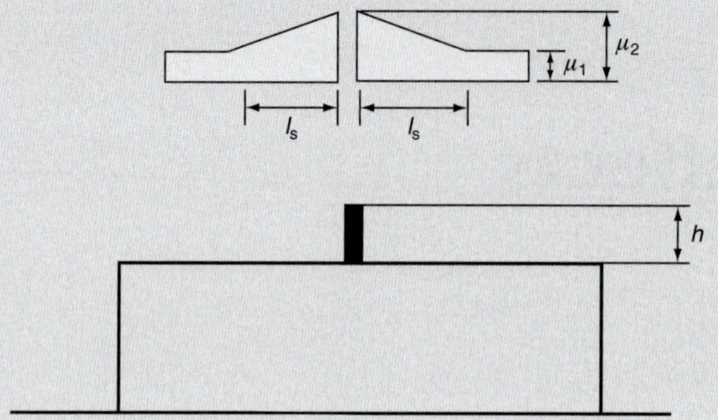

From the expression:

$\mu_1 = 0.8$, $\mu_2 = \gamma h/s_k$ (expression 6.1 EN 1991-1-3)
with the restriction: $0.8 \le \mu_2 \le 2.0$ (expression 6.2 EN 1991-1-3)

Taking for the weight density of snow the value $\gamma = 2.0\,\text{kN/m}^3$, we get the variation of the shape coefficient μ_2 illustrated in the graph below. Line A represents the variation of μ_2 within the imposed limits.

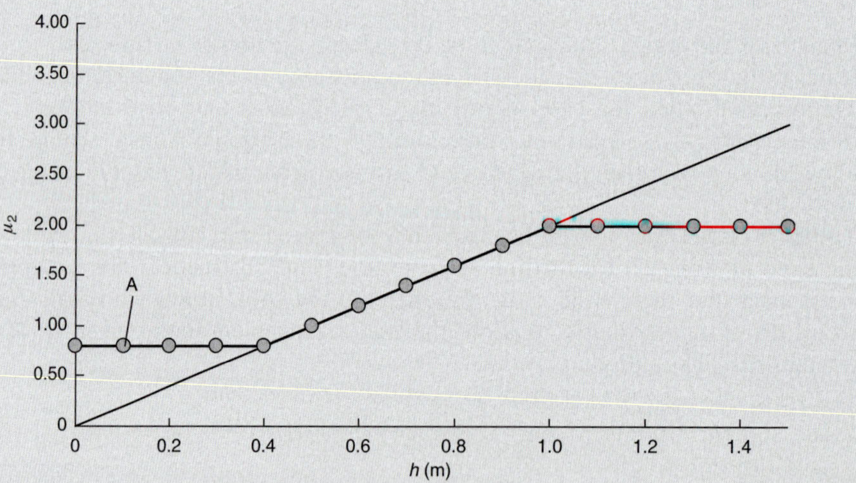

From the above graph it is evident that the accumulation of snow at the obstacle is significant, in comparison with the undrifted distribution μ_1 which is present on the rest of the roof surface. This is only if the obstacle height is greater than the undrifted roof snow cover, which depends on s_k. In this case the minimum height is $\mu_1 s_k/\gamma = 0.8 \times 1.0/2.0 = 0.4\,\text{m}$.

The drift length l_s, to be taken equal to $2h$, is limited within the range $5 \le l_s \le 15\,\text{m}$. For all cases examined in the graph the drift length is equal to 5.0 m.

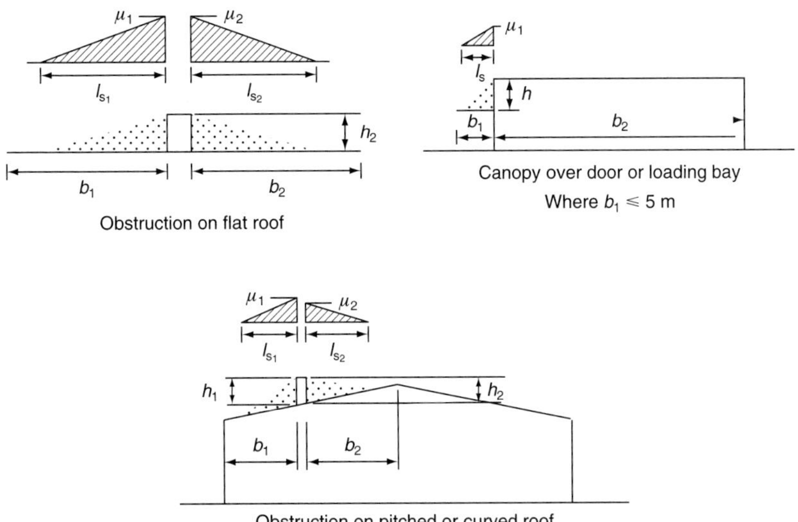

Fig. 6.1. Drifting at projections and obstructions allowed through the National Annex for specific local conditions

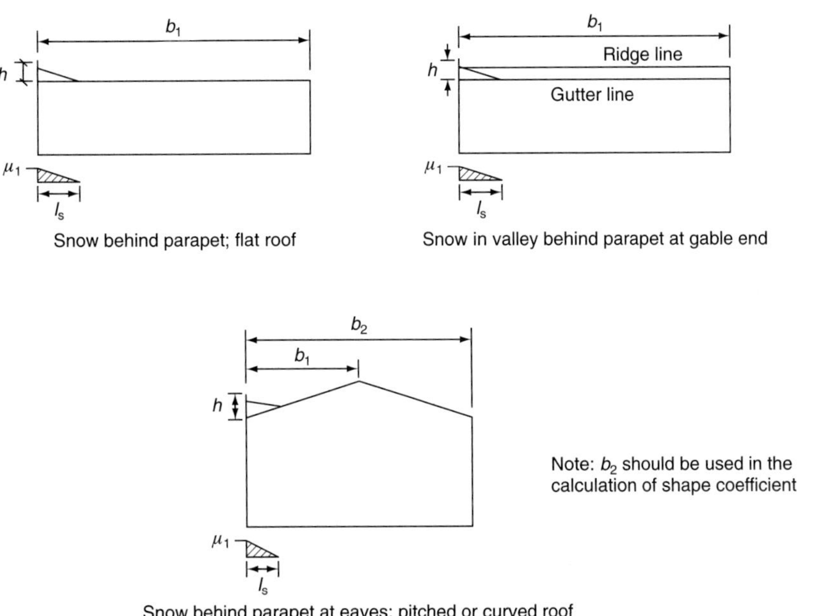

Fig. 6.2. Drifting at parapets allowed through the National Annex for specific local conditions

6.3. Snow overhanging the edge of a roof

For sites located at medium to high altitudes (recommended above 800 m) the design of the cantilevered parts of a roof, beyond the external walls, should take account of the possibility of snow overhanging; an example of this phenomenon is shown in Fig. 6.3. *cl.6.3(1): 1991-1-3* *Note to cl.6.3(1): 1991-1-3*

The load s_e per metre length of the roof's edge is given by the following formula: *cl.6.3(2): 1991-1-3*

$$s_e = ks^2/\gamma \qquad \text{(expression 6.4: EN 1991-1-3)}$$

where:

s is the most onerous undrifted load case appropriate for the roof under consideration
γ is the weight density of snow, which may be taken as $3\,\text{kN/m}^3$
k is a coefficient to take account of the irregular shape of the snow.

Fig. 6.3. Snow overhang

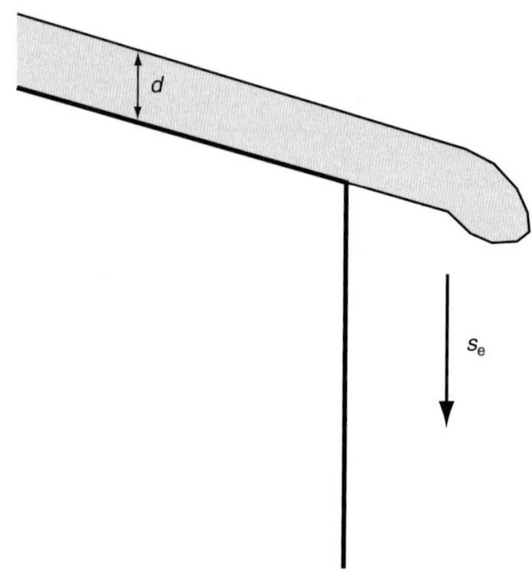

Fig. 6.4. Snow overhang

The values for the dimensionless coefficient k may be given by the National Annex. The recommended values are as follows:

Note to cl.6.3(2): 1991-1-3

$k = 3/d$, but $k \leq d\gamma$

where d is the roof snow cover depth, measured as shown in Fig. 6.4.

Example 6.2

Consider the edge of a roof illustrated in Fig. 6.4 and suppose the building being located at 1200 m a.s.l. in a zone where the characteristic ground snow load is $s_k = 1.50\,\text{kN/m}^2$. Suppose also that the edge considered is part of a duo-pitched roof with identical pitch angles $\alpha = 25°$.

From 5.3.1 and 5.3.3 we get the most onerous undrifted load value for the pitch terminating with the edge under consideration:

$\mu_1(\alpha = 25°) = 0.8$
$s = 1.50 \times 0.8 = 1.20\,\text{kN/m}^2$

The roof snow cover depth d is: $d = s/\gamma = 1.20/3.0 = 0.40\,\text{m}$.
The coefficient k in this case is given by the upper limit $k = d\gamma = 1.20$.
Finally we get the action due to snow overhang:

$s_e = ks^2/\gamma = 1.20 \times 1.20^2/3.0 = 0.58\,\text{kN/m}$

6.4. Snow loads on snowguards and other obstacles

cl.6.4(1): 1991-1-3

In accordance with Chapter 5 of this Part of this Designers' Guide, snow cover on the roof is assumed to remain in place to a pitch angle of 60°. The presence of snowguards or other obstacles on the roof will need special consideration, both for the design of the roof's structure, where the snow cover on the roof may not reduce below the obstacle's

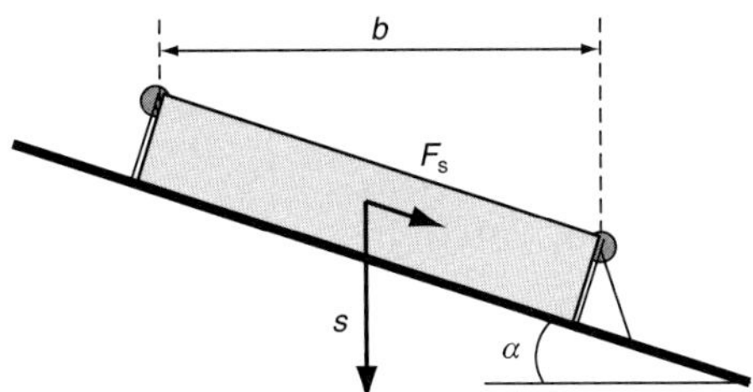

Fig. 6.5. Snow load on a snowguard

height independently of pitch angle, as well as for the design of the snowguard or of a more general obstacle.

Figure 6.5 shows the load pattern to be considered for the design of a snowguard.

The force F_s exerted by a sliding mass of snow, in the direction of slide, per unit length of the building is given by:

$$F_s = sb \sin \alpha \qquad \text{(expression 6.5: EN 1991-1-3)}$$

where:

s is the snow load on the roof relative to the most onerous undrifted load case appropriate for roof area from which snow could slide (see 5.2 and 5.3)

b is the width on plan (horizontal) from the guard or obstacle to the next guard or to the ridge

α is the pitch of the roof, measured from the horizontal.

Example 6.3

Consider the monopitched roof illustrated in Fig. 6.5, with pitch angle 30°, located in a place where the characteristic ground snow load is $s_k = 1.00\,\text{kN/m}^2$.

Let us calculate the force F_s exerted on the snowguard by the sliding mass of snow, assuming that the distance b between two consecutive snowguards is $b = 3\,\text{m}$.

From 5.3.1 and 5.3.2 we get the most onerous undrifted load value for the pitch under consideration:

$$\mu_1(\alpha = 30°) = 0.8$$

The corresponding roof snow load is $s = 0.8 \times 1.00 = 0.80\,\text{kN/m}^2$.

From expression 6.5 it results:

$$F_s = sb \sin \alpha = 0.80 \times 3.0 \times \sin(30°) = 1.20\,\text{kN/m}$$

CHAPTER 7

Annex A. Design situations and load arrangements to be used for different locations

This annex summarises, in a tabular form, the four possible cases for the definition of design situations and load arrangements to be used in accordance with the specifications contained in the National Annex.

For detailed description of the four cases see Chapter 3. *cl.A.1: EN 1991-1-3*

Table A.1: EN 1991-1-3. Design situations and load arrangements for different locations *Table A.1: 1991-1-3*

Normal	Exceptional conditions		
Case A	Case B1	Case B2	Case B3
No exceptional falls No exceptional drift	Exceptional falls No exceptional drift	No exceptional falls Exceptional drift	Exceptional falls Exceptional drift
3.2(1)	3.3(1)	3.3(2)	3.3(3)
Persistent/transient design situation: (1) Undrifted $\mu_i C_e C_t s_k$ (2) Drifted $\mu_i C_e C_t s_k$	Persistent/transient design situation: (1) Undrifted $\mu_i C_e C_t s_k$ (2) Drifted $\mu_i C_e C_t s_k$ Accidental design situation (where snow is the accidental action): (3) Undrifted $\mu_i C_e C_t C_{esl} s_k$ (4) Drifted $\mu_i C_e C_t C_{esl} s_k$	Persistent/transient design situation: (1) Undrifted $\mu_i C_e C_t s_k$ (2) Drifted $\mu_i C_e C_t s_k$ (except for roof shapes in Annex B) Accidental design situation (where snow is the accidental action): (3) Drifted $\mu_i s_k$ (for roof shapes in Annex B)	Persistent/transient design situation: (1) Undrifted $\mu_i C_e C_t s_k$ (2) Drifted $\mu_i C_e C_t s_k$ (except for roof shapes in Annex B) Accidental design situation (where snow is the accidental action): (3) Undrifted $\mu_i C_e C_t C_{esl} s_k$ (4) Drifted $\mu_i s_k$ (for roof shapes in Annex B)

Note 1: Exceptional conditions are defined according to the National Annex.
Note 2: For Cases B1 and B3 the National Annex may define design situations which apply for the particular local effects described in Section 6.

CHAPTER 8

Annex B. Snow load shape coefficients for exceptional snow drifts

This chapter is covered in Chapters 5 and 6 of this Part of this Designers' Guide as appropriate. Note that μ values given in the present annex are to be taken as non-dimensional values.

CHAPTER 9

Annex C. European ground snow load maps

Annex C of EN 1991-1-3 presents the European ground snow load maps, which are one of the main results of scientific work, carried out under contract to DGIII/D-3 of the European Commission, by a specifically formed research Group.[2,3]

The maps represent a significant innovation to the EN version of this Part of Eurocode 1. In the previous ENV 1991-2-3,[6] ground snow loads in different CEN countries, which at that time numbered 18, were given in the form of individual national maps.

This assembly of snow information, applicable in different CEN member states, encouraged comparative examinations, which revealed discrepancies along national boundaries. These were due to a large number of causes:

- National regulations were elaborated completely independently from one another.
- Differing procedures of measuring the amount of snow.
- The way of processing the data.
- The treatment of exceptional snow falls.
- The choice of an appropriate type of statistical distribution to calculate characteristic values.
- The choice of the return period.

In the research work reported in Reference 2 a homogeneous approach was set up for the elaboration of snow data and for the drawing of maps, to help National Competent Authorities to redraft their national maps, to eliminate or reduce the inconsistencies of snow load values in CEN member states at borders between countries.

The majority of maps presented in this annex were elaborated, starting from the collection of snow data in the 18 CEN member states, which were included in the previous ENV code. The snow database consists of approximately 2600 weather stations, spread all over Europe within a latitude ranging between 35° (Cyprus) and 70° North Cape Norway.

Maps presented in C(5) for Czech Republic, C(6) for Iceland and C(7) for Poland in Annex C of EN 1991-1-3, were added, but were not elaborated within the research work.

Maps are based upon a common procedure for the statistical treatment of data coming from the single weather station. The return period to calculate the characteristic value is fixed to 50 years, as indicated for climatic actions by EN 1990. The statistical analysis was performed on the yearly extreme values, using the Cumulative Distribution Function Gumbel Type I. The winters when there was no snow were taken into account.

Exceptional ground snow loads (see Sections 4.1 and 4.3 of this Part of this Designers' Guide) were excluded from the database.

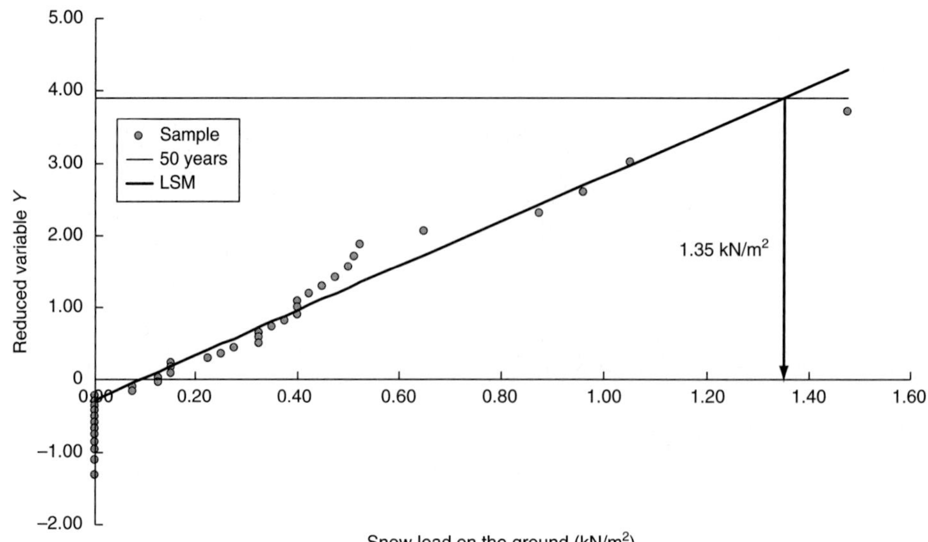

Fig. 9.1. Gumbel Type I probability paper, weather station in Parma, Italy; 50-year characteristic value = 1.35 kN/m^2

Figure 9.1 shows, as an example, the probability paper for the weather station in Parma, Italy, where the distinction between the years with zero and non-zero maximum snow loads is evident. It also represents the best-fitting line for the non-zero values and the characteristic value, obtained from the intersection of this line with the horizontal line located at 3.90 of the reduced probability scale (ordinate), corresponding to 0.02 probability of exceedance, which is approximately associated with a 50-year return period.

The characteristic values of ground snow load at the location of the weather station need to be interpolated with other values, to get a continuous geographical representation of the ground load variation.

At first sight, the simplest solution would be to set up a map giving directly the characteristic snow load at any place. However, in all mountainous areas of Europe such a map would have to be extremely detailed and would largely follow the topographic relief.

In the majority of regions with homogeneous climatic features, the best way to present the snow loads in a map for use by design engineers is to define areas in which a given altitude function can be applied, thus assuming that snow load varies with altitude only, disregarding other influences caused by local effects such as presence of great water masses (lakes) or concentration of snow in valleys etc. These effects were carefully analysed and data from weather stations which presented significant alterations with respect to the surrounding ones were excluded from the database.

The 18 CEN countries were divided into homogeneous climatic regions, through the analysis of the correlation presented by ground snow load characteristic values versus altitude of the relative weather station. The European territory was divided into ten climatic regions as illustrated in Fig. 9.2.

Borders of each region, defined through the correlation criteria illustrated above, are not necessarily coincident with national borders; an example is the Alpine region, to which belongs Switzerland, Austria and parts of Germany, France and Italy (see Fig. 9.3).

Figure 9.4 shows the scatterplot (ground snow load – altitude a.s.l.) for the Alpine region.

In the scatterplot each point represents a weather station and it is possible to calculate, by way of regression analysis, the best-fitting curve correlating the ground snow load with altitude. In the example, the curve is the parabola given by expression C.1.

$$s_k = (0.642Z + 0.009)\left[1 + \left(\frac{A}{728}\right)^2\right] \qquad \text{(C.1)}$$

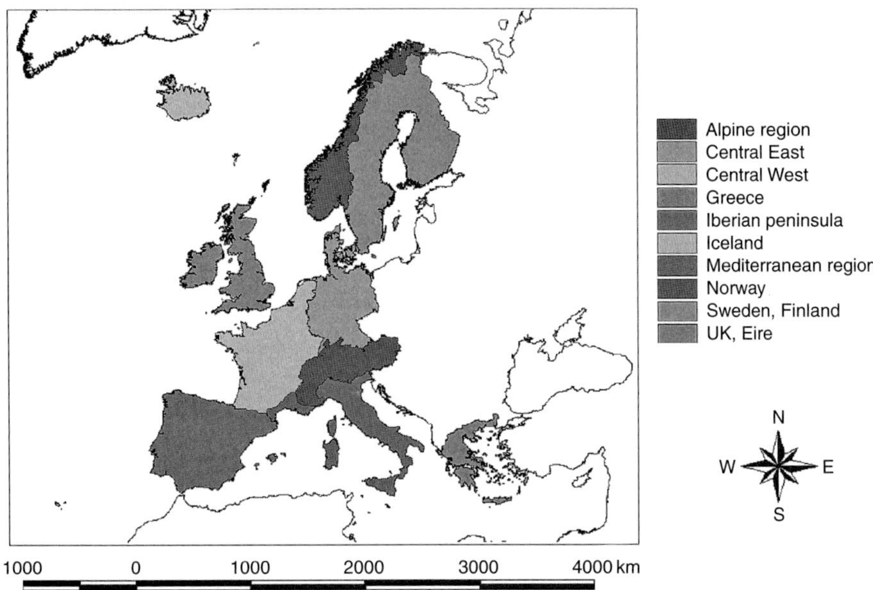

Fig. 9.2. European climatic regions

where:

A is the altitude a.s.l. of the site being considered in m
Z is the zone number given on the map for the site considered.

It is stressed that expression C.1 applies only for the Alpine region, and each climatic region (see Fig. 9.2) has its unique expression. These expressions are given in Table C.1 of EN 1991-1-3.

The variation of Z allows the coverage of the whole scatterplot in n intervals, which in the case under consideration is 5. Each interval is represented by the mean curve marked with a dashed line in Fig. 9.4.

Weather stations belonging to the same interval are geographically identified and grouped in the same zone. In Fig. 9.3 the five zones identified in the scatterplot are reduced to only four. The two upper zones, numbered 4 and 5, are grouped together here.

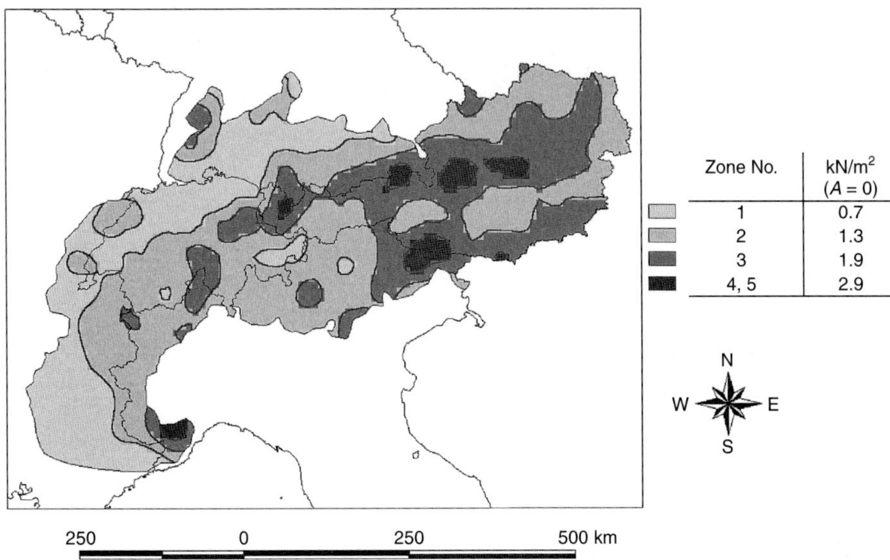

Fig. 9.3. Alpine region. Snow load at sea level

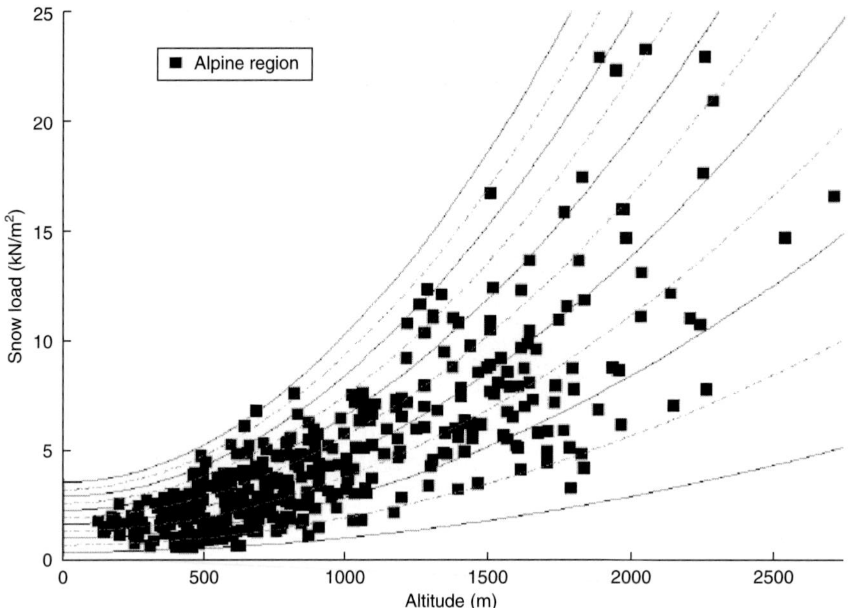

Fig. 9.4. Alpine region. Correlation between characteristic ground snow load and altitude a.s.l.

Example C.1

Consider the ground snow load map for the Central West climatic region in the figure below, to calculate the ground snow load in Paris, 75 m a.s.l.

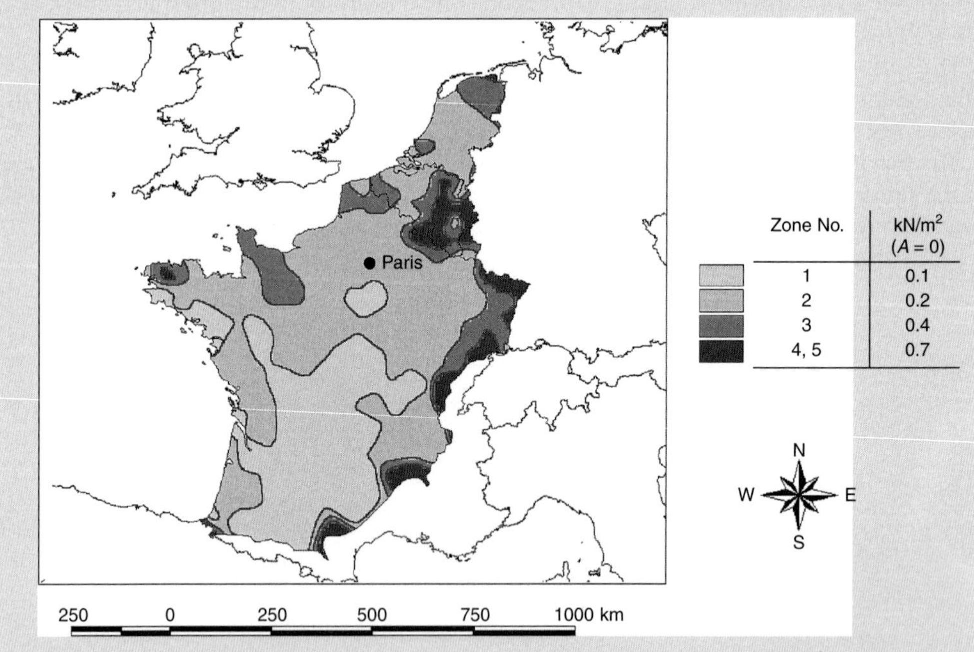

The zone number corresponding to the location of Paris is 2.

The altitude function for the Central West region is given by the following expression:

$$s_k = 0.164Z - 0.082 + \frac{A}{966}$$

with $Z = 2$ and $A = 75$ it gives:

$s_k = 0.32\,\text{kN/m}^2$

Data from weather stations from border regions, located within 100 km into the neighbouring regions, have also been included. In this way the discontinuities at borders between climatic regions are significantly limited.

The use of the maps is extremely easy for the design engineer; he or she needs to input the location of the site on the map and its altitude in metres only. With the zone number and the altitude function known, the characteristic ground snow load on the site is easily obtained. The following example shows the procedure.

Maps given in Annex C to EN 1991-1-3 are recommended to be considered as a reference with the aim of satisfying the objectives indicated in Clause C(2) of EN 1991-1-3, listed here below:

- to help National Competent Authorities to redraft their national maps
- to establish harmonised procedures to produce the maps.

The final goal of the implementation of the use of this European map will be the elimination or reduction of the inconsistencies of snow load values in CEN member states and at borders between countries.

Further work is still needed for the enlargement of the map to cover the 'new' CEN member states, which now total 28.

CHAPTER 10

Annex D. Adjustment of the ground snow load according to return period

Ground snow load values given by maps in Annex C and by National Annexes are characteristic values, and by definition are based on annual probability of exceedance of 0.02, corresponding to a mean recurrence interval of approximately 50 years (see Section 4.1 of this Part of this Designers' Guide).

To obtain ground snow loads associated with any mean recurrence interval different to 50 years it is recommended to apply the following formula: *cls.D(1) and D(2): 1991-1-3*

$$s_n = s_k \left\{ \frac{1 - V\frac{\sqrt{6}}{\pi}\{\ln[-\ln(1 - P_n)] + 0.57722\}}{(1 + 2.5923V)} \right\} \qquad \text{(Expression D.1)}$$

where:

s_k is the characteristic snow load on the ground (with a return period of 50 years, in accordance with EN 1990: 2002)

s_n is the ground snow load with a return period of n years

P_n is the annual probability of exceedance (equivalent to approximately $1/n$, where n is the corresponding recurrence interval in years)

V is the coefficient of variation of annual maximum snow load.

Expression D.1 should not be used for probability of exceedance below 0.20, which is associated with a return period approximately equal to five years.

The use of the above formulation is based on the assumption that the available recorded series of yearly maxima ground snow loads shows a good fit to a Gumbel Type I probability distribution function (PDF). This has been shown to be the case for the majority of European snow data samples analysed in Reference 2. Any other best-fitting statistical distribution (such as log-normal or Weibull) can be adopted where allowed by the National Annex. Obviously, in these cases expression D.1 has to be appropriately modified. *Note 1 to cl.D(2): 1991-1-3*

For the evaluation of the coefficient of variation V of recorded yearly maxima, the designer has to refer to specific information for the geographical location under consideration. Obviously values of V may vary significantly from site to site. Figure 10.1 shows an example where the coefficient of variation calculated for 331 weather stations in Germany is plotted against the mean ground snow load.[3]

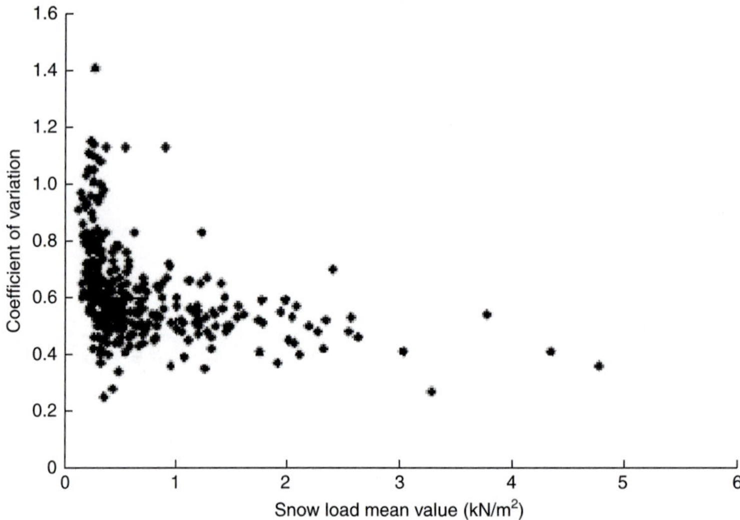

Fig. 10.1. Coefficient of variation plotted against mean ground snow loads for 331 weather stations in Germany

In Fig. 10.2 the coefficient of variation is plotted versus the altitude of the weather stations.

From Figs 10.1 and 10.2 it is evident that the scatter of the coefficient of variation may be very large (in this example ranging from 0.2 to 1.4), the wider scatter being associated with lower mean ground snow loads, which are normally registered at weather stations located at lower altitudes.

For initial calculations, and where no refined estimates are needed, it may be useful to refer to mean values of the coefficient of variation for a single climatic region or even for wider regions.

In Reference 3 the coefficient of variation has been calculated for 18 countries covered by the European map in Annex C, and elaborated within the research work; the resulting overall mean value is $V = 0.6$.

cl.D(3): 1991-1-3 Figure 10.3 shows the graphical representation of expression D.1, plotted for $0.2 \leq V \leq 0.6$.

For a return period of 10 years, which, for example, could be associated with a given execution phase of the structure with an expected nominal duration of one year (see Part

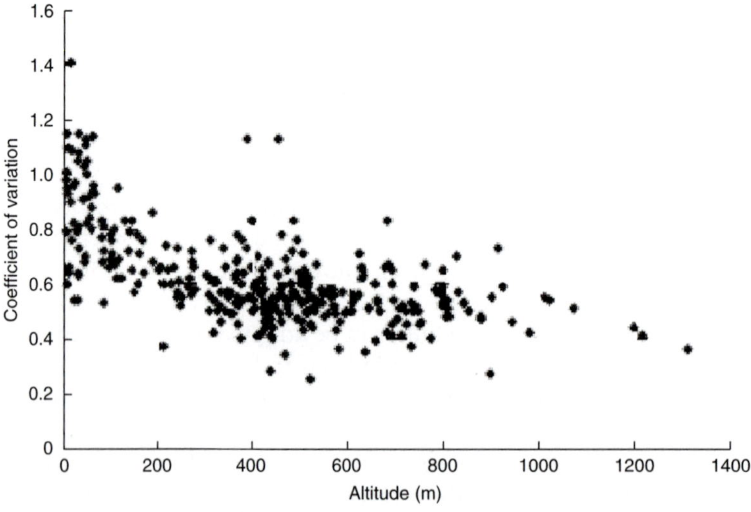

Fig. 10.2. Coefficient of variation plotted against altitude for 331 weather stations in Germany

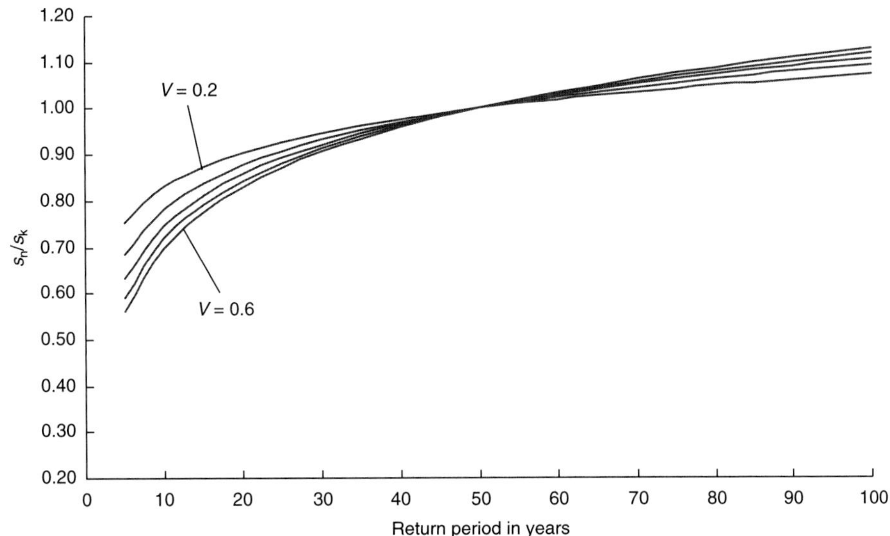

Fig. 10.3. Adjustment of the ground snow load according to return period

6 of this Designers' Guide and Chapter 2 of EN 1991-1-6), the ratio s_n/s_k varies from 0.83 for $V = 0.2$ to 0.70 for $V = 0.6$.

Expression D.1, where permitted by the National Annex, may also be used to calculate snow loads on the ground for other probabilities of exceedance, such as for structures where a higher risk of exceedance is deemed acceptable (e.g. greenhouses) or where greater than normal safety is required (e.g. strategic structures). *cl.D(4): 1991-1-3*

CHAPTER 11

Annex E. Bulk weight density of snow

The bulk weight density of snow is a function of many parameters among which are included:

- the duration of the snow cover on the ground (or on the roof)
- the location of the site
- the exposure to solar radiation
- the climate and the altitude of the site itself *cl.E(1): 1991-1-3*

Del Corso and Formichi[7] illustrated and discussed in detail the influences on the determination of snow bulk weight density registered in Italy, which are briefly summarised below as an example valid for regions in a Mediterranean climate.

First, the air temperature during the snow fall plays an extremely important role in the formation of crystals, on the definition of their size and shape and on the tendency to clump together, leading to a low snow density in the case of dry cold air and high densities in the case of humid relatively warm air. That is, wet snow particles (generally in humid, relatively warm air) are susceptible to clumping, due to cohesion, whereas dry snow particles (generally in dry, cold air) behave more akin to sand particles, governed by frictional effects.

Furthermore, it has been observed that the bulk weight density rapidly increases after a few hours from the snow event. This is mainly due to the effect of the modification of lower-layer crystals' shape, under pressure exerted by upper snow layers; this results in a reduction of the overall layer volume and in the increase of its mass.

Duration on the ground (or on the roof) of the snow layer is also another extremely important parameter, significantly affecting snow density. For long-lasting snow layers, the overall density varies due to the drifting–compression–erosion process, consequence of successive snow events in time, wind action and air temperature variations. Therefore 'younger' snow layers have a lower density than 'older' ones. Once again this is the effect of the compression of the lower and oldest snow layers, which may have incorporated water due to the melting of upper layers.

Measurements taken at several high-altitude weather stations in the Italian Alps are described by Zanon,[8] where snow lasts on the ground for periods up to several months; the data show a clear tendency to an increase in density with time. Specifically it has been observed that a rapid increase occurs at the beginning of winter, due to the drifting and of the successive compression of different fresh snow layers. This is followed by a 'central' period, during which, probably due to the crystals' stabilisation shape and structure, snow

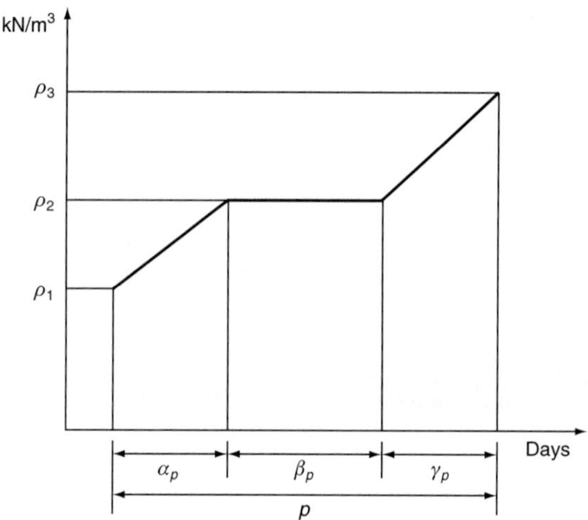

Fig. 11.1. Bulk weight density of snow as a function of layer duration, adopted in Italy

density can be assumed to remain constant. Finally, during the melting phase the density increases rapidly due to the presence of water.

Figure 11.1 shows the simplified bulk weight density law based upon the above studies, adopted in Italy for the conversion of ground snow depths into ground snow loads, where the duration of the snow layer (p) is greater than 10 days.

Values ρ_1, ρ_2, ρ_3 are the limits of variation of the snow density during the period 'p'. Sub-periods α_p and γ_p are the initial and the final ones, associated with the increase in snow density; β_p is the 'central' sub-period during which density may be assumed to remain constant. Values are calibrated through experimental measures[8–10] as follows:

$$\rho_1 = 2.15\,\text{kN/m}^3, \quad \alpha_p = 0.3$$
$$\rho_2 = 3.50\,\text{kN/m}^3, \quad \beta_p = 0.4$$
$$\rho_3 = 5.15\,\text{kN/m}^3, \quad \gamma_p = 0.3$$

For snow layers with a duration lower than 10 days, which is typical of low-altitude sites where the snow layer results from one single snow event, bulk weight density has been assumed to remain constant and equal to $2.50\,\text{kN/m}^3$.

Where no direct measures are available, and except where specified in Sections 1 to 6 of EN 1991-1-3, mean bulk weight density of snow on the ground may be assumed to vary within the limits given in Table 11.1 (***Clause E(2): EN 1991-1-3***).

cl.E(2): 1991-1-3

Table 11.1. Mean bulk weight density of snow

Type of snow	Bulk weight density (kN/m^3)
Fresh	1.0
Settled (several hours or days after its fall)	2.0
Old (several weeks or months after its fall)	2.5–3.5
Wet	4.0

References to Part 3

1. EN 1991-1-3: 2003. *Eurocode 1 – Actions on Structures – Part 1-3: General Actions – Snow loads*. European Committee for Standardisation, Brussels, 2003.
2. *Final Report to the European Commission, Scientific Support Activity in the Field of Structural Stability of Civil Engineering Works: Snow Loads – Phase 1*. Department of Structural Engineering, University of Pisa, March 1998. The document is available at the Commission of the European Communities DG III, D-3 Industry, Rue de la Loi, 200 B, 1049 Brussels, or at the Università degli Studi di Pisa Dipartimento di Ingegneria Strutturale, Via Diotisalvi, 2, 56100 Pisa, Italy. Or online at: http://www2.ing. unipi.it/dis/snowloads/
3. *Final Report to the European Commission, Scientific Support Activity in the Field of Structural Stability of Civil Engineering Works: Snow Loads – Phase 2*. Department of Structural Engineering, University of Pisa, September 1999. The document is available at the Commission of the European Communities DG III, D-3 Industry, Rue de la Loi, 200 B, 1049 Brussels, or at the Università degli Studi di Pisa Dipartimento di Ingegneria Strutturale, Via Diotisalvi, 2, 56100 Pisa, Italy. Or online at: http://www2.ing. unipi.it/dis/snowloads/
4. ISO 4355: 1998. *Bases for Design of Structures – Determination of Snow Loads on Roofs*. International Organisation for Standardisation, Geneva, 1998.
5. Irwin, P., Hochstenbach, F. and Gamble, L. *Wind and Snow Considerations for Wide Span Enclosures*. Rowan Williams Davies & Irwin, Ontario, Canada, 2004.
6. ENV 1991-2-3: 1996. *Eurocode 1. Basis of Design and Actions on Structures. Actions on Structures. Snow loads*. European Committee for Standardisation, Brussels, 1996.
7. Del Corso, R. and Formichi, P. A proposal for a new normative snow load map for the Italian territory in *Proceedings of the 5th International Conference on Snow Engineering, Davos, Switzerland*, 2004. A.A. Balkema, Rotterdam, 2004.
8. Zanon, G. Osservazioni sul manto nevoso alla stazione pilota della Fedaia nell'inverno 1962–1963. *Bollettino Comitato Glaciologico Italiano*, 1967, Series II, No. 13, Part 2.
9. Del Corso, R. and Formichi, P. Duration of the snow load on the ground in Italy. *Proceedings of the 4th International Conference on Snow Engineering, Trondheim, Norway*, June 2000. A.A. Balkema, Rotterdam, 2000.
10. Del Corso, R. and Formichi, P. Shape coefficients for conversion of ground snow loads to roof snow loads. *Proceedings of the 16th International Congress of the Precast Concrete Industry, Venice, Italy*, May 1999.

PART 4: EN 1991-1-4: Eurocode 1: Part 1-4: Wind Actions

This Part gives a brief introduction to EN 1991-1-4: *Eurocode 1 – Actions on Structures: Part 1.4 General Actions – Wind Actions*.[1] A comprehensive treatment for this Eurocode part is given in the Thomas Telford Designers' Guide on wind actions by N. J. Cook.[2]

The style of this brief introduction is different to the other Parts in this Designers' Guide which cover other Parts of EN 1991: *Actions on Structures*, in that it gives guidance on selected clauses only. This Part makes numerous references to the UK National Annex to EN 1991-1-4 and BS 6399: Part 2. In cases where the information is substantial, the text, graphs etc. have been boxed.

The contents of EN 1991-1-4

EN 1991-1-4 contains eight sections and six informative annexes as follows:

Section 1 General
Section 2 Design situations
Section 3 Modelling of wind actions
Section 4 Wind velocity and velocity pressure
Section 5 Wind actions
Section 6 Structural factor $c_s c_d$
Section 7 Pressure and force coefficients
Section 8 Wind actions on bridges
Annex A (informative) Terrain effects
Annex B (informative) Procedure 1 for structural factor $c_s c_d$
Annex C (informative) Procedure 2 for structural factor $c_s c_d$
Annex D (informative) Graphs of $c_s c_d$ for common building forms
Annex E (informative) Vortex shedding and aeroelastic instabilities
Annex F (informative) Dynamic characteristics of structures

CHAPTER 1

General

1.1. Scope

EN 1991-1-4 gives guidance on the determination of natural wind actions for the structural design of buildings and civil engineering works for each of the loaded areas under consideration. The wind actions are given for the whole or parts of the structure, e.g. components, cladding units and their fixings. *cl.1.1(1): 1991-1-4*

EN 1991-1-4 is applicable to: *cl.1.1(2): 1991-1-4*

- building and civil engineering works with heights up to 200 m
- bridges with spans of not more than 200 m (subject to certain limitations based on dynamic response criteria).

EN 1991-1-4 is intended to predict characteristic wind actions on land-based structures, their components and appendages. *cl.1.1(3): 1991-1-4*

The specific exclusions are: *cl.1.1(11): 1991-1-4*

- lattice towers with non-parallel chords
- guyed masts and guyed chimneys (covered in EN 1993)
- cable-supported bridges
- bridge deck vibration from transverse wind turbulence
- torsional vibrations of buildings
- modes of vibration higher than the fundamental mode.

The application range of EN 1991-1-4 is much wider than compared to some European National Wind Standards. For example, comparing EN 1991-1-4 to BS 6399-2,[3] the scope of EN 1991-1-4 is much wider than BS 6399-2, as it includes wind actions on other structures, which in the UK are given in a number of other British Standards and design guides. In some cases, for example dynamic response of buildings, there is no equivalent UK standard.

1.2. Definitions and symbols

Some of the terminology, symbols and definitions used in EN 1991-1-4 will be new for designers. *cls.1.6 and 1.7: 1991-1-4*

Regarding the UK, the major differences between EN 1991-1-4 and UK practice[4] are given below. The clause numbers indicate where the term is first mentioned in EN 1991-1-4.

- **Background response factor *B* (6.3.1(1)):** accounts for the lack of correlation of the wind gusts over the surface of the structure or element. There is no equivalent value in BS 6399-2. *cl.6.3.1(1): 1991-1-4*

cl.4.5(1): 1991-1-4

- **Basic velocity pressure q_b (4.5(1)):** derived from the basic wind velocity ($q_b = 0.613v_b^2$). There is no equivalent value in BS 6399-2.

cl.4.2(2)P: 1991-1-4

- **Basic wind velocity v_b (4.2(2)P):** is the fundamental basic wind velocity modified to account for seasonal and directional effects. There is no direct equivalent value in BS 6399-2.

cl.4.2(2)P: 1991-1-4

- **Directional factor c_{dir} (4.2(2)P):** used to modify the basic wind velocity to produce wind speeds with the same risk of being exceeded in any wind direction. c_{dir} is the same as S_d in BS 6399-2.

cl.4.5(1): 1991-1-4

- **Exposure factor $c_e(z)$ (4.5(1)):** accounts for the effect of terrain, orography and building height. Similar to S_b^2 in BS 6399-2. Note: orography is equivalent to topography in BS 6399-2.

cl.5.3(2): 1991-1-4

- **Force coefficient c_f (5.3(2)):** the ratio of the force acting on a structure or element to the peak velocity pressure multiplied by an appropriate area. There are no equivalent values in BS 6399-2.

cl.5.3(2): 1991-1-4

- **Friction coefficient c_{fr} (5.3(2)):** the ratio of the frictional drag on surfaces aligned parallel with the wind to the peak velocity pressure multiplied by an appropriate area. Corresponds to C_f in BS 6399-2.

cl.4.2(1)P: 1991-1-4

- **Fundamental basic wind velocity $v_{b,0}$ (4.2(1)P):** this is the mean wind velocity for a 10-minute averaging period with an annual risk of being exceeded of 0.02, at a height of 10 m above ground level in flat open country terrain (terrain category II). There is no direct equivalent value in the main body of BS 6399-2. For use in the UK, $v_{b,0} = v_{b,0}^* c_{alt}$ where $v_{b,0}^*$ is the mean wind velocity for a 10-minute averaging period with an annual risk of being exceeded of 0.02, at a height of 10 m above sea level in terrain category II and c_{alt} is the altitude factor which accounts for the effects of altitude on the fundamental basic wind velocity.

Note 4 to cl.4.2(2)P: 1991-1-4

- **Mean wind velocity $v_m(z)$ (4.2(2)P Note 4):** is the basic wind velocity modified to account for terrain roughness category and orography effects (defined below). There is no direct equivalent value in BS 6399-2.

cl.4.3.1(1): 1991-1-4

- **Orography factor $c_o(z)$ (4.3.1(1)):** used to account for the increase in wind speed due to topographic features such as hills, cliffs and escarpments. $c_o(z)$ is obtained using the same method for calculating topographic effects as included in the S_a factor in BS 6399-2.

cl.4.5(1): 1991-1-4

- **Peak velocity pressure $q_p(z)$ (4.5(1)):** is the site wind velocity taking account of the terrain and building size. $q_p(z)$ corresponds to q_s in BS 6399-2.

cl.5.2(1): 1991-1-4

- **Pressure coefficients c_{pe} and c_{pi} (5.2(1)):** the ratio of the pressure acting on the external or internal surfaces to the peak velocity pressure. Corresponds to C_{pe} and C_{pi} in BS 6399-2.

Note 4 to cl.4.2(2)P: 1991-1-4

- **Probability factor c_{prob} (4.2(2)P Note 4):** used to modify the basic wind velocity to change the risk of the wind speed being exceeded. c_{prob} is the same as S_p in BS 6399-2.

cl.6.3.1(1): 1991-1-4

- **Resonant response factor R (6.3.1(1)):** accounts for the effects of wind turbulence in resonance with the vibration of the structure in its fundamental mode of vibration. There is no equivalent value in BS 6399-2.

cl.4.3.1(1): 1991-1-4

- **Roughness factor $c_r(z)$ (4.3.1(1)):** used to modify the mean wind speed to account for the terrain roughness upwind of the site and the height of the building or structure under consideration. $c_r(z)$ corresponds to S_c in the BS 6399-2 directional method.

cl.4.2(2)P: 1991-1-4

- **Season factor c_{season} (4.2(2)P):** used to modify the basic wind velocity to produce wind speeds with the same risk of being exceeded in any specific sub-annual period. c_{season} is the same as S_s in BS 6399-2.

cl.5.3(2): 1991-1-4

- **Structural factor $c_s c_d$ (5.3(2)):** takes account of the effect of non-simultaneous wind action over the surfaces of the structure or element combined with the effect of dynamic response of the structure or element. c_s and c_d correspond to C_a and $(1 + C_r)$ respectively in BS 6399-2.

cl.4.4(1): 1991-1-4

- **Turbulence intensity $I_v(z)$ (4.4(1)):** is the standard deviation of the wind turbulence divided by the mean wind velocity and is a measure of the 'gustiness' of the wind. $I_v(z)$ corresponds to $S_t T_t$ in the BS 6399-2 directional method.

- **Wind force F_w, $F_{w,e}$, $F_{w,i}$, F_{fr} (5.3(2)):** the wind force acting on the overall structure or element (F_w), on the external surfaces ($F_{w,e}$), on the internal surfaces ($F_{w,i}$) or due to frictional forces (F_{fr}). Corresponds to P and P_f in BS 6399-2. *cl.5.3(2): 1991-1-4*
- **Wind pressure w_e and w_i (5.2(1)):** the wind pressure acting on external and internal building surfaces respectively. Corresponds to p_e and p_i in BS 6399-2. *cl.5.2(1): 1991-1-4*

CHAPTER 2

Design situations

The general requirement of this chapter is that the relevant wind actions be determined for each design situation identified in EN 1990 (*Basis of Structural Design*), such as persistent, transient and accidental design situations. *cl.2.1(1)P: 1991-1-4*

In EN 1991-1-4, the following specific design situations should also be taken into account:

- Modification to wind actions due to other actions (such as snow, traffic or ice) which will affect the wind loads. *cl.2.1(2): 1991-1-4*
- Changes to the structure during construction which modify the wind loads. *cl.2.1(3): 1991-1-4*
- Where doors and windows that are assumed to be shut are open under storm conditions then they should be treated as accidental actions. *cl.2.1(4): 1991-1-4*
- Fatigue due to wind actions should be considered for susceptible structures. *cl.2.1(4): 1991-1-4*

CHAPTER 3

Modelling of wind actions

This chapter describes wind actions and defines characteristic values of wind velocity and velocity pressure according to EN 1990. Wind actions need to be classified according to their variation in time, spatial variation, origin and their nature and/or structural response. Wind actions are classified according to EN 1990 as variable fixed actions; this means that the wind actions will not always be present, and the wind actions have, for each considered wind direction, a fixed distributions along the structure. The classification to the origin of wind actions can be direct as well as indirect: direct for external surfaces and internal surfaces of open structures; indirect for internal surfaces of enclosed structures (***Clause 3.1(1): EN 1991-1-4***) and (***Clause 3.3(1): EN 1991-1-4***). *cl.3.1(1): 1991-1-4* *cl.3.3(1): 1991-1-4*

Wind actions calculated using EN 1991-1-4 are characteristic values (see EN 1990 Clause 4.1.2) having annual probabilities of exceedance of 0.02, which is equivalent to a mean return period of 50 years (***Clause 3.4(1): EN 1991-1-4***). *cl.3.4(1): 1991-1-4*

CHAPTER 4

Wind velocity and velocity pressures

4.1. Basis for calculation

One of the main parameters in the determination of wind actions on structures is the characteristic peak velocity pressure q_p. This parameter is the characteristic pressure due to the wind velocity of the undisturbed wind field. The peak wind velocity accounts for the mean wind velocity and a turbulence component. The characteristic peak velocity pressure q_p is influenced by the regional wind climate, local factors (e.g. terrain roughness and orography/terrain topography), terrain categories, see Table 4.1 and Fig. 4.1, and the height above terrain. See UK National Annex information in 4.3. *cl.4.1(1): 1991-1-4*

Table 4.1. Terrain categories and terrain parameters

Terrain category		z_0 (m)	z_{min} (m)
0	Sea or coastal area exposed to the open sea	0.003	1
I	Lakes or flat and horizontal area with negligible vegetation and without obstacles	0.01	1
II	Area with low vegetation such as grass and isolated obstacles (trees, buildings) with separations of at least 20 obstacle heights	0.05	2
III	Area with regular cover of vegetation or buildings or with isolated obstacles with separations of maximum 20 obstacle heights (such as villages, suburban terrain, permanent forest)	0.3	5
IV	Area in which at least 15% of the surface is covered with buildings and their average height exceeds 15 m	1.0	10

Note: Examples of the terrain categories are illustrated in Fig. 4. *cl.A.1: 1991-1-4*

4.2. Basic values

The wind climate for different regions/countries in Europe is described by values related to the characteristic 10-minute mean wind velocity at 10 m above ground of a terrain with low vegetation (terrain category II, see Table 4.1). These characteristic values correspond to annual probabilities of exceedance of 0.02 which corresponds to a return period of 50 years. In EN 1991-1-4 this variable is denoted as the *fundamental value of the basic wind velocity* $v_{b,0}$ which may be given in the National Annex through a National wind map. *cl.4.2(1)P: 1991-1-4* *Note 2 to cl.4.2(1)P: 1991-1-4*

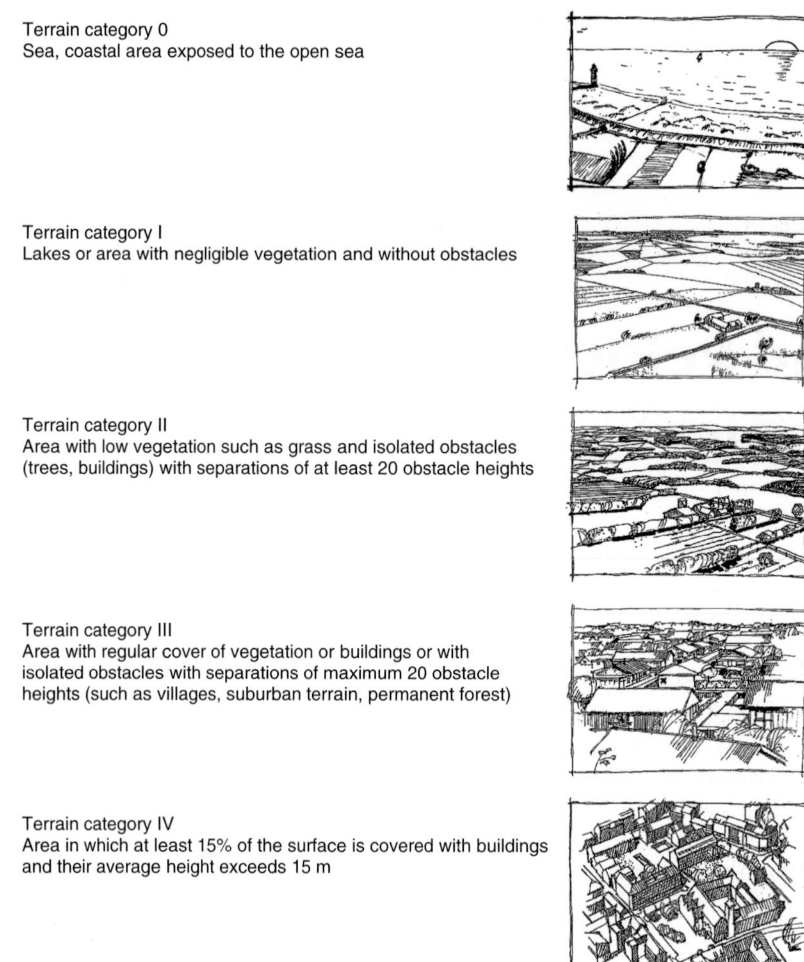
Terrain category 0
Sea, coastal area exposed to the open sea
Terrain category I
Lakes or area with negligible vegetation and without obstacles
Terrain category II
Area with low vegetation such as grass and isolated obstacles (trees, buildings) with separations of at least 20 obstacle heights
Terrain category III
Area with regular cover of vegetation or buildings or with isolated obstacles with separations of maximum 20 obstacle heights (such as villages, suburban terrain, permanent forest)
Terrain category IV
Area in which at least 15% of the surface is covered with buildings and their average height exceeds 15 m

Fig. 4.1. Illustrations of the upper roughness of each terrain category

cl.4.2(2)P: 1991-1-4

The *basic wind velocity* v_b having an annual probability of exceedance equal to 0.02 is determined with the expression:

$$v_b = c_{dir} c_{season} v_{b,0} \tag{4.1}$$

where:

$v_{b,0}$ is the fundamental value of basic wind velocity
v_b is the basic wind velocity
c_{dir} is the directional factor
c_{season} is the seasonal factor.

The following should be noted:

- The directional factor c_{dir} accounts for the fact that for particular wind directions the velocity v_b could be decreased.
- The seasonal factor c_{season} takes into account, for example the case of temporary structures for particular periods (e.g. in summer) where the probability of occurrence of high wind velocities is relatively low.

Note also that a transportable structure (such as light covering for exhibitions) may not be considered as a temporary structure, and the c_{season} factor should be taken equal to 1.0 (see also EN 1991-1-6 for the definition of temporary structures).

The recommended values for the directional factor c_{dir} and the seasonal factor c_{season} are in general equal to 1.0. *Notes 2 and 3 to cl.4.2(1)P: 1991-1-4*

EN 1991-1-4 expression (4.10) (given below as expression (4.2)) gives the following relationship between the basic wind velocity (v_b) and the basic wind velocity pressure (q_b):

$$q_b = \rho/2v_b^2 \tag{4.2}$$

where ρ is the density of air whose recommended value is 1.25 kg/m^3. This value represents the mean wind velocity pressure (averaging interval 10 min.) at a reference height of 10 m in open terrain of category II, with a return period of 50 years, and does not include the wind turbulence.

4.3. Mean wind velocity

The basic wind velocity pressure has to be transformed into the value at the reference height of the considered structure. The wind velocity at a relevant height (z) and the gustiness of the wind depend on the terrain roughness. The roughness factor ($c_r(z)$) describing the variation of the wind speed with height has to be determined in order to obtain the mean wind speed $v_m(z)$ at the relevant height z using expression (4.3): *cl.4.3.1.(1): 1991-1-4*

$$v_m(z) = c_r(z)c_o(z)v_b \tag{4.3}$$

where:

$v_m(z)$ is the mean velocity
$c_r(z)$ is the roughness factor
$c_o(z)$ is the orography factor (usually taken as 1.0).

The **roughness factor ($c_r(z)$)** accounts for the variability of the mean wind velocity at the site of the structure due to: *cl.4.3.2.(1): 1991-1-4*

- the height above ground level
- the ground roughness of the terrain upwind of the structure in the wind direction considered.

The recommended procedure for the determination of the roughness factor at height z is given by expression (4.4) and is based on a logarithmic velocity profile: *Note to cl.4.3.1.(1): 1991-1-4*

$$\begin{aligned} c_r(z) &= k_r \ln\left(\frac{z}{z_0}\right) \quad \text{for} \quad z_{min} \le z \le z_{max} \\ c_r(z) &= c_r(z_{min}) \quad \text{for} \quad z \le z_{min} \end{aligned} \tag{4.4}$$

where:

z_0 is the roughness length defined in Table 4.1
k_r is the terrain factor depending on the roughness length z_0 calculated using:
z_{min} is the minimum height defined in Table 4.1
z_{max} is to be taken as 200 m, unless otherwise specified in the National Annex

$$k_r = 0.19\left(\frac{z_0}{z_{0,II}}\right)^{0.07} \tag{4.5}$$

where:

$z_{0,II}$ = 0.05 m (terrain category II, Table 4.1)

With regard to the guidance in EN 1991-1-4, Fig. 4.2 shows the variation of C_r with height z, up to 200 m.

UK National Annex

Expressions (4.4) and (4.5) are not used in the BSI National Annex in which the classification of roughness categories has been simplified to give the following three terrain categories:

- Terrain category 0 is referred to as Sea terrain (Zone A).
- Terrain categories I and II have been considered together to give a single terrain category referred to as Country terrain (Zone B).
- Terrain categories III and IV have been considered together to give a single terrain category referred to as Town terrain (Zone C).

NA2-11: BSI NA: 1991-1-7 All inland lakes that extend more than 1 km in the direction of wind and closer than 1 km upwind of the site should be treated as Sea.

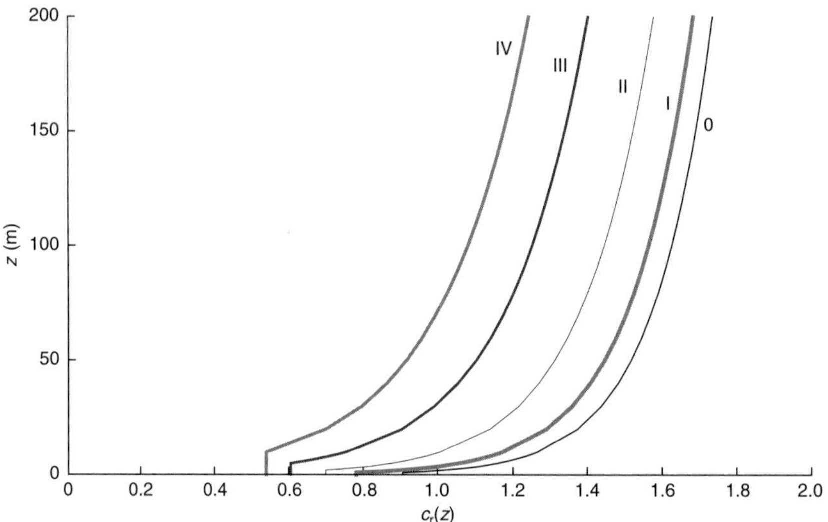

Fig. 4.2. Variation of the roughness factor $c_r(z)$ with the height above the terrain

4.4. Wind turbulence

cl.4.4. (1): 1991-1-4 The **turbulence intensity $I_v(z)$** at height z is defined as the standard deviation of the turbulence divided by the mean wind velocity.

Note 1 to cl.4.4(1): 1991-1-4 The turbulent component of wind velocity has a mean value of 0 and a standard deviation l_v. The standard deviation of the turbulence l_v is be determined using expression (4.6).

$$\sigma_v = k_r v_b k_I \tag{4.6}$$

where:

the terrain factor k_r is determined from expression (4.5)
the basic wind velocity v_b from expression (4.1)
for turbulence factor k_I see expression (4.7).

Note 2 to cl.4.4(1): 1991-1-4 The recommended rules for the determination of $I_v(z)$ are given in expression (4.7) (***Note 2 to Clause 4.4(1): EN 1991-1-4***):

$$I_v(z) = \frac{\sigma_v}{v_m(z)} = \frac{k_I}{c_o(z)\ln(z/z_0)} \quad \text{for} \quad z_{min} \le z \le z_{max}$$
$$I_v(z) = I_v(z_{min}) \quad \text{for} \quad z < z_{min} \tag{4.7}$$

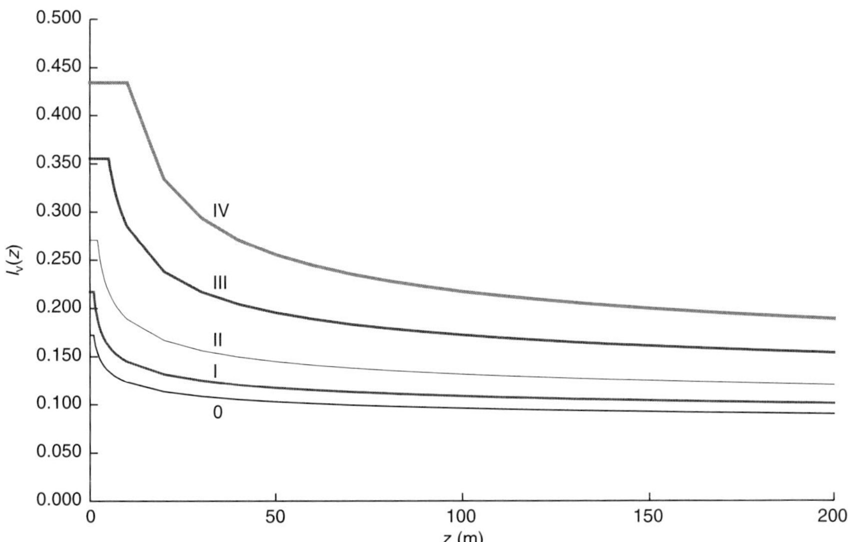

Fig. 4.3. Variation of the tubulence intensity factor $I_v(z)$ with the height above the terrain, for various terrain categories ($k_I = 1.0$, $c_0(z) = 1.0$)

where:

k_I is the turbulence factor. The value of k_I to be used in a country may be found in its National Annex. The recommended value is $k_I = 1.0$

c_o is the orography factor as described in Clause 4.3.3

z_0 is the roughness length, given in Table 4.1.

Figure 4.3 shows the variation of the turbulence intensity factor $I_v(z)$, for various terrain categories, having assumed both $k_I = 1.0$ and $c_0(z) = 1.0$.

4.5. Peak velocity pressure

The recommended method for determining the peak velocity pressure $q_p(z)$ at height z, which includes mean and short-term velocity fluctuations, is given below. *cl.4.5(1): 1991-1-4*

$$q_p(z) = [1 + 7I_v(z)]\tfrac{1}{2}\rho v_m^2(z) = c_e(z)q_b \tag{4.8}$$

where:

ρ is the air density, which depends on the altitude, temperature and barometric pressure to be expected in the region during wind storms; its value may be given by the National Annex. The recommended value is $1.25\,\text{kg/m}^3$.

$c_e(z)$ is the exposure factor given in expression (4.9):

$$c_e(z) = \frac{q_p(z)}{q_b} \tag{4.9}$$

q_b is the basic velocity pressure given in expression (4.10):

$$q_b = \tfrac{1}{2}\rho v_b^2 \tag{4.10}$$

The value 7 in expression (4.8) is based on a peak factor equal to 3.5 and is consistent with the values of the pressure and force coefficients given in Chapter 7.

For flat terrain where $c_0(z) = 1.0$ (see Section 4.3 above), the exposure factor $c_e(z)$ is illustrated in Fig. 4.4 as a function of height above terrain and a function of terrain category as defined in Table 4.1. *Note to cl.4.5(1): 1991-1-4*

In Fig. 4.4 the variation of the exposure factor $c_e(z)$ is separated into the two components: the mean part and the part due to the turbulence, in order to clarify the effect of the introduction of the I_v factor.

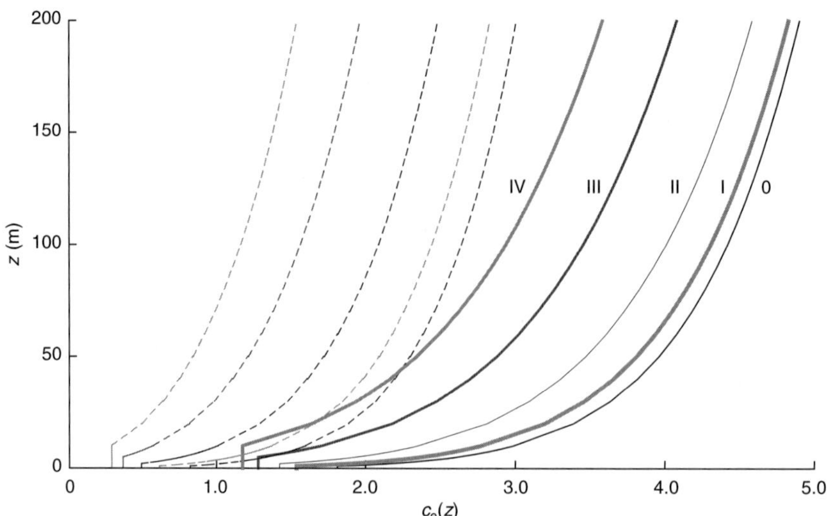

Fig. 4.4. Variation of the exposure factor $c_e(z)$ with height above the terrain for different terrain categories. The dashed lines refer to the mean wind pressure, the solid lines include the effect of turbulence effects

UK National Annex

The UK National Annex has a different expression for determining the peak velocity pressure q_p where $c_e(z)$ is the exposure factor which corresponds to S_b^2 in BS 6399-2; except that in EN 1991-1-4 a linearised version is used which disregards second-order turbulence effects and consequently can significantly underestimate the wind pressure, particularly in urban terrain where the turbulence is greatest. $c_{e,T}$ is an exposure correction factor for town terrain.

NA2-17: BSI NA: 1991-1-7

For this reason expression (4.8) in EN 1991-1-4 is not used in the UK and has been replaced by expressions NA.3a and NA.3b.

$$q_p(z) = c_e(z)q_b \quad \text{for sites in Country terrain} \qquad \text{(NA.3a)}$$

$$q_p(z) = c_e(z)c_{e,T}q_b \quad \text{for sites in Town terrain} \qquad \text{(NA.3b)}$$

Figure NA.7: BSI NA 1991-1-7
Figure NA.8: BSI NA 1991-1-7
cl.A.5: 1991-1-4

The values of exposure factor $c_e(z)$ for Country terrain are given in Fig. 4.5 (reproduced from Fig. NA.7 of the UK National Annex) and the values of exposure correction factor for Town terrain $c_{e,T}$ are given in Fig. 4.6 (reproduced from Figure NA.8 of the UK National Annex). In these figures h_{dis} is the displacement height. See Section 4.6 of this Designers' Guide and Clause A.5 of EN 1991-1-4.

When orography is significant (i.e. shaded areas of Figure 4.7):

$$q_p(z) = [q_p(z) \text{ from expression NA.3a or NA.3b } (c_o(z) + 0.6)/1.6]^2 \quad \text{for} \quad z \leq 50\,\text{m} \qquad \text{(NA.4a)}$$

or:

$$q_p(z) = [1 + 3.0I_v(z)]^2 \times 0.5\rho v_m^2 \quad \text{for} \quad z > 50\,\text{m} \qquad \text{(NA.4b)}$$

Annex A of BSI NA: 1991-1-4

Annex A to the UK National Annex shows flow diagrams for the determination of $q_p(z)$.

Figure NA.2: BSI NA 1991-1-7

Orography is not significant in the *unshaded* areas of Fig. 4.7 (reproduced from Fig. NA.2 of the UK National Annex) (c_o {orography factor} = 1.0). Where there is significant orography, as defined by the *shaded* zones in Fig. 4.7, *A* (altitude factor in the BSI NA) should be taken as the altitude of the upwind base of the orographic feature for each wind direction considered.

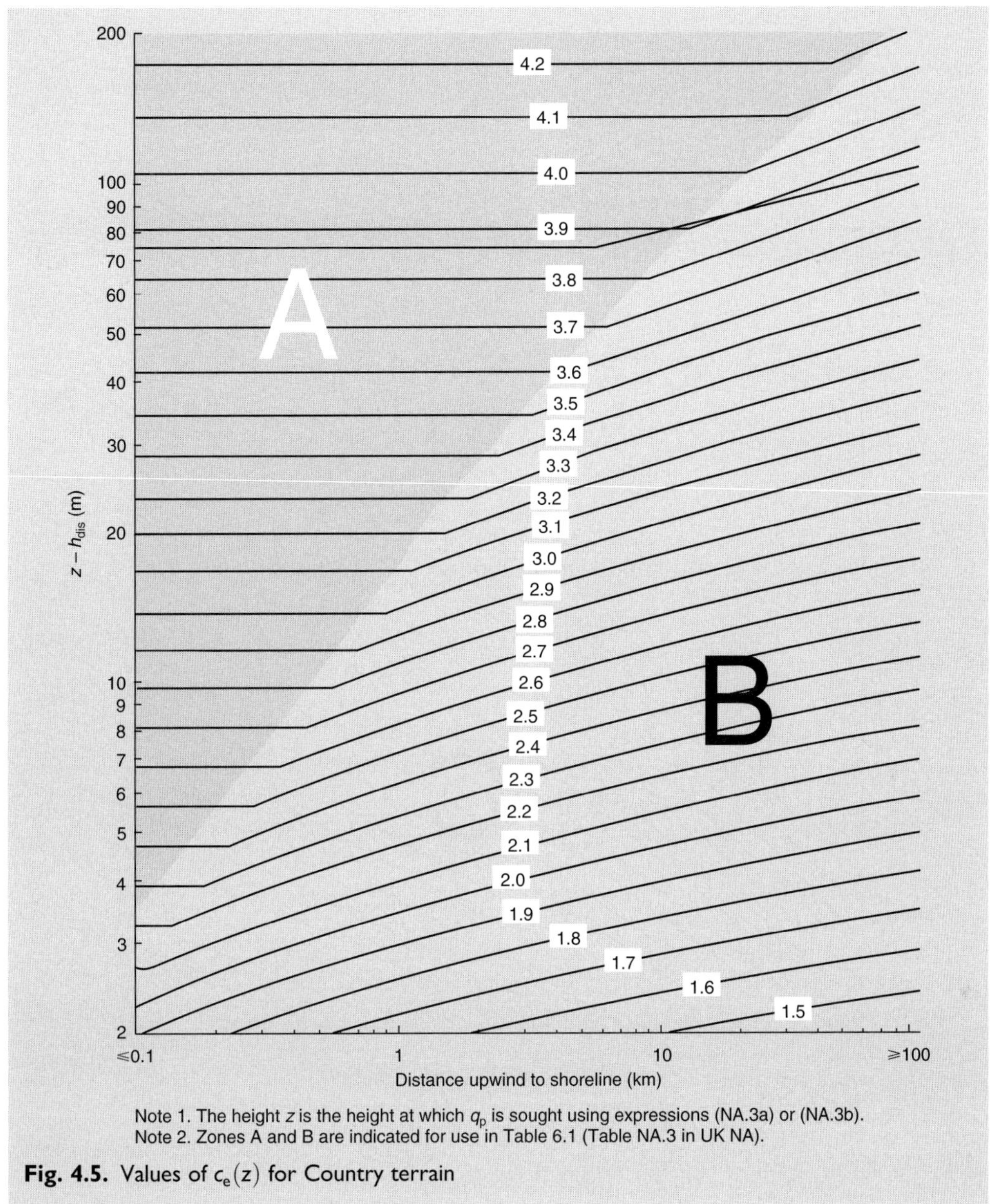

Note 1. The height z is the height at which q_p is sought using expressions (NA.3a) or (NA.3b).
Note 2. Zones A and B are indicated for use in Table 6.1 (Table NA.3 in UK NA).

Fig. 4.5. Values of $c_e(z)$ for Country terrain

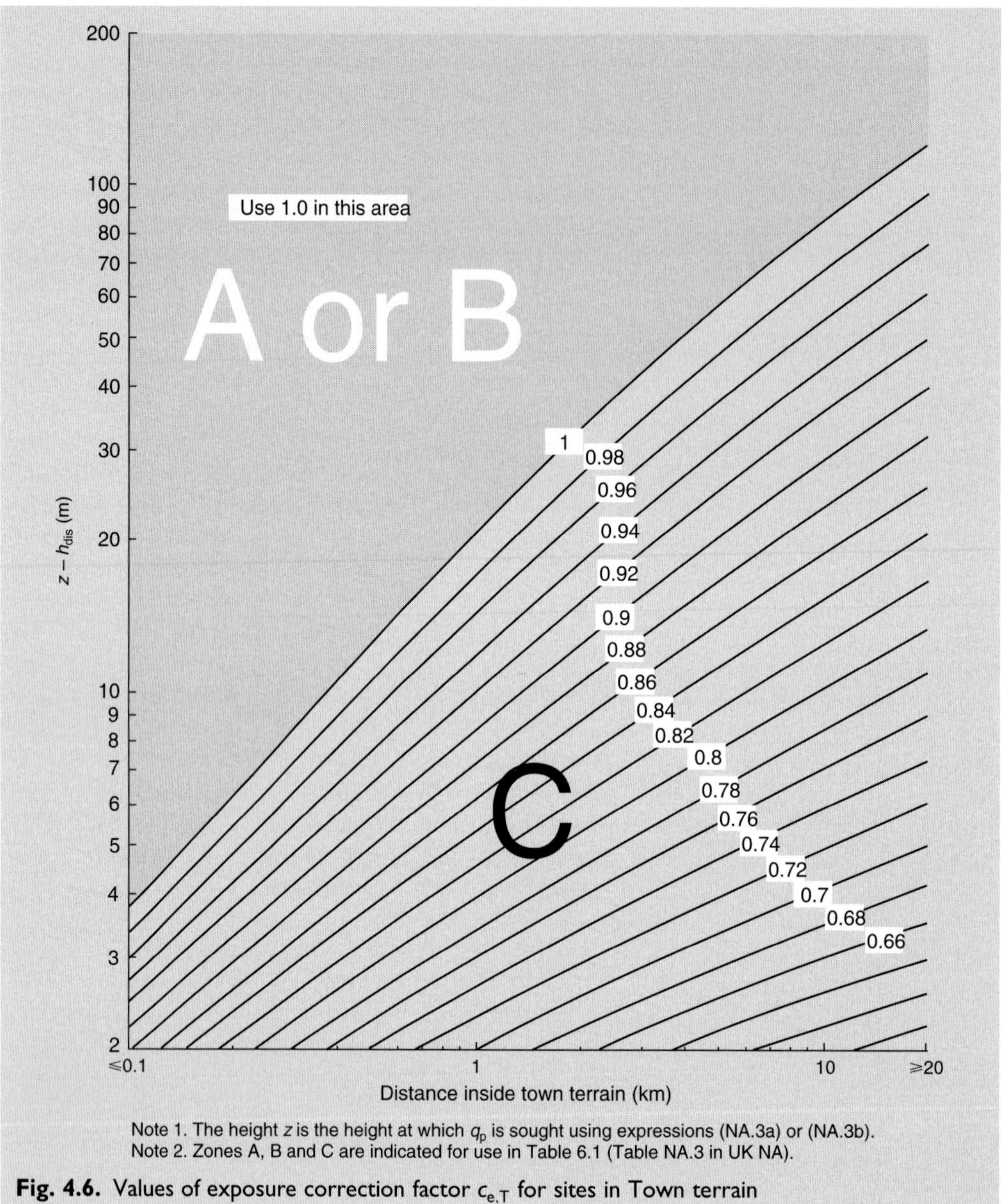

Note 1. The height z is the height at which q_p is sought using expressions (NA.3a) or (NA.3b).
Note 2. Zones A, B and C are indicated for use in Table 6.1 (Table NA.3 in UK NA).

Fig. 4.6. Values of exposure correction factor $c_{e,T}$ for sites in Town terrain

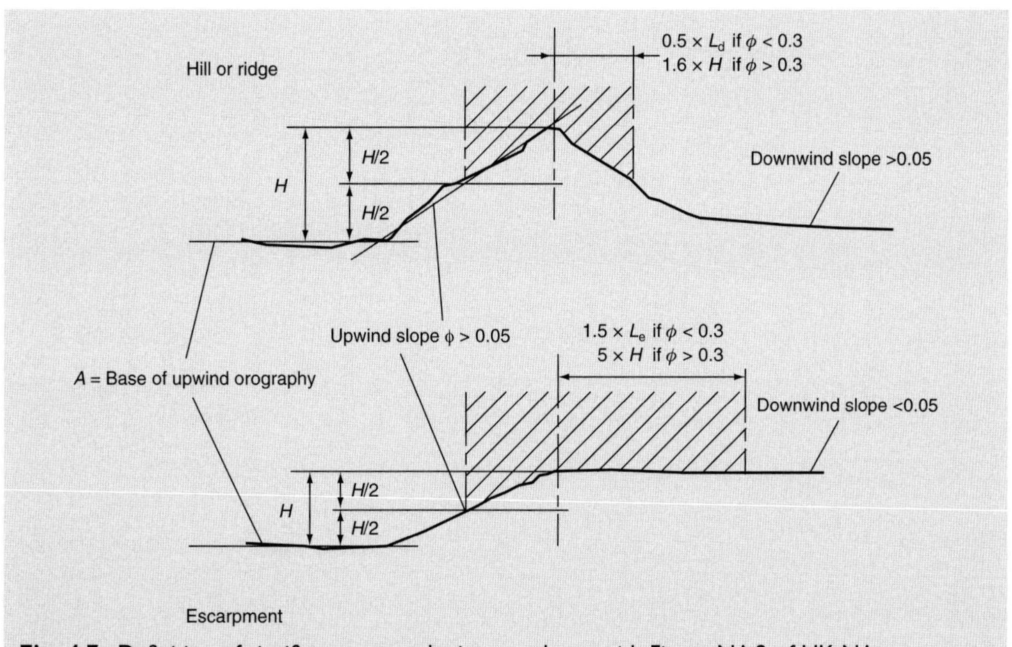

Fig. 4.7. Definition of significant orography in accordance with Figure NA.2 of UK NA

4.6. Explanation of h_{dis} displacement height

For buildings in terrain category IV (i.e. C in the BSI National Annex), closely spaced buildings and other obstructions cause the wind to behave as if the ground level were raised to a displacement height, h_{dis}. h_{dis} may be determined by expression (A.15) (see also Fig. 4.8).

$$x \leq 2h_{ave} \qquad h_{dis} \text{ is the lesser of } 0.8h_{ave} \text{ or } 0.6h$$

$$2h_{ave} < x < 6h_{ave} \qquad h_{dis} \text{ is the lesser of } 1.2h_{ave} - 0.2x \text{ or } 0.6h \qquad \text{(A.15)}$$

$$x \geq 6h_{ave} \qquad h_{dis} = 0$$

In the absence of more accurate information the obstruction height may be taken as $h_{ave} = 15\,\text{m}$ for terrain category IV.

These rules are direction dependent; the values of h_{ave} and x should be established for each 30° sector as described in Clause 4.3.2 of EN 1991-1-4. *cl.A.5(1): 1991-1-4*

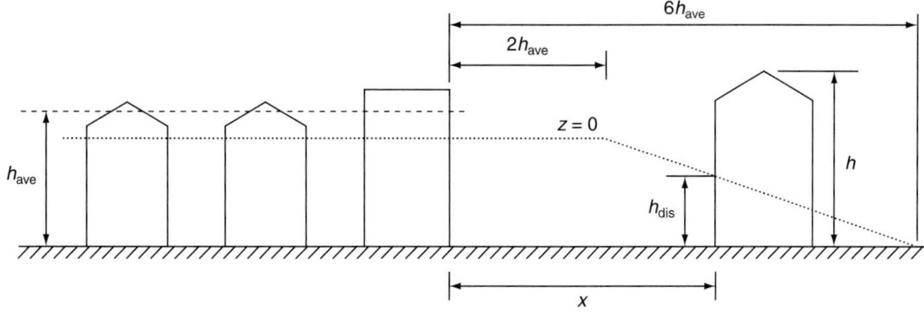

Fig. 4.8. Obstruction height and upwind spacing

CHAPTER 5

Wind actions

5.1. General

Section 5 of EN 1991-1-4 gives the rules for determining the wind pressures on internal and external surfaces, the wind forces and frictional forces. *cl.5.1. (1)P: 1991-1-4*

5.2. Wind pressures on surfaces

EN 1991-1-4 gives guidance for the determination of the external wind pressure w_e on the structure's cladding and the internal wind pressure w_i in case of openings in the cladding. Both types of wind pressure depend on the geometry of the considered structure. In addition the internal pressure varies with the permeability of the building.

Wind actions on a structure and/or on structural elements have to be determined taking into account both external and internal wind pressures.

The wind pressure acting on the external surfaces, w_e, is obtained from expression (5.1). *cl.5.2(1): 1991-1-4*

$$w_e = q_p(z_e)c_{pe} \tag{5.1}$$

where:

$q_p(z_e)$ is the peak velocity pressure (see Section 4.5)
z_e is the reference height for the external pressure given in Section 7 of EN 1991-1-4
c_{pe} is the pressure coefficient for the external pressure, see Section 7 of EN 1991-1-4.

The wind pressure acting on the internal surfaces of a structure, w_i, is obtained from expression (5.2). *cl.5.2(2): 1991-1-4*

$$w_i = q_p(z_i)c_{pi} \tag{5.2}$$

where:

z_i is the reference height for the internal pressure given in Section 7 of EN 1991-1-4
c_{pi} is the pressure coefficient for the internal pressure given in Section 7 of EN 1991-1-4.

The net pressure on a wall, roof or element is the difference between the pressures on the opposite surfaces taking account of their signs (+ve or −ve). Pressure, directed towards the surface is taken as positive, and suction, directed away from the surface as negative. Examples are given in Fig. 5.1. *cl.5.2(3): 1991-1-4*

5.3. Wind forces

The wind forces for the whole structure or of a structural component are determined **by integration of the wind pressure over the whole surface** either (***Clause 5.3(1): EN 1991-1-4***): *cl.5.3(1): 1991-1-4)*

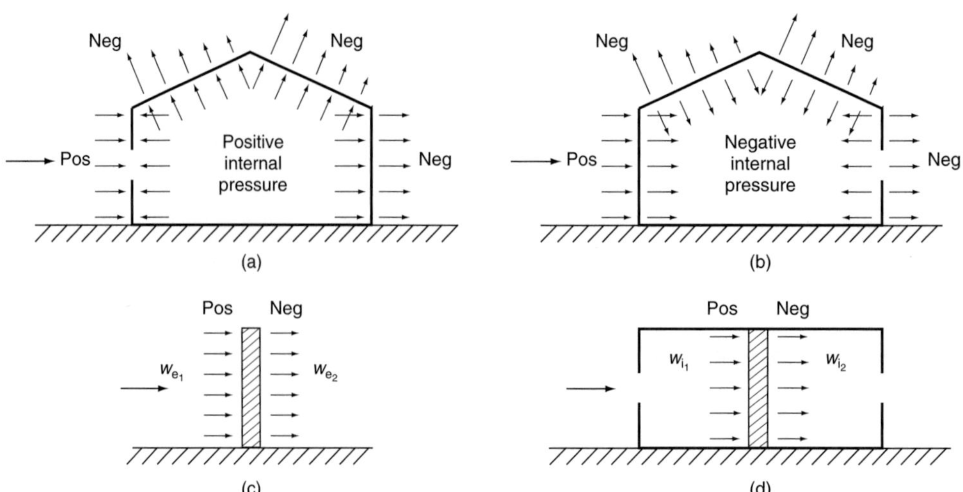

Fig. 5.1. Example of signs for pressure on surfaces

- by calculating forces using force coefficients given in Section 7 of EN 1991-1-4 (see 5.3(2)); or
- by calculating forces by vectorial summation surface pressures over the individual structural element (see 5.3.(3)).

The wind force F_w acting on a structure or a structural component may be determined directly by using expression (5.3):

$$F_w = c_s c_d c_f q_p(z_e) A_{ref} \tag{5.3}$$

cl.5.3(2): 1991-1-4

or by vectorial summation over the individual structural elements (as shown in Section 7.2.2 of EN 1991-1-4) by using expression (5.4):

$$F_w = c_s c_d \sum_{\text{elements}} c_f q_p(z_e) A_{ref} \tag{5.4}$$

where:

$c_s c_d$ is the structural factor as defined in Chapter 6
c_f is the force coefficient for the structure or structural element, given for buildings in Section 7 of EN 1991-1-4 (see also below)
$q_p(z_e)$ is the peak velocity pressure (defined in Section 4.5 above) at reference height z_e (defined for buildings in Section 7 of EN 1991-1-4)
A_{ref} is the reference area of the structure or structural element, given for buildings in Section 7 of EN 1991-1-4.

Wind forces may be determined for the calculation of wind actions on a limited set of structures or structural elements (see also Section 7 of EN 1991-1-4).

Note to cl.5.3(2): 1991-1-4

Chapter 7 gives c_f values for structures or structural elements such as prisms, cylinders, roofs, signboards, plates and lattice structures. These values include friction effects.

In accordance with EN 1991-1-4 rectangular/polygonal shapes force coefficients are determined by:

$$c_f = c_{f,0} \psi_r \psi_\lambda$$

where:

$c_{f,0}$ is the force coefficient for shapes with sharp corners
ψ_r is the reduction factor for rounded corners at rectangular structures

cl.7.13: 1991-1-4

ψ_λ is the end-effect factor for elements with free end flow as defined in Clause 7.13 of EN 1991-1-4

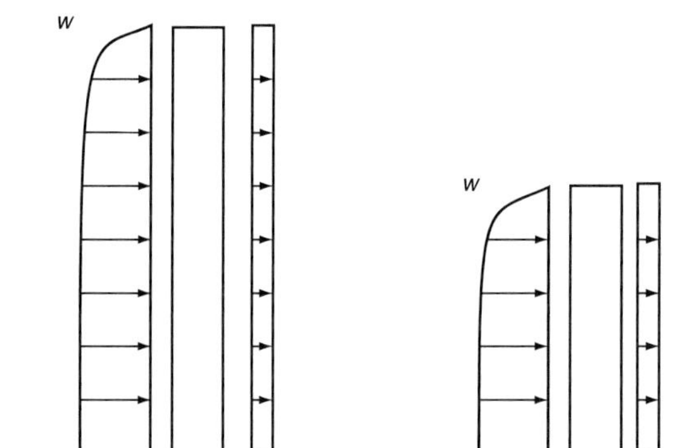

Fig. 5.2. Relative influence of the flow at the edge of the structure on the resulting wind force is smaller for slender structures than for compact structures (where *w* is the wind pressure)

The reduction factor for rounded corners at rectangular structure ψ_r makes allowance for the fact that the wind pressure for rounded corners is lower than for sharp corners which provide a higher obstacle to the flow. This leads to a decreased wind force at structures with rounded corners.

Furthermore, independent of the shape of the corner, the end-effect factor ψ_λ takes into account at the top of structures that the resulting wind pressure is lower than the average value at the inner surface. This effect, i.e. this reduction, decreases relatively with structures of higher slenderness. See Fig. 5.2.

Force coefficients for other shapes and geometries are given in Section 7 of EN 1991-1-4.

The wind force F_w acting on a structure or a structural element may be determined by vectorial summation of the forces $F_{w,e}$, $F_{w,i}$ and F_{fr} calculated from the external and internal pressures using expressions (5.5) and (5.6) and the frictional forces resulting from the friction of the wind parallel to the external surfaces, calculated using expression (5.7). *cl.5.3(3): 1991-1-4*

External forces:

$$F_{w,e} = c_s c_d \sum_{\text{surfaces}} w_e A_{ref} \qquad (5.5)$$

Internal forces:

$$F_{w,i} = \sum_{\text{surfaces}} w_i A_{ref} \qquad (5.6)$$

Friction forces:

$$F_{fr} = c_{fr} q_p(z_e) A_{fr} \qquad (5.7)$$

where:

$c_s c_d$ is the structural factor as defined in Chapter 6 below
w_e is the external pressure on the individual surface at height z_e, given in expression (5.1)
w_i is the internal pressure on the individual surface at height z_i, given in expression (5.2)
A_{ref} is the reference area of the individual surface
c_{fr} is the friction coefficient derived from Clause 7.5 of EN 1991-1-4
A_{fr} is the area of external surface parallel to the wind, given in Clause 7.5 of EN 1991-1-4.

For elements (e.g. walls, roofs), the wind force is equal to the difference between the external and internal resulting forces. *Note 1 to cl.5.3(3): 1991-1-4*

The friction forces F_{fr} act in the direction of the wind components parallel to external surfaces. These can be disregarded when the total area of all surfaces parallel with (or at a small angle to) the wind is equal to or less than four times the total area of all external surfaces perpendicular to the wind (windward and leeward). *Note 2 to cl.5.3(3): 1991-1-4*

CHAPTER 6

Structural factor $c_s c_d$

Section 6 of EN 1991-1-4 gives rules for determining the structural factor $c_s c_d$ which accounts for the effect of non-simultaneous occurrence of peak pressures over the surface of the structure (c_s) or element combined with the effect of dynamic response of the structure or element due to turbulence (c_d). *cl.6.1(1): 1991-1-4*

EN 1991-1-4 allows the separation of c_s and c_d through the National Annex. *Note to cl.6.1(1): 1991-1-4*

For the majority of traditional low-rise or framed buildings $c_s c_d$ may conservatively be taken as 1.0. Clause 6.2(1) of EN 1991-1-4 gives the following list of building and element types for which $c_s c_d$ may be taken as 1.0:

- For buildings with a height less than 15 m. *cl.6.2(1): 1991-1-4*
- For façade and roof elements having a natural frequency greater than 5 Hz.
- For framed buildings which have structural walls and which are less than 100 m high and whose height is less than four times the in-wind depth.
- For chimneys with circular cross-sections whose height is less than 60 m and 6.5 times the diameter.

For other building types, or where a more precise value is required, $c_s c_d$ is determined using the detailed procedure (i.e. expression 6.1) given in Clause 6.3.1. *cl.6.3(1): 1991-1-4*

The procedure using expression 6.1 can only be used if the conditions given in Clause 6.3.1(2) apply.

$$c_s c_d = \frac{1 + 2k_p I_v(z_e)\sqrt{B^2 + R^2}}{1 + 7I_v(z_e)} \tag{6.1}$$

where:

z_e is the reference height, see Fig. 6.2; for structures where Fig. 6.2 does not apply z_e may be equal to h, the height of the structure

k_p is the peak factor defined as the ratio of the maximum value of the fluctuating part of the response to its standard deviation

I_v is the turbulence intensity defined in Clause 4.4 of EN 1991-1-4 and this Designers' Guide

B^2 is the background factor, allowing for the lack of full correlation of the pressure on the structure surface

R^2 is the resonance response factor, allowing for turbulence in resonance with the vibration mode.

Note 1 to cl.6.3(1): 1991-1-4

The size factor c_s, which takes into account the reduction effect on the wind action due to the non-simultaneity of occurrence of the peak wind pressures on the surface, may be obtained from expression (6.2):

$$c_s = \frac{1 + 7I_v(z_e)\sqrt{B^2}}{1 + 7I_v(z_e)} \tag{6.2}$$

Note 1 to cl.6.3(1): 1991-1-4

The dynamic factor c_d, which takes into account the increasing effect from vibrations due to turbulence in resonance with the structure, may be obtained from expression (6.3):

$$c_d = \frac{1 + 2k_p I_v(z_e)\sqrt{B^2 + R^2}}{1 + 7I_v(z_e)\sqrt{B^2}} \tag{6.3}$$

Note 3 to cl.6.3(1): 1991-1-4

The procedure to be used in a Country terrain to determine k_p, B and R may be given in its National Annex. A recommended procedure is given in Annex B. An alternative procedure is given in Annex C. As an indication to the users, the differences in $c_s c_d$ using Annex C compared to Annex B does not exceed approximately 5%.

Annex B: 1991-1-4

In the BSI National Annex it has been decided to separate $c_s c_d$ into a size factor c_s and a dynamic factor c_d. The BSI NA gives Table NA3 (Table 6.1 in this Designers' Guide) for determining c_s values and Fig. NA9 (Fig. 6.2 in this Designers' Guide) for determining c_d. These have been derived using the detailed procedure in Annex B of EN 1991-1-4 (***Annex B: EN 1991-1-4***).

In Fig. 6.1 graphs are given for four classes of structure which correspond to various values of logarithmic decrement of structural damping, δ_s. The benefits of separating c_s and c_d are greatest for large plan area low-rise buildings and are shown in Example 6.1 below.

Table 6.1. Size factor c_s for zones A, B and C indicated in Figs 4.6 and 4.7 (from Table NA3 UK NA)

$b + h$ (m)	$z - h_{dis} = 6$ m			$z - h_{dis} = 10$ m			$z - h_{dis} = 30$ m			$z - h_{dis} = 50$ m			$z - h_{dis} = 200$ m		
	A	B	C	A	B	C	A	B	C	A	B	C	A	B	C
1	0.99	0.98	0.97	0.99	0.99	0.97	0.99	0.99	0.98	0.99	0.99	0.99	0.99	0.99	0.99
5	0.96	0.96	0.92	0.97	0.96	0.93	0.98	0.97	0.95	0.98	0.98	0.96	0.98	0.98	0.98
10	0.95	0.94	0.88	0.95	0.95	0.90	0.96	0.96	0.93	0.97	0.96	0.94	0.98	0.97	0.97
20	0.93	0.91	0.84	0.93	0.92	0.87	0.95	0.94	0.90	0.95	0.95	0.92	0.96	0.96	0.95
30	0.91	0.89	0.81	0.92	0.91	0.84	0.94	0.93	0.88	0.94	0.93	0.90	0.96	0.95	0.93
40	0.90	0.88	0.79	0.91	0.89	0.82	0.93	0.91	0.86	0.93	0.92	0.88	0.95	0.94	0.92
50	0.89	0.86	0.77	0.90	0.88	0.80	0.92	0.90	0.85	0.92	0.91	0.87	0.94	0.94	0.91
70	0.87	0.84	0.74	0.88	0.86	0.77	0.90	0.89	0.83	0.91	0.90	0.85	0.93	0.92	0.90
100	0.85	0.82	0.71	0.86	0.84	0.74	0.89	0.87	0.80	0.90	0.88	0.82	0.92	0.91	0.88
150	0.83	0.80	0.67	0.84	0.82	0.71	0.87	0.85	0.77	0.88	0.86	0.79	0.90	0.89	0.85
200	0.81	0.78	0.65	0.83	0.80	0.69	0.85	0.83	0.74	0.86	0.84	0.77	0.89	0.88	0.83
300	0.79	0.75	0.62	0.80	0.77	0.65	0.83	0.80	0.71	0.84	0.82	0.73	0.87	0.85	0.80

b = cross-wind breadth of building or building part or width of element.

h = height of building or building part or length of element.

cl.7.2.2(1): 1991-1-4

z = height of building or height to top of element (or height of building part; subject to BS EN 1991-1-4: 2005 7.2.2(1)) interpolation may be used (***Clause 7.2.2(1): EN 1991-1-4***).

The zone A, B or C to be used for a building can be determined as follows:

- For sites in country, it is determined with respect to distance from shore and $(z - h_{dis})$ using Fig. 4.5.
- For sites in town, using the distance into town and $(z - h_{dis})$ in Fig. 4.6, it is first determined whether zone C applies. If not, zone A or B will apply depending on the distance of the site from shore and $(z - h_{dis})$ as shown in Fig. 4.5.

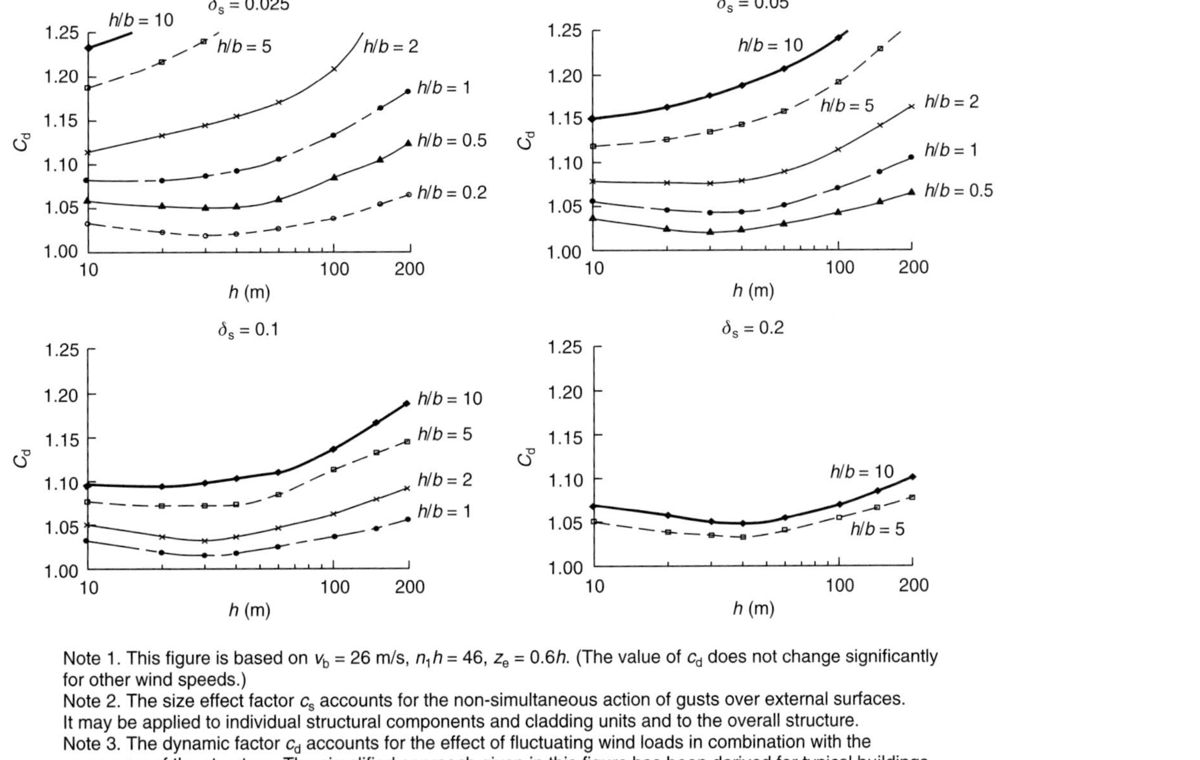

Note 1. This figure is based on v_b = 26 m/s, $n_1 h$ = 46, $z_e = 0.6h$. (The value of c_d does not change significantly for other wind speeds.)
Note 2. The size effect factor c_s accounts for the non-simultaneous action of gusts over external surfaces. It may be applied to individual structural components and cladding units and to the overall structure.
Note 3. The dynamic factor c_d accounts for the effect of fluctuating wind loads in combination with the resonance of the structure. The simplified approach given in this figure has been derived for typical buildings with typical damping and natural frequency characteristics. More accurate values will be given using the procedure in Clause 6.3 of BS EN 1991-1-4: 2005. *cl.6.3: 1991-1-4*
Note 4. The dynamic factor c_d may be taken as 1.0 for framed buildings with structural walls and masonry internal walls and for cladding panels and elements.
Note 5. Values of δ_s for typical classes of structure are given in Annex F.5 of BS EN 1991-1-4: 2005. *Annex A5: 1991-1-4*
(N.B. for reinforced concrete buildings δ_s = 0.1 and for steel buildings δ_s = 0.05).

Fig. 6.1. Dynamic factor c_d for various values of logarithmic decrement of structural damping, δ_s

Expression (6.1) can be used provided the following requirements are satisfied:

- The structure corresponds to one of the general shapes shown in Fig. 6.2. *cl.6.3(1P): 1991-1-4*
- Only the along-wind vibration in the fundamental mode is significant, and this mode shape has a constant sign and the contribution to the response from the second or higher along-wind vibration modes is negligible.
- The height of the building is less than 200 m.

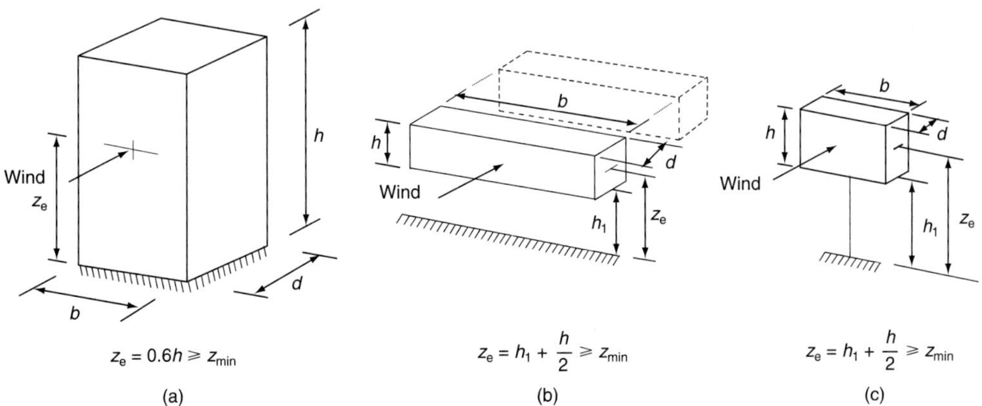

Fig. 6.2. General shapes of structures covered by the design procedure: (a) vertical structures such as buildings; (b) parallel oscillator, i.e. horizontal structures such as beams; (c) pointlike structure such as signboards. The structural dimensions and the reference height used are also shown

Example 6.1. Comparing c_sc_d values between EN 1991-1-4 and the BSI NA
Consider wind loads on the long face of an office building 10 m high with plan dimensions of 60 m × 20 m in Town terrain.

Using the BSI NA $c_d = 1.0$ from Table 6.2 (for $h/b = 0.167$) and $c_s = 0.77$ from Table 6.2 (for $b + h = 70$ and $z - h_{dis} = 10$) thus giving a c_sc_d value of 0.77.

This will give a 23% reduction in wind load on this face compared with the c_sc_d value of 1.0 recommended in Clause 6.3.1 of EN 1991-1-4.

Wake buffeting

cl.6.3.3(1): 1991-1-4

For slender buildings where $h/d > 4$ and chimneys ($h/d > 6.5$) in tandem or grouped arrangement, the effect of increased turbulence in the wake of nearby structures (wake buffeting) needs to be taken into account.

cl.6.3.3(2): 1991-1-4

Wake buffeting effects may be assumed to be negligible if at least one of the following conditions applies:

- the distance between two buildings or chimneys is larger than 25 times the cross-wind dimension of the upstream building or chimney
- the natural frequency of the downstream building or chimney is higher than 1 Hz.

CHAPTER 7

Pressure and force coefficients

General

Section 7 gives pressure and force coefficients for a wide range of buildings, structures and elements.

Depending on the structure, EN 1991-1-4 uses four separate types of aerodynamic coefficient:

- Internal and external pressure coefficients, see paragraph (1) in 'Choice of aerodynamic coefficient' below. *cl.7.1(1): 1991-1-4*
- Net pressure coefficients, see paragraph (2) in 'Choice of aerodynamic coefficient'.
- Friction coefficients, see paragraph (3) in 'Choice of aerodynamic coefficient'.
- Force coefficients, see paragraph (3) in 'Choice of aerodynamic coefficient'.

Choice of aerodynamic coefficient

(1) **The pressure coefficients** should be determined for:

- buildings, using Clause 7.2 of EN 1991-1-4, for both internal and external pressures; and for *cl.7.1.1(1): 1991-1-4*
- circular cylinders, using Clause 7.2.9 of EN 1991-1-4 for the internal pressures and Clause 7.9.1 for the external pressures.

External pressure coefficients give the effect of the wind on the external surfaces of buildings; internal pressure coefficients give the effect of the wind on the internal surfaces of buildings. *Note 1 to cl.7.1(1): 1991-1-4*

In EN 1991-1-4, the external pressure coefficients are divided into overall coefficients and local coefficients. Local coefficients give the pressure coefficients for loaded areas of $1\,m^2$ or less, e.g. for the design of small elements and fixings; overall coefficients give the pressure coefficients for loaded areas larger than $10\,m^2$. *Note 2 to cl.7.1(1): 1991-1-4*

(2) **The net pressure coefficients** should be determined for:

- canopy roofs, using Clause 7.3 of EN 1991-1-4 *cl.7.1(2): 1991-1-4*
- free-standing walls, parapets and fences using Clause 7.4 of EN 1991-1-4.

Net pressure coefficients give the resulting effect of the wind on a structure, structural element or component per unit area. *Note to cl.7.1(2): 1991-1-4*

(3) **Friction coefficients** should be determined for walls and surfaces defined in Clause 5.2(3) of EN 1991-1-4 and the Designers' Guide using Clause 7.5 of EN 1991-1-4. *cl.7.1(3): 1991-1-4*

(4) **Force coefficients** should be determined for:

- signboards, using Clause 7.4.3 of EN 1991-1-4
- structural elements with rectangular cross-section, using Clause 7.6 of EN 1991-1-4
- structural elements with sharp-edged section, using Clause 7.7 of EN 1991-1-4
- structural elements with regular polygonal section, using Clause 7.8 of EN 1991-1-4
- circular cylinders, using Clause 7.9.2 and 7.9.3 of EN 1991-1-4
- spheres, using Clause 7.10 of EN 1991-1-4
- lattice structures and scaffoldings, using Clause 7.11 of EN 1991-1-4
- flags, using Clause 7.12 of EN 1991-1-4.

cl.7.1(4): 1991-1-4 A reduction factor depending on the effective slenderness of the structure may be applied, using Clause 7.13 of EN 1991-1-4.

Note to cl.7.1(4): 1991-1-4 Force coefficients give the overall effect of the wind on a structure, structural element or component as a whole, including friction, if not specifically excluded.

CHAPTER 8

Annexes to EN 1991-1-4

There are six informative annexes in EN 1991-1-4:

- Annex A – Terrain effects
- Annex B – Procedure 1 for determining the structural factor $c_s c_d$
- Annex C – Procedure 2 for determining the structural factor $c_s c_d$
- Annex D – $c_s c_d$ values for different types of structure
- Annex E – Vortex shedding and aeroelastic instabilities
- Annex F – Dynamic characteristics of structures

Annex A gives guidance on the following: *Annex A: 1991-1-4*

- Description of terrain types, see Fig. 4.1 of this Designers' Guide.
- Fetch factors which are defined as *upwind extent of each kind of ground roughness.*
- Orography.
- Effect of neighbouring structures.
- Displacement height (*for buildings in terrain category IV, closely spaced buildings cause the wind to behave as if the ground level were raised to a displacement height*). These are explained in Chapter 4 of this Part of this Designers' Guide.

Annexes B, C and D give procedures for determining $c_s c_d$ factor as follows: *Annex B, C and D: 1991-1-4*

- Annex B – Procedure 1, which is the recommended procedure.
- Annex C – Procedure 2.
- Annex D – gives charts for determining $c_s c_d$ for common building forms.

Annex E gives guidance on the following: *Annex E: 1991-1-4*

- Vortex shedding which *occurs when vortices are shed alternatively from opposite sides of the structure.*
- Galloping which is the *self-induced vibration of a flexible structure in cross-wind bending mode.*
- Interference galloping for free-standing cylinders which is a *self-excited oscillation which may occur if two or more cylinders are close together but not connected.*
- Divergence and flutter which are *instabilities that occur for flexible plate-like structures, e.g. signboards.*

Annex F gives guidance on: *Annex F: 1991-1-4*

- dynamic characteristics of structures
- natural frequency
- damping
- mode shapes.

References to Part 4

1. EN 1991-1-4: 2005. *Eurocode 1: Actions on Structures – Part 1-4: General Actions – Wind actions*. European Committee for Standardisation, Brussels, 2005.
2. Cook, N. J. *Designers' Guide to EN 1991-1-4. Eurocode 1: Actions on Structures, General Actions Part 1-4. Wind actions*. Thomas Telford, London, 2007, ISBN 9 7807 2773 1524.
3. BS 6399-2: 1997. *Loadings for Buildings: Part 2: Code of Practice for Wind Loads*. British Standards Institution, London, 1997.
4. Blackmore, P. *The Application of EN 1991: Eurocode 1 – Actions on Structures: Part 1: General Actions: Guide to the Use of EN 1991-1-4 – Wind actions*. BRE, Watford, 2005. Available on the CLG website at: http://www.communities.gov.uk/index.asp?id = 1502963

PART 5: EN 1991-1-5: Eurocode 1: Part 1-5: Thermal Actions

CHAPTER 1

General

This chapter is concerned with the general aspects of EN 1991-1-5: *Eurocode 1 – Actions on Structures: Part 1.5: General Actions – Thermal actions.*[1] The material described in this chapter is covered in the following clauses:

- Scope *Clause 1.1: 1991-1-5*
- Normative references *Clause 1.2: 1991-1-5*
- Assumptions *Clause 1.3: 1991-1-5*
- Distinction between Principles and Application Rules *Clause 1.4: 1991-1-5*
- Terms and definitions *Clause 1.5: 1991-1-5*
- Symbols *Clause 1.6: 1991-1-5*

1.1. Scope

EN 1991-1-5: *Eurocode 1 – Actions on Structures: Part 1.5: General Actions – Thermal actions* is one of the ten Parts of EN 1991. It gives design guidance and rules for calculating thermal actions on buildings, bridges and other structures including their structural elements. Principles needed for cladding and other appendages of the buildings are also provided in EN 1991-1-5. *cl.1.1(1): 1991-1-5*

EN 1991-1-5 also describes the changes in the temperature of structural elements. It gives characteristic values of thermal actions for use in the design of structures which are exposed to daily and seasonal climatic changes. This clause further states that '*Structures not so exposed may not need to be considered for thermal actions.*' *cl.1.1(2): 1991-1-5*

EN 1991-1-5 also gives guidance for structures in which thermal actions are mainly a function of their use (e.g. cooling towers, silos, tanks, warm and cold storage facilities, hot and cold services). *cl.1.1(3): 1991-1-5*

This Designers' Guide will only give guidance on the parts of EN 1991-1-5 dealing with buildings. Bridges are covered by The Thomas Telford Guide: *EN 1991: Actions on Structures: Bridges.*

Before discussing each clause relating to buildings in EN 1991-1-5, a brief introductory advice for using this EN 1991-1-5 for the design of buildings is given below.

1.2. Normative references

No comment is necessary on the quoted references with the exception of EN 1991-1-6 dealing with actions during execution (see also appropriate Part in this Guide). EN 1991-1-6 provides guidance on the **return periods for the determination of the characteristic values of climatic actions** for structures with design working lives different than 50 years and also for the duration of an execution phase. *cls.1.2: 1991-1-5*

Introductory advice for using this EN 1991-1-5 for the design of buildings

Introduction

There are buildings which can be characterised as sensitive to the effects of thermal actions which need to be designed to resist their effects. The designer should use experience and the guidance offered by EN 1991-1-5 and this manual to determine whether a building is sensitive to the effects of thermal actions.

The four criteria that are important and need to be considered by the designer are:

1. Material
2. Geometry
3. Restraint
4. Temperature range and frequency.

These criteria are described below.

Materials

Coefficients of linear expansion to determine temperature-induced strains are given in Table 4.1 of this Part of the Designers' Guide for a selection of common building materials.

Geometry

Buildings that have abrupt changes in geometry such as wings and courtyards will be prone to differential movement at junctions of architectural massing. This movement will arise from thermal action and may be accompanied by movement from other actions such as the differential settlement of foundations or seismic effects.

Restraints

Most buildings rely on bracing or framing systems to provide overall structural stability. These stability systems have the potential to create points of restraint against temperature movement which in turn leads to temperature-induced stresses. The combination of changes in plan geometry and the proximity of such systems can cause particular problems.

Long buildings (normally greater that 50 m) with stability systems located at the ends of the building are likely to cause significant temperature-induced stresses in both the stability system and connected elements of structure. It will be preferable to locate stability systems in the middle of the external face of the building or distributed evenly along it provided that each stability system only works in one direction.

Multi-storey buildings may exhibit significant differential temperature movements between the frame and its cladding. Cladding support systems should be designed to accommodate differential movement of the structural frame and the cladding while still catering for normal vertical and horizontal loads from the cladding.

Consideration should be given to the construction conditions which may expose elements of structure to temporary temperature effects.

Temperature range and frequency

Methods for determining an appropriate temperature range are given in EN 1991-1-5. Frequency should be determined for the project under consideration and the following information may be helpful:

- Most occupied buildings are either heated and/or cooled and therefore tend to have a daily temperature cycle which, in the majority of cases, will not be significant.
- Unheated buildings and car parks will have daily and seasonal temperature cycles much larger than normal buildings. Buildings like these should be designed for thermal action.

- Temperatures can be increased by solar gain so roofs and top decks of car parks should be checked. Car parks with surfacing are particularly susceptible to solar gain.

Extreme temperature changes due to equipment failures should be taken into account in the design.

At the construction stage normal thermal action may need to be added to other strain-related effects such as:

- concrete curing temperatures (for large or thick pours)
- shrinkage.

1.3. Assumptions

The statements and assumptions given in EN 1990 *Clause 1.3* apply to all the Eurocode parts and designers' guidance for these are given in the Designers' Guide to EN 1990. *cl.1.3: 1991-1-5*

1.4. Distinction between Principles and Application Rules

The statements and assumptions given in EN 1990 *Clause 1.4* apply to all the Eurocode parts and designers' guidance for these are given in the Designers' Guide to EN 1990.

1.5. Terms and definitions

Most of the definitions given in EN 1991-1-5 derive from ISO 2394, ISO 3898 and ISO 8930. In addition reference should be made to EN 1990 which provides a basic list of terms and definitions which are applicable to EN 1990 to EN 1999, thus ensuring a common basis for the Eurocode suite. *cl.1.5: 1991-1-5*

For the structural Eurocode suite, attention is drawn to the following key definitions, which may be different from current national practices:

- '*Action*' means a load, or an imposed deformation (e.g. temperature effects or settlement)
- '*Effects of Actions*' or '*Action effects*' are internal moments and forces, bending moments, shear forces and deformations caused by actions.

From the many definitions provided in EN 1990, those that apply for use with EN 1991 are described in Chapter 1 *Clause 1.4(a), (b), (c) and (d): 1991-1-1.*

The following comments are made to help the understanding of particular definitions in EN 1991-1-5.

'*Maximum shade air temperature* T_{max} *and maximum shade air temperature* T_{max}' have an annual probability of being exceeded of 0.02 (equivalent to a mean return period of 50 years), based on the maximum or minimum hourly values recorded respectively. *cl.1.5.3 and 1.5.4: 1991-1-5*

1.6. Symbols

The notation in *Clause 1.6: EN 1991-1-5* is based on ISO 3898. *cl.1.7(1): 1991-1-3*

EN 1990 *Clause 1.6* provides a comprehensive list of symbols, some of which may be appropriate for use with EN 1991-1-5. The symbols given in *Clause 1.6(2): 1991-1-5* are additional notations specific to this part of EN 1991-1-5.

CHAPTER 2

Classification of actions

This chapter is concerned with the classification of the actions in EN 1991-1-5: *Eurocode 1 – Actions on structures: Part 1.5: General Actions – Thermal actions.*

EN 1991-1-5 classifies thermal actions as variable and indirect actions. *cl.2.1(1)P: 1991-1-5*

The values of thermal actions, and the maximum and minimum shade air temperatures T_{max} and T_{min} given in EN 1991-1-5 are characteristic values unless it is stated otherwise in the code. The characteristic values of thermal actions given in EN 1991-1-5 are values with an annual probability of being exceeded of 0.02 (equivalent to a mean return period of 50 years) unless otherwise stated, e.g. for transient design situations. *cl.2.1(2): 1991-1-5*

For determining the maximum and minimum temperatures for shorter return periods ($T_{max,p}$ and $T_{min,p}$) relating to transient design situations, the related values of thermal actions may be derived using the calculation method given in Annex 1 of EN 1991-1-5 and the characteristic values T_{max} and T_{min} are given in National maps showing isotherms for the minimum and maximum temperatures. These values have been based on a return period of 50 years, but formulae are given in Annex A of EN 1991-1-5, based on Gumbel law (i.e. law of extreme values of type I), for the assessment of temperatures based on a return period other than 50 years. As a simplification, the results of these formulae are given diagrammatically (see Fig. 2.1) as ratios between the maximum (minimum) for a *cl.2.1(3): 1991-1-5* *Note to cl.2.1(3): 1991-1-5*

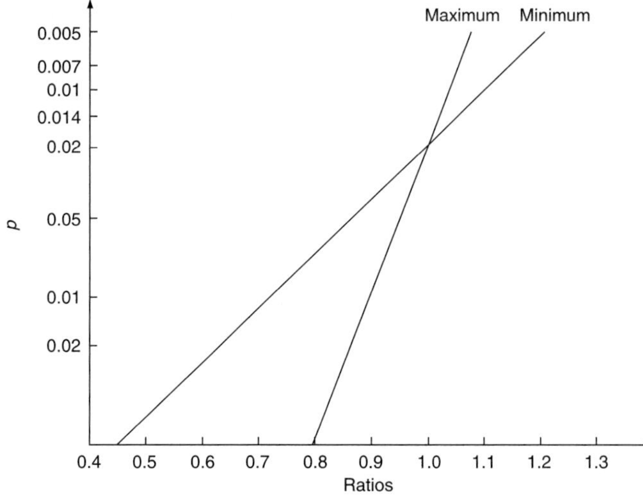

Fig. 2.1. (From Figure A1 in EN 1991-1-5.) Ratios $T_{max,p}/T_{max}$ and $T_{min,p}/T_{min}$

A2: 1991-1-5 probability p and the maximum (minimum) for a return period of 50 years (probability $p = 0.02$) (***A2: 1991-1-5***).

Example 2.1

Using Fig. 2.1 determine the ratios for $T_{max,p}/T_{max}$ and $T_{min,p}/T_{min}$ for a return period of 10 years.

For a return period of 50 years $p = 1/50 = 0.02$
For a return period of 10 years $p = 1/10 = 0.01$

Thus for maximum temperature the ratio for $T_{max,p}/T_{max}$ is equal to 0.9 and for minimum temperatures the ratio $T_{min,p}/T_{min}$ is equal to 0.72.

CHAPTER 3

Design situations

This chapter is concerned with the general concepts of design situations relating to EN 1991-1-3: *Eurocode 1 – Actions on Structures: Part 1.5: General Actions – Thermal Actions*.

EN 1990 Clause 3.2 identifies the following design situations for the verification of ultimate limit states. *cl.3.1(1)P: 1991-1-5*

- persistent design situations, which refer to the conditions of normal use
- transient design situations, which refer to temporary conditions applicable to the structure, e.g. during execution or repair
- accidental design situations, which refer to exceptional conditions applicable to the structure or to its exposure, e.g. explosions
- seismic design situations.

Each of these design situations is linked to a particular expression for the combination of action effects as follows:

- persistent and transient design situations, which refer to *expressions (6.10), or (6.10a) and (6.10b) in EN 1990*
- accidental design situations, which refers to *expression (6.11b) in EN 1990*
- seismic design situations, which refers to *expression (6.12b) in EN 1990.*

In addition, thermal actions need to be determined for the verification of serviceability limit states and the following expressions for the combination of action effects given in EN 1990:

- the characteristic combination which refers to *expression (6.14b) of EN 1990*
- the frequent combination which refers to *expression (6.15b) of EN 1990*
- the quasi-permanent combination which refers to *expression (6.16b) of EN 1990.*

The thermal actions that need to be identified for each of the above design situations for ultimate and serviceability limit state verifications include the following:

- A characteristic value (Q_k) which usually corresponds to an upper value with an intended probability of not being exceeded during a specific reference period (normally 50 years for buildings). Depending upon the design situation being considered for the ultimate or serviceability limit states verifications (described in *EN 1990 Clauses 6.4 and 6.5* and its Designers' Guide), other representative values of a variable action need to be determined as follows:
- The combination value, represented as a product $\psi_0 Q_k$.
- The frequent value, represented as a product $\psi_1 Q_k$. For buildings, the frequent value is generally chosen so that the time it is exceeded is 0.01 of the reference period (i.e. 50 years).

- The quasi-permanent value, represented as a product $\psi_2 Q_k$. Quasi-permanent values are also used for the calculation of long-term effects. This does not occur in the case of thermal actions.

EN 1990 Table A1

ψ_0 and ψ_1 are factors for the combination and frequent value, respectively of a variable action. For thermal actions $\psi_0 = 0.6$ and $\psi_1 = 0.5$. As mentioned above, the quasi-permanent value $\psi_2 Q_k$ does not apply to thermal actions.

cl.3.1(2)P: 1991-1-5

The elements of load-bearing structures which can be characterised as sensitive to the effects of thermal actions will need to be designed so that thermal movements will not cause overstressing of the structure. This can be either by the provision of movement joints or by including the effects of the thermal actions in the design.

CHAPTER 4

Representation of actions

This chapter is concerned with representation of actions in EN 1991-1-1: *Eurocode 1 – Actions on Structures: Part 1.1: General Actions – Thermal actions.*

The magnitude of the thermal actions and their distribution throughout individual elements of a structure are a function of numerous parameters, some of which are difficult to interpret numerically.

Daily and seasonal changes in shade air temperature, solar radiation, re-radiation, etc. give rise to variations of the temperature distribution within individual elements of a structure. *cl.4(1): 1991-1-5*

To determine the effects on a building due to thermal actions, the following parameters, some linked to the conditions of the particular building in question, have to be considered: *cl.4(2): 1991-1-5*

- local climatic conditions
- the presence of nearby structures that act as solar radiation screens
- the orientation of the building
- the building's total mass (and consequent thermal inertia)
- finishes (e.g. cladding in buildings) for determining the degree of solar energy absorption
- heating, air-conditioning, ventilation regimes and thermal insulation of the building
- the structural form and the nature of connections and movement joints.

The temperature profile within an individual structural element is made up of the four separate components as illustrated in Fig. 4.1: *cl.4(3): 1991-1-5*

(a) a uniform temperature component, designated ΔT_u (see Fig. 4.1(a))
(b) a linearly varying temperature difference component about the z–z axis, designated ΔT_{MY} (see Fig. 4.1(b))

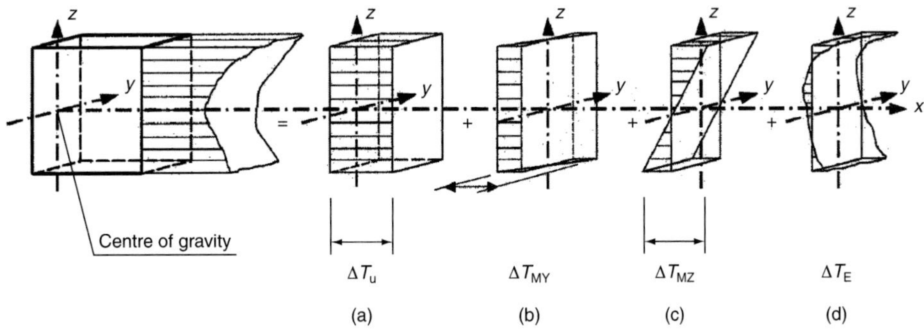

Fig. 4.1. Diagrammatic representation of constituent components of a temperature profile

(c) a linearly varying temperature difference component about the y–y axis, designated ΔT_{MZ} (see Fig. 4.1(c))
(d) a non-linear temperature difference component, designated ΔT_E (see Fig. 4.1(d)), resulting in a system of self-equilibrated stresses which produce no net load effect on the element.

Of the four thermal components, the self-equilibrating component ΔT_E (see Fig. 4.1(d)) causes the least effect on the structure or element under consideration.

cl.4(4): 1991-1-5 The strains, deformations and the resulting stresses induced by the thermal actions are dependent upon the geometry and boundary conditions of the element being considered and on the physical properties of the material employed in the construction. When materials with different coefficients of linear expansion are used compositely the thermal effect should be taken into account.

For determining the thermal effects, the coefficient of linear expansion for a material
cl.4(5): 1991-1-5 will be required (***Clause 4(5): 1991-1-5***). Annex C of EN 1991-1-5 gives a comprehensive list of the coefficient of linear expansion for a selection of commonly used construction
Annex C of 1991-1-5 materials (***Annex C of EN 1991-1-5***). A sample from the UK National Annex to EN 1991-1-5 is reproduced in Table 4.1.

Table 4.1. Coefficients of linear expansion

Material	α_T ($\times 10^{-6}$/°C)
Aluminium, aluminium alloy	–
Stainless steel	–
Structural steel, wrought or cast iron	12[1]
Concrete, normal density	12[2]
Concrete, lightweight aggregate	–
Masonry with concrete units	–
Masonry with clay units	–
Glass	–
Timber, along grain	–
Timber, across grain	–

[1] For composite structures the coefficient of linear expansion of the steel component may be taken as equal to 10×10^{-6}/°C to neglect restraining effects from different α_T values.
[2] When limestone aggregates are used in concrete, α_T, may be taken as 12×10^{-6}/°C; however, the coefficient may be as low as 9×10^{-6}/°C, therefore both values should be considered.

CHAPTER 5

Temperature changes in buildings

This chapter is concerned with the temperature changes in buildings in EN 1991-1-5: *Eurocode 1 – Actions on Structures: Part 1.5: General Actions – Thermal actions*. The material described in this chapter is covered in the following clauses:

- General *Clause 5.1: 1991-1-5*
- Determination of temperatures *Clause 5.2: 1991-1-5*
- Determination of temperature profiles *Clause 5.3: 1991-1-5*

5.1. General

Thermal actions on buildings due to:

- climatic temperature changes, and
- operational temperature changes inside the building

need to be considered in the design of buildings where there is a possibility of the ultimate or serviceability limit states being exceeded due to thermal movement and/or stresses (***Clause 5.1 (1)P: 1991-1-5***). *cl.5.1(1)P: 1991-1-5*

In addition thermal movement and stresses within the structure caused by temperature changes can also be influenced by:

- the shading from nearby structures
- the use of different materials in the structure with different thermal expansion coefficients and heat transfer
- the different geometric shapes of member cross-sections.

(***Note 1 to Clause 5.1(1)P: 1991-1-5***). *Note 1 to cl.5.1(1)P: 1991-1-5*

Moisture and other environmental factors (orientation of the structure) may also affect the volume changes of elements (***Note 2 to Clause 5.1(1)P: 1991-1-5***). *Note 2 to cl.5.1(1)P: 1991-1-5*

5.2. Determination of temperatures

This Section 5.2 of EN 1991-1-5 gives principles and rules for the determination of thermal actions on buildings due to climatic and operational temperature changes. Regional data and experience should also be taken into account (***Clause 5.2(1): 1991-1-5***). *cl.5.2(1): 1991-1-5* In the event that a building is not exposed to significant daily or seasonal temperature variations caused by activities within the building, the effects of short-term thermal actions can be neglected in the structural analysis.

cl.5.2(2)P: 1991-1-5 Climatic effects need to be determined by considering the variation of shade air temperature and solar radiation as described later in this section. The influence of activities (i.e. operations) carried out in the building's interior (due to heating, cooling, technological or industrial processes) also need to be considered for the particular project if appropriate.

cl.5.2(3)P: 1991-1-5 In accordance with the temperature components given in Chapter 4 of this Part, thermal actions caused by climatic and operational influences on a structural element need to be specified using the following basic quantities:

cl.5.2(5)P: 1991-1-5

(a) A uniform temperature component ΔT_u given by the difference between the average temperature T of an element and its initial temperature T_0. ΔT_u is defined as:

$$\Delta T_u = T - T_0 \qquad (5.1\text{: 1991-1-5})$$

where:

T is an average temperature of a structural element due to climatic temperatures in the winter or summer season and due to operational temperatures, and

cl.1.5.5: 1991-1-5 T_0 is defined as the temperature of a structural element at the relevant stage of its restraint (completion). Normally the value for T_0 may not be known at the time of the design. The designer should make a safe assessment for the value of T_0, or national guidance may be available. In Italy a value of $T_0 = 15°C$ is recommended.[2]

(b) A linearly varying temperature component given by the difference ΔT_M between the temperatures on the outer and inner surfaces of a cross-section, or on the surfaces of individual layers.

(c) A temperature difference ΔT_p of different parts of a structure given by the difference of average temperatures of these parts.

The values of ΔT_M and ΔT_p are normally determined for the particular project.

cl.5.2(4): 1991-1-5 In addition, local effects of thermal actions should be considered where relevant (e.g. at supports or fixings of structural and cladding elements). Adequate representation of thermal actions should be defined taking into account the location of the building and structural detailing.

The quantities ΔT_u, ΔT_M, ΔT_p and T are determined in accordance with the guidance provided in Section 5.3 below using regional data. When regional data are not available, the rules in Section 5.3 may be applied.

5.3. Determination of temperature profiles

cl.5.3(1): 1991-1-5 The temperature T in expression (5.1: 1991-1-5) is determined as the value of the average winter or summer temperature of the structural element in question by adopting a specific profile that defines the temperature distribution throughout the element's thickness. In the case of a sandwich element, T is the average temperature of a particular layer.

Annex D: 1991-1-5 Temperature profiles in elements for buildings may be obtained using the thermal transmission theory given in Annex D of EN 1991-1-5.

Note 1 to cl.5.3(1): 1991-1-5 For elements of one layer and when the environmental conditions on both sides are similar, a simplified procedure is given by EN 1991-1-5 and T may be approximately determined as the average of inner and outer environment temperature T_{in} and T_{out}. Thus:

$$T = (T_{out} + T_{in})/2 \qquad \textit{(Clause 5.3(1): 1991-1-5)}$$

cl.5.3(2): 1991-1-5 For the determination of the temperature of the inner environment, T_{in}, and the outer environment T_{out}, EN 1991-1-5 provides three tables. For determining the temperature for the inner environment Table 5.1 is provided. To determine the temperature of the outer environment, T_{out}, Table 5.2 should be used for parts located above ground level, and Table 5.3 for the parts below ground level.

Table 5.1. Indicative temperatures of inner environment T_{in}

Season	Temperature T_{in}
Summer	T_1
Winter	T_2

EN 1991-1-5 recommends the following values for T_1, and T_2 where more precise information on the temperature of the internal environment of a building is not available.

- $T_1 = 20°C$, and
- $T_2 = 25°C$.

See also Fig. 5.1 for a diagrammatical representation of external and internal temperatures.

(a) Determination of the temperature of the inner environment, T_{in}

The temperature of the inner environment, T_{in}, is determined in accordance with Table 5.1 of EN 1991-1-5 and Fig. 5.1.

As additional guidance the CIBSE Guide A Enviromental Design, 2006[3] in its Table 1.5 sets internal building environment comfort criteria for some typical building uses as follows.

- education buildings 21–23°C
- dwellings 23–25°C
- hospitals 21–25°C
- hotels 21–25°C
- libraries 21–25°C
- museums, art galleries 21–23°C
- offices 22–24°C
- retail 21–25°C
- sports halls 14–16°C
- swimming pools 23–26°C

(b) Determination of the temperature of the outer environment, T_{out}

The temperature of the outer environment, T_{out}, is determined in accordance with:

(i) Table 5.2 for parts located above ground level
(ii) Table 5.3 for parts located below ground level,

and Fig. 5.1.

The temperatures T_{out} for the summer season as given in Table 5.2 (for the parts of buildings above ground level) are dependent on the surface absorptivity and the orientation of the building. Table 5.2 gives:

- the maximum temperature T_{out} which is usually reached in Europe for latitudes between 45° and 55° (and also in accordance with the UK National Annex) for surfaces facing west, south-west or for horizontal surfaces
- the minimum T_{min} for surfaces facing north or north-east.

With reference to Table 5.2, Fig. 5.2 shows the recommended values for T_3, T_4 and T_5 for different orientations of a building.

The values given on absorptivity depend on both colour (which EN 1991-1-5 notes) and finish (which EN 1991-1-5 does not note).

In reality one may find surfaces with absorptivity from 5% to 95%. Generally 0.5 means a light grey while 0.7 (which means reflectivity 0.3) is a medium-dark surface which

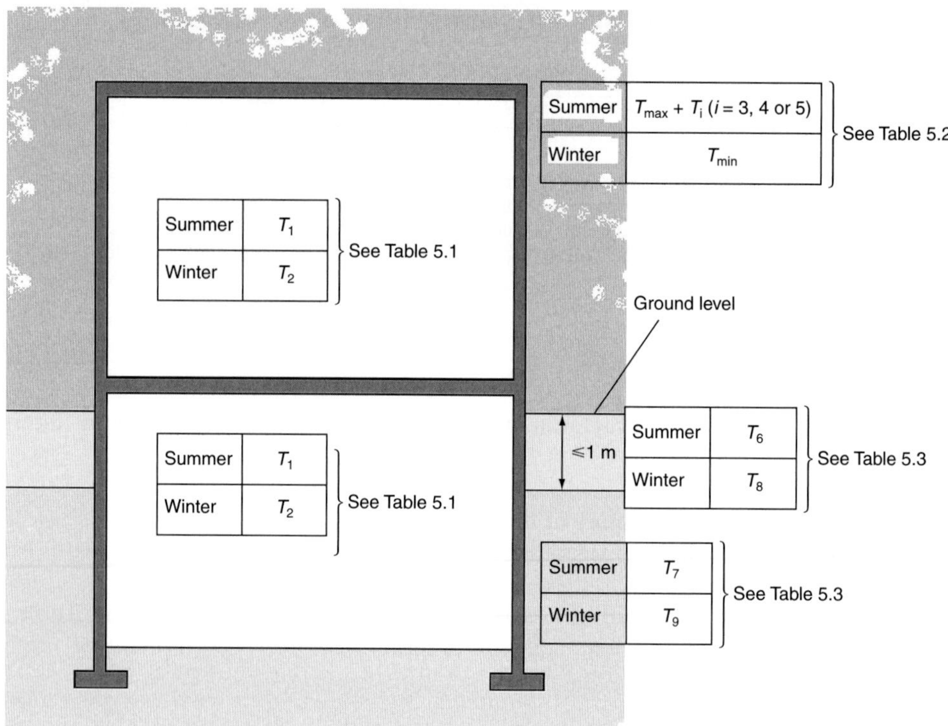

Fig. 5.1. Diagrammatic representations of internal and external temperatures

could be a medium grey or a glossy darker surface, and 0.9 (0.1 reflectivity) is very dark, like matt black.

The temperatures T_{out} for both the summer and winter seasons are given in Table 5.3 for the parts of buildings below ground level for parts of Europe between latitudes 45° and 55°; these are dependent upon the depth below ground level.

Table 5.2. Indicative temperatures T_{out} for buildings above ground level for latitudes between 45° and 55°

Season	Significant factor		Temperature T_{out} in °C
Summer	Relative absorptivity depending on surface colour	0.5 e.g. bright light surface (e.g. steel, light-coloured bricks, plaster, whitewash, light-coloured paints)	$T_{max} + T_3$
		0.7 e.g. light-coloured surface (e.g. yellow or buff bricks, red bricks, stone, concrete, dark paints)	$T_{max} + T_4$
		0.9 e.g. dark surface (e.g. black, non-metallic surfaces)	$T_{max} + T_5$
Winter			T_{min}

The values of the:

- maximum shade air temperature T_{max}, and
- minimum shade air shade temperature T_{min}

are given in maps in the National Annexes for mean return periods of 50 years.

The recommended solar radiation effects T_3, T_4, and T_5 are given by the following values:

- $T_3 = 0°C$, $T_4 = 2°C$ and $T_5 = 4°C$ should be used for north or north-east facing elements, and
- $T_3 = 18°C$, $T_4 = 30°C$ and $T_5 = 42°C$ for south, south-west or horizontal facing elements.

Table 5.3. Indicative temperatures T_{out} for parts of buildings below ground for latitudes between 45° and 55°

Season	Depth below the ground level	Temperature T_{out} in °C
Summer	Less than 1 m	T_6
	More than 1 m	T_7
Winter	Less than 1 m	T_8
	More than 1 m	T_9

The following values for T_6, T_7, T_8 and T_9 are recommended in EN 1991-1-5 and also used, for example, in the UK and Italian National Annexes:

$T_6 = 8°C$
$T_7 = 5°C$
$T_8 = -5°C$
$T_9 = -3°C$

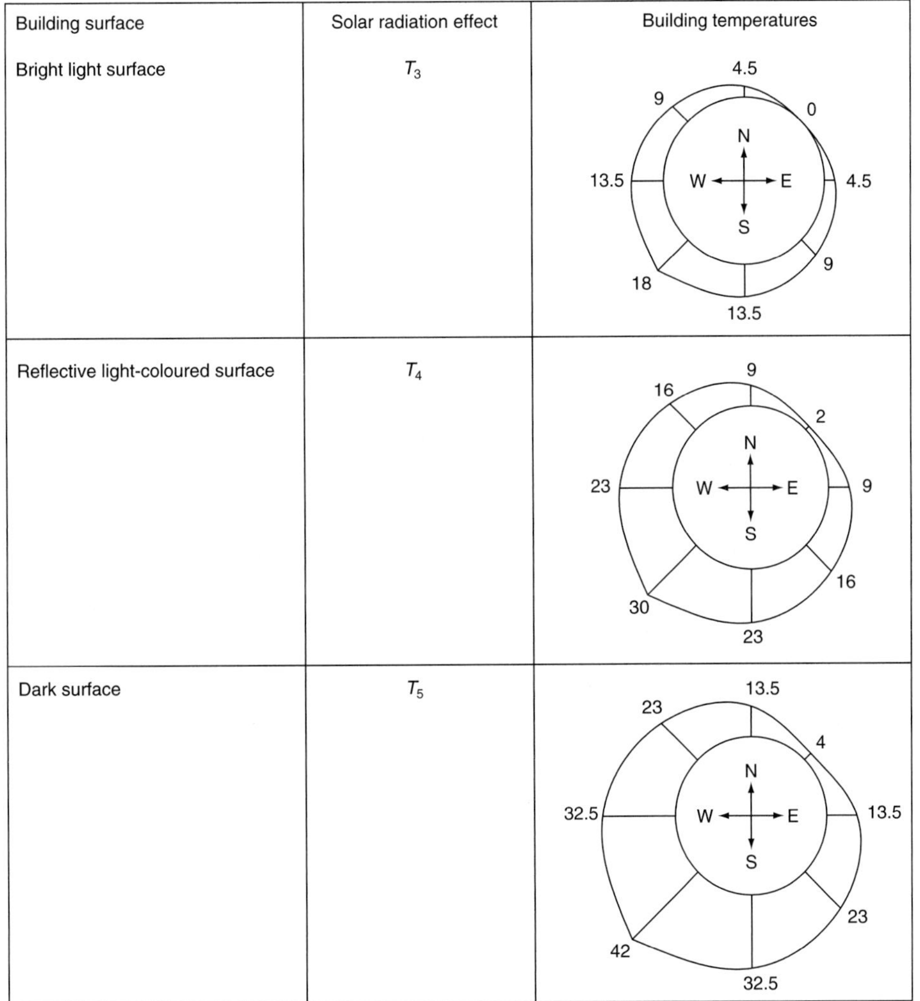

Fig. 5.2. Solar radiation effects T_3, T_4 and T_5 for different building orientations

CHAPTER 6

Annex A to Thermal Actions Part of the Manual

Section 6 of EN 1991-1-5 'Temperature changes in bridges' is covered in the *Designers' Guide for Actions on Road and Rail Bridges*. *Section 6: 1991-1-5*

Section 7 of EN 1991-1-5 'Temperature changes in industrial chimneys, pipelines, silos, tanks and cooling towers' is outside the scope of this Designers' Guide. *Section 7: 1991-1-5*

Guidance on information given in Annex A 'Isotherms of National minimum and maximum shade air temperatures' is given in Figure 2.1 of this Designers' Guide. *Annex A: 1991-1-5*

Guidance on information given in Annex B 'Temperature differences for various surfacing depths' is covered in the *Designers' Guide for Actions on Road and Rail Bridges*. *Annex B: 1991-1-5*

Guidance on information given in Annex C 'Coefficients of linear expansion' are given in Table 2.1 of this Designers' Guide. *Annex C: 1991-1-5*

CHAPTER 7

Annex B to Thermal Actions Part of the Manual

Temperature Profiles in Buildings and other Construction Works covered in Annex D of EN 1991-1-5

This Annex B gives information on Annex D of EN 1991-1-5 'Temperature Profiles in Buildings and other Construction Works'. It is referred to in the main text of EN 1991-1-5 in Clause 5.3(1). *Annex D: 1991-1-5* *cl.5.3(1): 1991-1-5*

This Annex D of EN 1991-1-5 gives the thermal transmission theory for determining the temperature profile for a simple sandwich element (e.g. slab, wall, shell). Assuming that local thermal bridges do not exist, a temperature $T(x)$ at a distance x from the inner surface of the cross-section can be determined assuming steady thermal state as: *D(1): 1991-1-5*

$$T(x) = T_{\text{in}} - \frac{R(x)}{R_{\text{tot}}}(T_{\text{in}} - T_{\text{out}})$$

where:

T_{in} is the air temperature of the inner environment
T_{out} is the temperature of the outer environment
R_{tot} is the total thermal resistance of the element including resistance of both surfaces
$R(x)$ is the thermal resistance at the inner surface and of the element from the inner surface up to the point X. *Figure D.1/1991-1-5*

The resistance values R_{tot}, [m^{2}K/W] is determined using the coefficient of heat transfer given in EN ISO 6946[4] as follows: *D(2): 1991-1-5*

$$R_{\text{tot}} = R_{\text{in}} + \sum_i \frac{h_i}{\lambda_i} + R_{\text{out}}$$

where:

R_{in} [m^{2}K/W] is the thermal resistance at the inner surface
R_{out} [m^{2}K/W] is the thermal resistance at the outer surface
λ_i [W/(mK)] is the thermal conductivity and h_i [m] is the thickness of the layer i.

The resistance $R(x)$ [m^{2}K/W] is determined using the coefficients of thermal conductivity given in EN ISO 13370[5]: *D(2): 1991-1-5*

$$R(x) = R_{\text{in}} + \sum_i \frac{h_i}{\lambda_i}$$

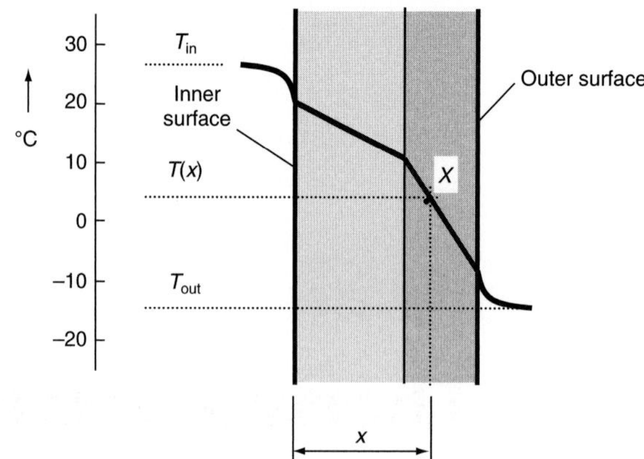

Fig. D.1. Thermal profile of a two-layer element

where layers (or part of a layer) from the inner surface up to point x (see Fig. D.1/EN 1991-1-5) are considered only.

Note to D(2): 1991-1-5

For buildings:

- the thermal resistance $R_{in} = 0.10$ to $0.17\,[m^2K/W]$ (depending on the orientation of the heat flow)
- $R_{out} = 0.04$ (for all orientations)
- the thermal conductivity λ_i for concrete (of volume weight from 21 to $25\,kN/m^3$) varies from $\lambda_i = 1.16$ to $1.71\,[W/(mK)]$.

References to Part 5

1. EN 1991-1-5: 2003. *Eurocode 1: Actions on Structures – Part 1-5: General Actions – Thermal actions*. European Committee for Standardisation, Brussels, 2003.
2. Norme Tecniche per le Costruzioni, DM 14.01.08. *Gazzetta Ufficiale della Repubblica Italiana*, n. 29 del 4 febbraio 2008 – Suppl. Ordinario n. 30.
3. *CIBSE Guide A: Environmental Design*, Chartered Institution of Building Services Engineers, 2006.
4. ISO 6949: 2007. *Building components and building elements – Thermal resistance and thermal transmittance – Calculation method*. International Organisation for Standardisation, Geneva, Switzerland, 2007.
5. ISO 13370: 2007. *Thermal performance of buildings – Heat transfer via the ground – Calculation methods*. International Organisation for Standardisation, Geneva, Switzerland, 2007.

PART 6: EN 1991-1-6: Eurocode 1: Part 1-6: Actions during execution

CHAPTER 1

General

This chapter is concerned with the general aspects of EN 1991-1-6: *Eurocode 1 – Actions on Structures: Part 1.6: General Actions – Actions during execution.*[1] The material described in this chapter is covered in the following clauses:

- Scope *Clause 1.1: 1991-1-6*
- Normative references *Clause 1.2: 1991-1-6*
- Assumptions *Clause 1.3: 1991-1-6*
- Distinction between Principles and Application Rules *Clause 1.4: 1991-1-6*
- Terms and definitions *Clause 1.5: 1991-1-6*
- Symbols *Clause 1.6: 1991-1-6*

1.1. Scope

EN 1991-1-6: *Eurocode 1 – Actions on Structures: Part 1.6: General Actions – Actions during execution* is one of the ten Parts of EN 1991. It gives design guidance and rules for the determination of actions which should be taken into account during execution of buildings and civil engineering works. *cl.1.1(1): 1991-1-6*

The guidance given in EN 1991-1-6 can also be used for the determination of actions to be taken into account for the design of:

- auxiliary construction works (auxiliary construction works include those which are sometimes known as temporary structures, which are typically structures that are not required after use when the related construction works are completed), and
- different types of construction works (e.g. structural alterations such as refurbishment and/or partial or full demolition). *Note 1 to cl.1.1(1): 1991-1-6*

It is stressed that EN 1991-1-6 does not provide any guidance concerning the safety of people (other than due to structural failure) in and around the construction site. Such rules may be defined for the individual project in accordance with National requirements (e.g. from the Health and Safety Executive in the UK). *cl.1.1(1): 1991-1-6*

A brief introduction is given below to the sections of this Part of EN 1991 which are discussed in greater detail in Chapters 2 to 4 of this Part of this Designers' Guide. *cl.1.1(1): 1991-1-6*

- *Section 1: General.* The scope of EN 1991-1-6 is wide, and it is the first code in Europe on this topic, and will benefit and be applicable to many types of users, including clients, designers and contractors of different disciplines.
- *Section 2: Classification of actions.* Among the many defined actions, particular attention is paid to the introduction of a number of different types of construction loads, which

may typically be present during the execution stages but which are unlikely to be present after completion of the structure.

- *Section 3: Design situations and limit states.* Principles and general rules on the choice of design situations and combinations of actions are given in accordance with EN 1990: Eurocode: Basis of structural design. Additionally, the planned duration of a relevant phase of execution can be associated with a theoretical nominal duration which may be used as a basis for the determination of characteristic values of climatic actions for return periods shorter than 50 years.
- *Section 4: Representation of actions.* EN 1991-1-6 describes Principles and Application rules for the determination of actions to be considered during execution of buildings and civil engineering works, including the following aspects. It should be noted that not all the actions and effects of actions below apply to buildings. In Chapter 4 only those actions relating to buildings will be discussed:
  - actions on structural and non-structural members during handling
  - geotechnical actions
  - actions due to prestressing effects
  - predeformations
  - temperature, shrinkage, hydration effects
  - wind actions
  - snow loads
  - actions caused by water
  - actions due to atmospheric icing
  - construction loads
  - accidental actions
  - seismic actions.
- *Annex A1: Supplementary rules for buildings*, which is a normative annex, where additional Basis of structural design information is given relating to Construction loads.
- *Annex A2: Supplementary rules for bridges*, which is a normative annex. This annex is covered by the Thomas Telford *Designers' Guide for Actions on Bridges.*
- *Annex B: Actions on structures during alteration, reconstruction or demolition*, which is an informative annex where general introductory guidance is given by EN 1991-1-6.

cl. 1.6.2.1: 1991-1-6 EN 1991-1-6 also gives rules for the determination of actions which may be used for the design of auxiliary construction works which are defined as follows.

Auxiliary construction works
Any works associated with the construction processes that are not required after use when the related execution activities are completed and they can be removed (e.g. falsework, scaffolding, propping systems, bracing).

This Designers' Guide will only give guidance on the parts of EN 1991-1-6 dealing with buildings. Bridges are covered by the Thomas Telford *Designers' Guide for Actions on Bridges.*

Before discussing each clause relating to buildings in EN 1991-1-6, a brief introductory advice for using this EN 1991-1-6 for the design of buildings is given in the following 1.3 Assumptions.

1.2. Normative references

cls. 1.2: 1991-1-6 No comment is necessary on the quoted references (***Clauses 1.2: 1991-1-6***).

1.3. Assumptions

cl. 1.3: 1991-1-6 The statements and assumptions given in EN 1990 *Clause 1.3* apply to all the Eurocode Parts and designers' guidance, for these are given in the Designers' Guide to EN 1990 (***Clause 1.3: 1991-1-6***).

Introductory advice for using this EN 1991-1-6 for the design of buildings
The guidance given in this part of the Designers' Guide is based on EN 1991-1-6 which gives guidance for the determination of actions for the design or verification of structures during their execution stages. It gives rules also for the determination of actions to be used for the design of auxiliary construction works, as well as for the design of structures in transient design situations, as defined in EN 1990, such as refurbishment, reconstruction and partial or total demolition. Auxiliary construction works include those that are sometimes known as temporary structures. The scope of EN 1991-1-6 is wide and it is the first such code in Europe and will benefit many, including clients, designers and contractors of different disciplines. Among all the many defined actions, particular attention is paid to the introduction of a number of different types of construction loads, which may typically be present during execution stages but which are unlikely to be present after completion of the structure.

1.4. Distinction between Principles and Application Rules

The statements and assumptions given in EN 1990 *Clause 1.4* apply to all the Eurocode Parts and designers' guidance, for these are given in the Designers' Guide to EN 1990. *cl.1.3: 1991-1-6*

1.5. Terms and definitions

Most of the definitions given in EN 1991-1-6 derive from ISO 2394, ISO 3898 and ISO 8930. In addition reference should be made to EN 1990 which provides a basic list of terms and definitions which are applicable to EN 1990 and to EN 1999, thus ensuring a common basis for the Eurocode suite. *cl.1.5: 1991-1-6*

For the structural Eurocode suite, attention is drawn to the following key definitions, which may differ from current national practices:

- '*Action*' means a load, or an imposed deformation (e.g. temperature effects or settlement).
- '*Effects of actions*' or '*Action effects*' are internal moments and forces, bending moments, shear forces and deformations caused by actions.

From the many definitions provided in EN 1990, those that apply for use with EN 1991 are described in Chapter 1 of this Designers' Guide. *cl.1.4(a), (b), (c) and (d): 1991-1-1*

The following comments are made to help the understanding of particular definitions in EN 1991-1-6:

- *construction loads* which are loads that can be present due to execution activities, but are not present when the execution activities are completed. *cl.1.5.2.2: 1991-1-6*

1.6. Symbols

cl.1.6: EN 1991-1-6

The notation in ***Clause 1.6: EN1991-1-6*** is based on ISO 3898. *cl.1.6(1): 1991-1-6*

EN 1990 *Clause 1.6* provides a comprehensive list of symbols, some of which may be appropriate for use with EN 1991-1-6. The symbols given in ***Clause 1.6(2): 1991-1-6*** are additional notations specific to this Part of EN 1991-1-6. *cl.1.6(2): 1991-1-6*

cl.1.6(2): 1991-1-6

CHAPTER 2

Classification of actions

This chapter is concerned with the classification of the actions in EN 1991-1-6: Eurocode 1 – *Actions on Structures: Part 1.6: General Actions – Actions during execution*. The material described in this chapter is covered in the following clauses:

- General *Clause 2.1: 1991-1-6*
- Construction loads *Clause 2.2: 1991-1-6*

2.1. General

EN 1991-1-6 classifies two sets of action types that occur during execution. The first set of actions are the general actions (e.g. snow loads, wind actions, thermal actions, self-weight) that occur during the executing process. The second set of actions are the construction loads (due to the operations of the execution process) which may typically be present during execution stages but which are unlikely to be present after completion of the structure. *cl.2.1(1): 1991-1-6*

The general actions that occur during the execution process are classified in accordance with EN 1990: 2002, Clause 4.1.1, and explained in the Designers' Guide to EN 1990. *cl.2.1(1): 1991-1-6*

Table 2.1 of EN 1991-1-6 gives the recommended classifications of actions for the general actions, appropriate for buildings with respect to their variation in time (permanent/variable/accidental), origin (direct/indirect), their spatial variation (fixed/free), by their nature or structural response (static/dynamic). *Table 2.1 of EN 1991-1-6*

Actions not necessarily relevant for buildings are shown shaded in Table 2.1. *Note 1 of cl.2.1(1): 1991-1-6*

Account should be taken of the fact that during the execution stages, as well as after completion of a structure, some of the actions may have to be reclassified. A typical example is the prestressing action applied to a concrete beam, which is to be regarded as a permanent action for the completed structure, but may have to be taken into account as a variable action for the design of anchorage areas during the execution phase for the jacking process.

2.2. Construction loads

Construction loads are represented in EN 1991-1-6, by a unique symbol Q_c. Construction loads, which can act on structures during execution activities, and which are not present after the completion of the works, are classified as direct variable actions. *cl.2.2(1): 1991-1-6*

Depending on their nature, construction loads are generally classified free actions. Table 2.2, which reproduces Table 2.2 of EN 1991-1-6, gives a general overview of the classification of construction loads. *Note 1 of cl.2.2(1): 1991-1-6*

Table 2.1. Classification of actions (other than construction loads) during execution stages (from ***EN 1991-1-6 Table 2.1***)

Action	Classification				Remarks	Source
	Variation in time	Classification/ origin	Spatial variation	Nature (static/dynamic)		
Self-weight	Permanent	Direct	Fixed with tolerance/free	Static	Free during transportation/storage Dynamic if dropped	EN 1991-1-1
Soil movement	Permanent	Indirect	Free	Static		EN 1997
Earth pressure	Permanent/ variable	Direct	Free	Static		EN 1997
Prestressing	Permanent/ variable	Direct	Fixed	Static	Variable for local design (anchorage)	EN 1990, EN 1992 to EN 1999
Pre-deformations	Permanent/ variable	Indirect	Free	Static		EN 1990
Temperature	Variable	Indirect	Free	Static	To be used with the National Annex	EN 1991-1-5
Shrinkage/hydration effects	Permanent/ variable	Indirect	Free	Static		EN 1992, EN 1993, EN 1994
Wind actions	Variable/ accidental	Direct	Fixed/free	Static/dynamic	To be used with the National Annex	EN 1991-1-4
Snow loads	Variable/ accidental	Direct	Fixed/free	Static/dynamic	To be used with the National Annex	EN 1991-1-3
Actions due to water	Permanent/ variable/ accidental	Direct	Fixed/free	Static/dynamic	Permanent/variable according to project specifications Dynamic for water currents if relevant	EN 1990
Atmospheric ice loads	Variable	Direct	Free	Static/dynamic		ISO 12494[2]
Accidental	Accidental	Direct/indirect	Free	Static/dynamic	To be used with the National Annex	EN 1990, EN 1991-1-7
Seismic	Variable/ accidental	Direct	Free	Dynamic	To be used with the National Annex	EN 1990, EN 1998

Note 1 of Table 2.2: 1991-1-6 Section 4.11.1 and Table 4.2 of this Designers' Guide give a comprehensive description of the various types of construction loads.

Note 1 of Table 2.2: 1991-1-6 Construction loads, which are caused by cranes, equipment, auxiliary construction works or structures, may be classified as fixed or free actions depending on their possible positions for use. See Section 4.11 below.

cl.2.2(3): 1991-1-6 Where the construction loads are classified as fixed, then tolerances for possible deviations from the theoretical assumed position should be defined. This will normally be for the individual project.

cl.2.2(4): 1991-1-6 Where the construction loads are classified as free, then the limits of the area where they may be moved or positioned should be determined. EN 1991-1-6 allows these limits to be *Note 1 to cl.2.2(4): 1991-1-6* defined in the National Annex and also for the individual project. In the UK National Annex the limits of movement for construction loads classified as 'free', need to be defined for the individual project.

In accordance with EN 1990: 2002, 1.3(2), and Annex B of EN 1990, control measures should be adopted, in particular to structures that fall within Consequence Classes CC2 and CC3 in accordance with Table B1 from Annex B of EN 1990 to verify the conformity of the position and moving of construction loads with the design assumptions.

cl. 1.3(2): EN 1990
Table B1: EN 1990
Note 2 to cl.2.2(4): 1991-1-6

Table 2.2. Classification of construction loads (***EN 1991-1-6 Table 2.2***)

	Classification					
Action (short description)	Variation in time	Classification/ origin	Spatial variation	Nature (static/dynamic)	Remarks	Source
Personnel and hand tools	Variable	Direct	Free	Static		
Storage movable items	Variable	Direct	Free	Static/dynamic	Dynamic in case of dropped loads	EN 1991-1-1
Non-permanent equipment	Variable	Direct	Fixed/free	Static/dynamic		EN 1991-3
Movable heavy machinery and equipment	Variable	Direct	Free	Static/dynamic		EN 1991-2, EN 1991-3
Accumulation of waste materials	Variable	Direct	Free	Static/dynamic	Can impose loads on for example vertical surfaces also	EN 1991-1-1
Loads from parts of structure in temporary states	Variable	Direct	Free	Static	Dynamic effects are excluded	EN 1991-1-1

CHAPTER 3

Design situations and limit states

This chapter is concerned with the design situations and limit states in EN 1991-1-6: *Eurocode 1 – Actions on Structures: Part 1.6: General Actions – Actions during execution.* The material described in this chapter is covered in the following clauses:

- General – Identification of design situations *Clause 3.1: 1991-1-6*
- Ultimate limit states *Clause 3.2: 1991-1-6*
- Serviceability limit states *Clause 3.3: 1991-1-6*

3.1. General – Identification of design situations

The process of the execution of a building is primarily a transient design situation. Accidental actions and therefore accidental design situations, for example loss of static equilibrium due to the fall of a member, earthquake, storm conditions, etc., can also occur. Therefore, the transient, accidental or seismic design situations need to be selected and taken into account as appropriate. *cl.3.1(1)P: 1991-1-6*

When designing for wind actions during storm conditions (e.g. a hurricane) EN 1991-1-6 allows the National Annex to select the design situation to be used, recommending the accidental design situation. The UK National Annex specifies the use of the accidental design situation for wind actions during storm conditions together with the BCSA Publication No. 39/05 *Guide to steel erection in windy conditions for steel structures.*[3] *Note to cl.3.1(1)P: 1991-1-6*

For verification of the execution stages, the design situations should be selected for:

- the structure as a whole
- the structural members
- the partially completed structure, and
- also for auxiliary construction works and equipment *cl.3.1(2)P: 1991-1-6*

as appropriate.

The selected design situations need to take into account the conditions that apply from stage to stage during execution in accordance with EN 1990: 2002, 3.2(3)P, which states: '*The selected design situations shall be sufficiently severe and varied so as to encompass all conditions that can reasonably be foreseen to occur during the execution and use of the structure*'. Therefore for the verification it is necessary to consider the conditions that can change during execution, including, for example, the shape of the structure, the structural system and especially the extent and degree of structural completeness. *cl.3.2(3)P: 1990* *cl.3.1(3)P: 1991-1-6 cl.3.1(4)P: 1991-1-6* Additionally the selected design situations need to be in accordance with the execution processes anticipated and need to take account of any revisions to the execution processes.

Choice of characteristic values of variable actions for transient design situations

The major problem concerning the choice of characteristic values of variable actions, especially climatic actions, for the transient design situations is related with the possibility of defining these characteristic values on the basis of return periods shorter than those accepted for persistent design situations. For the persistent design situation, in particular to climatic actions, a return period equal to the design working life of a structure is assumed.

The question to be discussed is:

(a) is it acceptable or not, and
(b) by how much,

to reduce the characteristic values of variable actions during execution and, more generally, during transient design situations?

This question is often posed, for practical reasons, because common sense considers it unlikely that rather high values of these actions (e.g. their characteristic value corresponding to the design working life of the completed structure) are reached during short periods (which is often the case for design situations during execution). Taking these high values into account may in some cases be very uneconomical.

EN 1990 states: '*The characteristic value of climatic actions is based upon the probability of 0.02 of its time-varying part being exceeded for a reference period of one year. This is equivalent to a mean return period of 50 years for the time-varying part.* ***However in some cases the character of the action and/or the selected design situation makes another fractile and/or return period more appropriate.***' (Note 2 to cl.4.1.2(7)P: 1990.) The emboldened part of the clause refers for example to transient design situations which correspond to shorter return periods. In addition, CEB Bulletin 191[4] states in Clause 6.2.1 that for variable actions the characteristic value is 'chosen with regard to the design situation under consideration'.

This reduction at the execution phase has been used by some National codes. For example, the French code for road bridges (dated 1971) specifies that in common cases wind pressure may be reduced during execution to 50% for phases of less than three months, 62.5% for longer phases.

EN 1991-1-6, through Clause 3.1(5), allows this reduction and this is discussed below.

cl.3.1(5): 1991-1-6 The expected time period or duration of a particular stage of execution may be associated with a nominal duration of the selected design situation, thus enabling different return periods of climatic actions to be taken into account. The nominal duration is intended to be equal to or greater than the anticipated duration of the stage of execution under consideration. Four ranges of return periods are recommended in EN 1991-1-6 as indicated in Table 3.1 below.

Regarding the nominal duration of three days, that is chosen for short execution phases, this corresponds to the extent in time of reliable meteorological predictions for the location of the site. Therefore in determining on the actions and their values, reliable meteorological predictions can be used instead in accordance with EN 1991-1-6. This choice may be kept for a slightly longer execution phase if appropriate organisational measures are taken. Generally, the concept of mean return period is generally not

Table 3.1. Recommended return periods for the assessment of the characteristic values of climatic actions Q_k depending on nominal duration of execution phase

Nominal duration of execution phase t	$t \leq 3$ days	Return period R	2 years	$p = 0.5$
	3 days $< t \leq 3$ months		5 years	$p = 0.2$
	3 months $< t \leq 1$ year		10 years	$p = 0.1$
	$t > 1$ year		50 years	$p = 0.02$

appropriate for short-term duration. However there may be situations when the time of the short execution phase cannot be predicted, and in those cases a return period of two years will give a reasonable value to be used. *Note (a) to Table 3.1: 1991-1-6*

Regarding the nominal duration of up to three months, actions may be determined taking into account appropriate seasonal and shorter-term meteorological climatic variations. For example, if for 3 days $< t \leq 3$ months, the period of the year under consideration is between June and the end of August, the likelihood of snow is very low. However there may be situations when the time of the execution phase cannot be predicted, and in those cases a return period of five years will give a reasonable value to be used. *Note (a) to Table 3.1: 1991-1-6*

EN 1991-1-6 allows for the return periods for the determination of characteristic values of variable actions during execution to be defined in the National Annex or for the individual project. The UK National Annex, for example, specifies that these are defined for the individual project, using the recommended values as a minimum. *Note to cl.3.1(5): 1991-1-6*

Determination of characteristic value of a climatic action Q_k for different return periods

The characteristic value of a climatic action Q_k may be determined on the basis of assumed probability distribution and selected return period R related to the probability p of its possible exceedance. The probability distribution of the basic variable Q_k may be derived on the basis of known data from the locality of site. According to recommendations given in EN 1991-1-6 and other Parts of EN 1991: *Eurocode 1: Actions on Structures*, the characteristic value $Q_{k,R}$ of a variable action for the return period of R years may be determined on the basis of the characteristic value $Q_{k,50}$ for a variable action for a 50-year return period. This may be determined from the general relationship given as:

$$Q_{k,R} = kQ_{k,50} \tag{3.1}$$

where k is the reduction coefficient of a variable action based on the extreme-value distributions as explained below.

The following relationships for thermal, snow and wind actions, respectively are recommended in the appropriate Parts of EN 1991. *Note 3 to cl.3.1(5)P: 1991-1-6*

(a) Thermal actions – in EN 1991-1-5

$$T_{max,R} = kT_{max,50} \quad \text{for } k = \{k_1 - k_2 \ln[-\ln(1 - 1/R)]\} \tag{3.2}$$

$$T_{min,R} = kT_{min,50} \quad \text{for } k = \{k_3 + k_4 \ln[-\ln(1 - 1/R)]\} \tag{3.3}$$

where:

$T_{max,50}/T_{min,50}$ is the maximum/minimum shade air temperature for 50 years return period and

$T_{max,R}/T_{min,R}$ for n years return period, and the coefficients $k_1 = 0.781$, $k_2 = 0.056$, $k_3 = 0.393$, $k_4 = -0.156$ may be used in case of lack of specific information on the coefficient of variation of temperature records.

See also the parts of this Designers' Guide dealing with EN 1991-1-5.

Table 3.2. Reduction coefficient k of actions $Q_{k,R}$ for different return periods R

Return period (years)	p	Reduction coefficient k			
		For thermal $T_{max,R}$	For thermal $T_{min,R}$	For snow $s_{n,R}$*	For wind $v_{b,R}$
2	0.50	0.80	0.45	0.64	0.77
5	0.20	0.86	0.63	0.75	0.85
10	0.10	0.91	0.74	0.83	0.90
50	0.02	1.00	1.00	1.00	1.00

* A coefficient of variation $v = 0.2$ has been assumed

(b) Snow actions – in EN 1991-1-3

According to the Gumbel formulation:

$$s_{k,R} = ks_{k,50} \quad \text{for } k = \left(\frac{1 - V\frac{\sqrt{6}}{\pi}\{\ln[-\ln(1-p)] + 0.57722\}}{(1 + 2.5923V)} \right) \tag{3.4}$$

where:

$s_{k,50}$ is the characteristic snow load on the ground for a 50-year return period, and
$s_{k,R}$ for n years return period, and
V is the coefficient of variation of annual maximum snow load,
p is the probability of exceedance during the reference period of n years, and it may be taken approximately equal to $1/n$.

See also the Parts of this Designers' Guide dealing with EN 1991-1-3.

(c) Wind actions – EN 1991-1-4

$$v_{b,R} = kv_{b,50} \quad \text{for } k = \left\{ \frac{1 - K\ln[-\ln(1-p)]}{1 - K\ln[-\ln(0.98)]} \right\}^{n} \tag{3.5}$$

where:

$v_{b,R}$ is the basic wind velocity for n years return period
$v_{b,50}$ is for 50-year return period
K is the shape parameter depending on the coefficient of variation of the extreme-value distribution, and n is the exponent. Recommended values are $K = 0.2$ and $n = 0.5$.

The calculated reduction coefficients k indicating the amount of reduction of the characteristic values of climatic actions $Q_{k,R}$ for different return periods R is shown in Table 3.2.

As mentioned above, the reduction of the characteristic value of a climatic action $Q_{k,R}$ associated with short expected durations of the execution activities under consideration, may not be appropriate. In such cases short-term meteorological predictions may be more appropriate to serve as a basis for estimation of climatic actions.

Note 2 to cl.3.1(5)P: 1991-1-6

Recommended values to limit inadequate reductions of characteristic values of climatic actions may be given in the National Annex, and examples are given in EN 1991-1-6 (e.g. a minimum wind speed of 20 m/s for up to durations of three months). The UK National Annex specifies that the value is defined for the individual project.

cl.3.1(6): 1991-1-6

When for the estimation of the characteristic climatic actions the design prescribes limiting climatic conditions, or weather window, the following should be taken into account:

- the anticipated duration of the particular execution stage
- the reliability of meteorological predictions for the area of the construction works
- the time to organise protection measures.

Example 3.1

The determination of characteristic values of variable actions $Q_{k,R}$ for four different return periods (R) can be seen below in Table 3.3 based on the k values given in Table 3.2.

This is for considered selected (2, 5 and 10 years) characteristic values of:

(a) For thermal actions a maximum shade air temperature $T_{max,50} = 32°C$.
(b) For thermal actions a minimum shade air temperature $T_{min,50} = -30°C$.
(c) For snow load on the ground $s_{n,50} = 1.5\,kN/m^2$.
(d) For basic wind velocity $v_b = 26\,m/s$ for a 50-year return period.

Table 3.3. Example of determination of climatic actions $Q_{k,R}$ for return periods R (2, 5 and 10 years)

Return period R (years)	p	Characteristic values of actions $Q_{k,R}$ (a) For thermal $T_{max,R}$ (°C)	(b) For thermal $T_{min,R}$ (°C)	(c) For snow $s_{n,R}$ (kN/m²)	(d) For wind $v_{b,R}$ (m/s)
2	0.50	25.6	−13.5	0.96	20.2
5	0.20	27.7	−18.8	1.13	22.2
10	0.10	29.0	−22.3	1.25	23.5
50	0.02	32.0	−30.0	1.50	26.0

When there is a likelihood of simultaneity of occurrence of climatic actions (snow loads, wind actions etc.) with the construction loads Q_c, the combinations of actions have to be in accordance with the rules given in EN 1990. See also Section 4.11.1 and Annex A of this Designers' Guide where advice is given on the safety factors (γ) and combination coefficients (ψ) to be used. *cl.3.1(7)P: 1991-1-6*

Clause 3.1(7): 1991-1-6 regarding the rules for the combination of snow loads and wind actions with construction loads Q_c is unhelpful, as is the UK National Annex which simply states that rules for combination should be defined for the individual project (***Clause 3.1(7): 1991-1-6***) and (***NA 2.6: BS EN 1991-1-6***). A reasonable assumption, to be used in the combination of action expressions given in EN 1990, would be to use the ψ_0 values given in EN 1990, Table A1.1 when either wind or snow loads are accompanying. For situations where the construction load Q_c is accompanying, Annex A of EN 1991-1-6 recommends a value for ψ_0 of between 0.6 and 1.0, with a recommended value of 1.0, which is also recommended by the UK National Annex. *cl.3.1(7): 1991-1-6* *NA 2.6: BS EN 1991-1-6*

In accordance with EN 1990 Clauses 3.5(3) and (7) imperfections in the geometry of the structure and of structural members should be defined for the individual project considering relevant CEN products (e.g. for concrete structures, these include CEN 'Precast Concrete Products' standards) and CEN execution standards. *cl.3.1(8): 1991-1-6*

Clause 3.1(9): 1991-1-6 is generally not applicable for building structures. Information may be obtained from the *Designers' Guide: Actions on Bridges*. *cl.3.1(9): 1991-1-6*

Where the structure or parts of it are subjected to accelerations that may give rise to dynamic or inertia effects, these effects should be taken into account. *cl.3.1(10): 1991-1-6*

Significant accelerations may be excluded where possible movements are strictly controlled by appropriate devices. *Note to cl.3.1(10): 1991-1-6*

Examples of situations that may cause accelerations are impact from a lorry, a dropped object, and reciprocating machinery left on, or beside a partially completed structure, etc. *cl.3.1(11): 1991-1-6*

Actions caused by water, including for example uplift due to groundwater, should be considered for the individual project, as appropriate.

Clause 3.1(12): 1991-1-6 is generally not applicable for building structures. Information may be obtained from the *Designers' Guide: Actions on Bridges*. *cl.3.1(12): 1991-1-6*

Actions due to creep and shrinkage and elastic shortening in concrete construction works should be determined on the basis of the expected dates and duration associated with the design situations, where appropriate. *cl.3.1(13): 1991-1-6*

Actions due to post-tensioning operations should be determined for short-term effects on the structure (elastic shortening) and load patterns on formwork (parasitic effects).

3.2. Ultimate limit states

cl.3.2(1)P: 1991-1-6

EN 1991-1-6 requires that ultimate limit states are verified in accordance with EN 1990, for all selected transient (i.e. expression 6.10 or 6.10a/6.10b from EN 1990), accidental (expression 6.11b) and seismic (expression 6.12b) design situations as appropriate during execution. The persistent design situation, which correspond to the conditions of normal use for the completed structure, is not taken into account.

For transient and accidental design situations the ultimate limit state verifications should be based on combinations of actions expressions referenced above and applied with the partial factors for actions γ_F and the appropriate ψ factors. Values of γ_G, γ_Q and ψ factors may be taken from EN 1990 Tables A1.1 to A1.4 for the actions given in Table A1.1 of EN 1990 (i.e. imposed loads, climatic actions).

The combinations of actions for accidental design situations, with expression 6.11b of EN 1990, are given below:

Note 1 to cl.3.2(1)P: 1991-1-6

$$\sum_{j\geq 1} G_{k,j} \text{"+"} P \text{"+"} A_d \text{"+"} (\psi_{1,1} \text{ or } \psi_{2,1}) Q_{k,1} \text{"+"} \sum_{i>1} \psi_{2,i} Q_{k,i}$$

can either include the accidental action explicitly or refer to a situation after an accidental event. The above expression 6.11b has to be used for a situation after an accidental event when the value for A_d becomes zero. It should be noted that the choice of factor ψ_1 or ψ_2 to be used with the leading variable action Q_k is given in the appropriate National Annex. The UK National Annex to EN 1990 specifies the use of ψ_1.

Generally, accidental design situations refer to exceptional conditions applicable to the structure or its exposure. During execution of the works these conditions may include:

Note 2 to cl.3.2(1)P: 1991-1-6

- impact
- local failure and subsequent progressive collapse
- fall of structural or non-structural parts, and
- in the case of buildings, abnormal concentrations of building equipment and/or building materials
- water accumulation on steel roofs
- fire, etc.

Note 3 to cl.3.2(1)P: 1991-1-6

Fire during construction should be treated as an accidental action.

No additional guidance is required to *Note 3 to Clause 3.2(1)P: 1991-1-6.*

cl.3.2(1)P: 1991-1-6

The verifications of the structure may need to take into account the appropriate geometry and resistance of the partially completed structure corresponding to the selected design situations. Patterns of temporary restraint should be taken into account.

3.3. Serviceability limit states

cl.3.3(1)P: 1991-1-6
cl.3.3(5): 1991-1-6

EN 1991-1-6 requires that serviceability limit states have to be verified as appropriate during execution in accordance with EN 1990 for the design situations concerning functioning (i.e. expression 6.10b from EN 1990) and appearance (expression 6.11b).

Note to cl.3.3(5): 1991-1-6

The frequent design situation is not required by EN 1991-1-6 to be taken into account unless specifically required for the individual project (***Note to Clause 3.3(5): 1991-1-6***).

cl.3.3(3)P: 1991-1-6

The objectives of these serviceability limit state verifications during execution is to limit cracking and/or early deflections which may adversely affect the durability, fitness for use and aesthetic appearance of the structure in the final stage.

cl.3.3(2): 1991-1-6

Hence the serviceability limit states during execution need to take into account the requirements for the completed structure.

The criteria associated with the serviceability limit states to be checked during execution may be defined in the National Annex and EN 1992 to EN 1999, but in many cases will need to be specified for the individual project. *Note to cl.3.3(2): 1991-1-6*

Load effects due to shrinkage, elastic shortening (post-tensioned concrete) and temperature need to be taken into account, generally by using the quasi-permanent load combination, in the design where appropriate, and should be minimised in the design by appropriate detailing. *cl.3.3(5): 1991-1-6*

Where relevant, serviceability requirements for auxiliary construction works need to be defined in order to avoid any unintentional deformations and displacements which affect the appearance or effective use of the structure or cause damage to finishes or non-structural members. *cl.3.3(6): 1991-1-6*

These requirements may be defined in the National Annex or for the individual project. Further guidance may be obtained from EN 12810,[5] EN 12811,[6] BS EN 12812,[7] and BS EN 12813.[8] *Note to cl.3.3(6): 1991-1-6*

CHAPTER 4

Representation of actions

This chapter is concerned with the representation of actions in EN 1991-1-6: *Eurocode 1 – Actions on Structures: Part 1.6: General Actions – Actions during execution*. The material described in this chapter is covered in the following clauses:

- General *Clause 4.1: 1991-1-6*
- Actions on structural and non-structural members during handling *Clause 4.2: 1991-1-6*
- Geotechnical actions *Clause 4.3: 1991-1-6*
- Actions due to prestresssing *Clause 4.4: 1991-1-6*
- Predeformations *Clause 4.5: 1991-1-6*
- Temperature, shrinkage, hydration effects *Clause 4.6: 1991-1-6*
- Wind actions *Clause 4.7: 1991-1-6*
- Snow loads *Clause 4.8: 1991-1-6*
- Actions caused by water *Clause 4.9: 1991-1-6*
- Actions due to atmospheric icing *Clause 4.10: 1991-1-6*
- Construction loads *Clause 4.11: 1991-1-6*
- Accidental actions *Clause 4.12: 1991-1-6*
- Seismic actions *Clause 4.13: 1991-1-6*

4.1. General

The determination of representative values of actions during execution need to be determined in accordance with EN 1990, EN 1991, EN 1997 and EN 1998 as appropriate. *cl.4.1(1)P: 1991-1-6*

The characteristic value F_k for the design situations during execution is a main representative value of a permanent or variable action. In accordance with EN 1990, the value of F_k needs to be specified as a mean, an upper or lower value, or a nominal value. If these values are specified for the individual project it must be ensured that consistency is achieved with methods given in the Eurocodes.

For variable actions, the two other representative values of actions F_{rep} that in common cases need to be considered for design for the execution phase are:

1. the combination value, represented as a product $\psi_0 Q_k$, used for the verification of the ultimate limit states and irreversible serviceability limit states
2. the quasi-permanent value, represented as a product $\psi_2 Q_k$, used for the verification of ultimate limit states, involving accidental actions and for verification of reversible serviceability limit states, for appearance etc.

Note 1 to cl.4.1(1)P: 1991-1-6
Note 2 and 3 to cl.4.1(1)P: 1991-1-6

The representative values of the actions (that also apply for the completed structure) during execution may be different from those used in the design of the completed structure. As explained in Section 3.1 of this guide, this may be due to the shorter duration of actions compared to the design working life of the structure.

No further guidance is necessary to Notes 2 and 3 of Clause 4.1(1)P (***Note 1 to Clause 4.1(1)P: 1991-1-6***) and (***Note 2 to Clause 4.1(1)P: 1991-1-6***).

cl.4.1(2): 1991-1-6

The determination of the representative values of construction loads will often depend on the specific process and period of execution as described in Section 3.1 of this Designers' Guide. In some cases it may be more appropriate to specify the values for the ψ factors, in some cases, for the individual project.

Regarding the ψ values for construction loads, Annex A1 of EN 1991-1-6 recommends ψ factors of construction loads for:

- $\psi_0 = 0.6$ to 1.0, and
- $\psi_2 = 0.2$

for building structures. It should be noted that ψ_1 does not normally apply to construction loads during execution.

Determination of γ and ψ values for construction loads Q_c

For construction loads the design value of an action F_d can be determined from the following expression:

$$F_d = \gamma_F F_{rep}$$

where γ_F is the partial factor for the action.

In EN 1990 three different sets of γ factors are recommended for variable loads in buildings depending on the ultimate limit state verification (EQU, STR and GEO) under consideration. For example, a value of $\gamma_f = 1.5$ is recommended (for verification of EQU and STR: the values of partial factors γ_f for Q_c may be altered nationally). (For an explanation of EQU, STR and GEO see Section 6.4 of the Designers' Guide to EN 1990.[9])

The design value of construction loads should take into account uncertainties in the modelling of the action and its effects. As characteristic values of construction loads are in many cases specified by the nominal value given in technical specifications for the individual project, the uncertainties in the determination of the action effect may be considered as low, provided they are not contravened during the execution of the works on site.

Thus, it might be reasonable to reduce the above mentioned γ_F factors for construction loads Q_c from 1.5 to 1.35 for building structures. However, it is also necessary to take into account other reductions described elsewhere in this Designers' Guide, e.g. magnitude of ψ factors influencing representative values of construction loads, and also reduction of characteristic values of other variable actions Q based on different return periods.

The γ and ψ values may also be altered in the National Annex or for the individual project, depending upon the allowed limits in the modelling of actions, on fixed tolerances in the specific project and the degree of approximation with which construction loads are defined.

cl.4.1(3): 1991-1-6

Interaction effects between structures and parts of structures need to be taken into account during execution. Such structures also include structures that form part of the auxiliary construction works.

cl.4.1(4)P: 1991-1-6

When parts of a structure are braced or supported by other parts of a structure (e.g. by propping floor beams for concreting), the actions on these parts resulting from bracing or supporting have to be taken into account.

Fig. 4.1. The slab of the lower level is supporting the casting weight of the upper one

Depending on the construction procedures, the supporting parts of the structure may be subjected to loads greater than the imposed loads for which they are designed for the persistent design situation, and this has to be checked. Additionally, concrete elements (e.g. supporting slabs) may not have developed their full strength capacities. *Note 1 to cl.4.1(4)P: 1991-1-6*

Figure 4.1 shows an example where the already constructed lower level is supporting the casting weight of the floor above.

Clause 4.1(4)P of EN 1991-1-6 is of particular relevance. Many collapses have occurred for reasons directly linked to the underestimation of the resistance of lower floors during the casting phases. As an example, Fig. 4.2 shows the collapse that occurred in the US[10] during the casting of concrete in the building where the lower supporting slabs had not developed their full strength. In the construction of the 26-storey building, concrete was being placed on the 24th floor and shorings were simultaneously being removed at the 22nd floor cast two weeks before and insufficient shear resistance of concrete caused progressive collapse.

Note 2 to cl.4.1(4)P: 1991-1-6

No further guidance is necessary to Note 2 of Clause 4.1(4)P.

Clause 4.1(5) may not generally have an application for buildings. However, where appropriate, any horizontal actions from friction effects need to be determined and based on the use of appropriate values of friction coefficients. *cl.4.1(5): 1991-1-6*

In the assessment, the lower and upper bounds of friction coefficients may have to be taken into account. Friction coefficients may be defined for the individual project. *Note 1 to cl.4.1(5): 1991-1-6*

4.2. Actions on structural and non-structural members during handling

The source for determining the self-weight of structural and non-structural members during handling is EN 1991-1-1. *cl.4.2(1): 1991-1-6*

Fig. 4.2. Progressive collapse during casting of concrete

Casting in in-situ concrete

During execution, non-uniformly distributed self-weight may occur when casting-in in-situ concrete and in order to avoid undesired effects the intended organisation of concreting and its control are essential (see also Section 4.11.2 below).

Regarding the weight of in-situ concrete for the assessment of vertical forces only, the characteristic density of fresh concrete is assessed by increasing the design characteristic value of the density of concrete by 1.00 kN/m^3, in accordance with EN 1991-1-1. After drying, this increased density is decreased to its normal design characteristic value.

For the execution of building structures, special attention should be paid to the early-aged material properties of concrete floors (e.g. lower strength, cracking, deflection). When the load imposed on a partially cured slab is higher than the capacity of the slab, the construction procedure must be changed to either reduce the loads on the slab, or to design accordingly so that the concrete strength at the time when the load is applied is increased.

cl.4.2(2): 1991-1-6 Where necessary, the dynamic or inertia effects of self-weight of structural and non-structural members will need to be taken into account. This will depend upon the individual project but further information may be found in EN 1991-1-1.

Clauses 4.2(3) and 4.2(4) regarding:

- *cl.4.2(3): 1991-1-6* actions on attachments for hoisting elements and materials and
- *cl.4.2(4): 1991-1-6* actions on structural and non-structural members due to support positions and conditions during hoisting, transporting or storage for which account needs to be taken, where appropriate of the actual support conditions and dynamic or inertia effects due to vertical and horizontal accelerations.

are generally more applicable to bridge structures, and reference should be made to the *Designers' Guide on Actions on Bridges*. *Note 1 to cl.4.2(2): 1991-1-6*

4.3. Geotechnical actions

The characteristic values of geotechnical parameters, soil and earth pressures, and limiting values for movements of foundations during execution, have to be determined using EN 1997. Further guidance may also be found in the *Designers' Guide to EN 1997-1*. *cl.4.3(1)P: 1991-1-6*

The soil movements of the foundations of both the structure and of auxiliary construction works (e.g. temporary supports during execution) need to be assessed from the results of geotechnical investigations. EN 1991-1-6 requires that the investigations should give information on both absolute and relative values of movements of foundations, their time dependency and possible scatter. *cl.4.3(2): 1991-1-6*

It should be noted that movements of auxiliary construction works may cause displacements and additional stresses on other parts of the structure. *Note 1 to cl.4.3(2): 1991-1-6*

The characteristic values of soil movements determined from statistical methods from geotechnical investigation data can be used as nominal values for determining imposed deformations of the structure. *cl.4.3(3): 1991-1-6*

If required, a more accurate determination for imposed deformations may be made by making a soil–structure interaction analysis. *Note 1 to cl.4.3(3): 1991-1-6*

4.4. Actions due to prestressing

No further guidance is necessary to Clause 4.4(1) and its note. *cl.4.4(1): 1991-1-6 and Note 1 to cl.4.4(1)*

In the case of prestressed concrete floors with unbonded tendons, special attention should be paid to the early-aged properties of the concrete, and to the load-balancing forces in combination with minimum variable floor loads during execution. These loads may be selected from the construction loads which are likely to be present during such phases, such as working personnel. For longer execution phases the loading related to the completion phases of the floors, such as storage of construction material etc., should be considered.

In general, the design situations to be checked are those concerned with the prevention of unacceptable cracks or crack widths (i.e. the characteristic and quasi-permanent load combinations for serviceability limit state verifications).

During the execution stages, as well as after completion of a structure, some of the actions may have to be reclassified. A typical example is the prestressing action applied to a concrete beam, which is to be regarded as a permanent action for the completed structure but may have to be taken into account as a variable action for the design of anchorage areas during the execution phase for the jacking process. *cl.4.4(3): 1991-1-6* *cl.4.4(2): 1991-1-6*

4.5. Predeformations

The treatments of the effects of predeformations need to be in conformity with the relevant design Eurocode (from EN 1992 to EN 1999). *cl.4.5(1)P: 1991-1-6*

Predeformations can result from displacements of supports, for example hangers and supports for floors. *Note to cl.4.5(1)P: 1991-1-6*

Action effects from execution processes should be taken into account, especially where predeformations (precamber or preset) are applied to a particular structural element in order to generate action effects for improving its final behaviour, particularly for structural safety and serviceability requirements. *cl.4.5(2): 1991-1-6*

A typical example is the self-weight distribution in concrete slabs poured over precast beams or floor elements with precamber, where the final appearance of the floor should be flat, see Fig. 4.3.

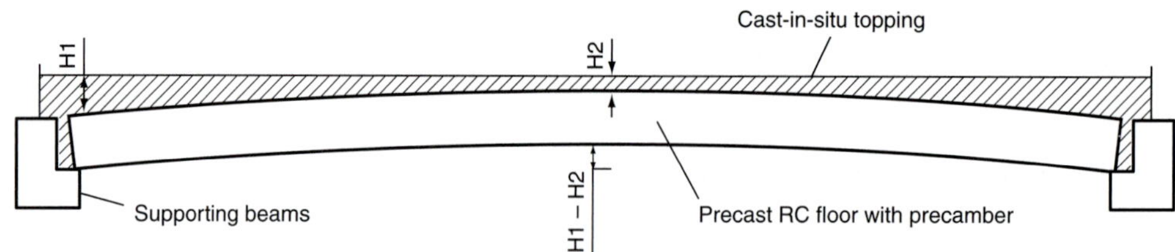

Fig. 4.3. Cast-in-situ topping over a precast prestressed reinforced concrete floor, with precamber

Another example is the case of a simply supported long-span prestressed reinforced concrete beam, with a significative depth and precamber. During the loading process the free movements of its end sections have to be assured, against the faces of the supporting elements.

cl.4.5(3): 1991-1-6 In critical cases, the action effects from predeformations should be checked against design criteria by measuring forces and deformations during execution.

4.6. Temperature, shrinkage, hydration effects

cl.4.6(1)P: 1991-1-6 The effects of temperature, shrinkage and hydration may have to be taken into account in each construction phase, as appropriate.

Note to cl.4.6(1)P: 1991-1-6 For buildings, the actions due to temperature and shrinkage are not generally significant if appropriate detailing has been provided for the persistent design situation.

Attention is given however to the effects due to a difference in time between casting one concrete element to another element that has already hardened. In general, the limit state to be checked is the prevention of unacceptable cracks or crack widths, especially in the case of composite steel and concrete structures.

Note 2 to cl.4.6(1)P: 1991-1-6 Note 2 to Clause 4.6(1)P generally applies to bridges and no further guidance is given. In addition, temperature can rise significantly in a massive concrete structure after casting, with consequent thermal effects.

cl.4.6(2): 1991-1-6 No further guidance is necessary for Clause 4.6(2).

If considered necessary by the designer thermal actions due to hydration, and shrinkage *cl.4.6(3): 1991-1-6* *Note to cl.4.6(3): 1991-1-6* effects in building materials should be determined according to EN 1992, to EN 1996 as appropriate. In addition, temperature can rise significantly in a massive concrete structure after casting, with consequent thermal effects.

cl.4.6(4): 1991-1-6 *cl.4.6(5): 1991-1-6* *cl.4.6(6): 1991-1-6* No further information is given on Clauses 4.6(4), 4.6(5) and 4.6(6) as these clauses do not normally apply for building structures.

4.7. Wind actions

Treatment of wind actions during execution
The programme for execution of the works should be phased so that it does not take place when the characteristic wind force assumed in the design is exceeded. Information on wind speeds should be obtained ahead of execution phases. Such information can be obtained from weather forecasts of the nearest meteorological station and local wind measurements.

cl.4.7(1): 1991-1-6 *Note to cl.4.7(1): 1991-1-6* Clause 4.7(1) will not apply to the majority of building structures. However, for dynamically sensitive buildings and structural elements such as slender columns or chimneys, the need for a dynamic response design procedure for wind actions should be determined for the execution stages, taking into account the degree of completeness and stability of the structure and its various elements. The criteria and procedures have to be defined for the individual project.

cl.4.7(2): 1991-1-6
Note to cl.4.7(2): 1991-1-6

The characteristic values of static wind forces Q_W should be determined according to EN 1991-1-4 and its Designers' Guide and the chapter on wind actions in this manual, for the appropriate return period. The appropriate return periods have been explained and specified in Section 3.1 of this Part of this Designers' Guide.

cl.4.7(3): 1991-1-6
Note to cl.4.7(3): 1991-1-6

For lifting and moving operations or other construction phases that are of short duration, the maximum acceptable wind speed for the operations should be specified, in accordance with Section 3.1 of this Part of this Designers' Guide.

cl.4.7(4): 1991-1-6

Clause 4.7(1) will not apply to the majority of building structures and no further guidance is provided here. The reader is referred to EN 1991-1-4 and its Designers' Guide.

cl.4.7(5): 1991-1-6
Note to cl.4.7(5): 1991-1-6

Wind actions on parts of the structure that are intended to be internal parts after its completion (e.g. internal walls) should be taken into account for execution processes. In such cases, the external pressure coefficients c_{pe} may have to be applied (e.g. for free-standing walls).

cl.4.7(6): 1991-1-6

When determining wind forces, the areas of equipment, falsework and other auxiliary construction works that are loaded should be taken into account.

4.8. Snow loads

cl.4.8(1)P: 1991-1-6

Snow loads need to be determined according to EN 1991-1-3 and its appropriate National Annex for the conditions of the site and the required return period.

Note 2 to cl.4.8(1)P: 1991-1-6

The appropriate return periods have been explained and specified in Section 3.1 of this Part of this Designers' Guide.

Note 1 to cl.4.8(1)P: 1991-1-6

Note 1 of Clause 4.8(1)P is not applicable to buildings.

4.9. Actions caused by water

cl.4.9(1): 1991-1-6

In general for buildings, actions due to water, including groundwater (Q_{wa}) should be represented as static pressures. For hydrodynamic effects which will generally not apply to buildings, see the *Designers' Guide on Actions on Bridges*.

Note to cl.4.9(1): 1991-1-6

No further guidance is given to the Note to Clause 4.9(1) which concerns hydrodynamic effects.

cl.4.9(2): 1991-1-6

Actions caused by water may be taken into account in combinations as permanent or variable actions. The classification of actions caused by water as permanent or variable may be defined for the individual project, taking account of the specific environmental conditions.

For example groundwater may be considered as a permanent action if the water level is considered to be always above the upper level of foundation and if the variations of the water level are relatively minor.

cl.4.9(3): 1991-1-6

No further guidance is given to the Note to Clause 4.9(3) which concerns, for example, dynamic effects, exerted by currents on immersed structures. Guidance may be found in the *Designers' Guide on Actions on Bridges*.

cl.4.9(4): 1991-1-6

No further guidance is given to Clause 4.9(4), which concerns the determination of the total horizontal force exerted by currents on a vertical surface. Guidance may be found in the *Designers' Guide on Actions on Bridges*.

cl.4.9(5): 1991-1-6

No further guidance is given to Clause 4.9(5), which concerns the possible accumulation of debris in rivers. Guidance may be found in the *Designers' Guide on Actions on Bridges*.

cl.4.9(6): 1991-1-6

No further guidance is given to Clause 4.9(5), which concerns actions due to ice and floating ice.

cl.4.9(7): 1991-1-6

Actions from rainwater need to be taken into account for conditions where there may be collection of water such as ponding effects from, for example, inadequate drainage, imperfections of surfaces, deflections and/or failure of dewatering devices. Normally the frequent design situation for serviceability limit state verification will apply.

4.10. Actions due to atmospheric icing

cl.4.10(1)P: 1991-1-6
Note to cl.4.10(1)P: 1991-1-6

No further guidance is given to Clause 4.9(10), which concerns actions due to atmospheric ice loads. Guidance on atmospheric ice loads may be found in EN 1993-3[11] and ISO 12494[2].

4.11. Construction loads

Defining construction loads

Actions that can be present due to the execution activities but are not present when the execution activities are completed are called construction loads and are an innovative part of the code.

For consistency with this definition, construction loads are classified as variable actions. A construction load may have vertical as well as horizontal components, static as well as dynamic effects.

Background on construction loads Q_c for buildings

Numerical data from Chapter 2 of *Cast-in-Place Concrete in Tall Building Design and Construction*[12] shows results from measures and investigations on construction sites. During the visits, data on equipment, material storage and worker loads were collected for two stages of construction:

1. before the concrete slab has been poured and
2. after the slab has been poured and preparations for the construction of the next floor.

To analyse the collected data, the surveyed floors were divided into successive sets of grids of various sizes. Table 4.1 shows the observed mean structural loads, transformed into an equivalent uniformly distributed load, for 10%, 1% and 0.5% fractiles.

Table 4.1. Observed mean structural loads, transformed into an equivalent uniformly distributed load, for 10%, 1% and 0.5% fractiles

Grid size (m^2)	Mean load (kN/m^2)	10% fractile Load (kN/m^2)	1% fractile Load (kN/m^2)	0.5% fractile Load (kN/m^2)
2.32	0.31	1.08	2.93	3.34
5.95	0.30	0.92	2.00	2.39
9.25	0.29	0.80	2.18	2.68
20.90	0.30	0.73	1.58	1.94
37.16	0.28	0.72	1.43	1.46

As an example, the 5% fractile value for the $9.25\,m^2$ grid size is $1.23\,kN/m^2$ (Gumbel distribution of the random variable is assumed).

From Table 4.1 it can be seen that the mean load is independent of the loaded area and its magnitude is low. But the coefficient of variation is very high, ranging from about 1.00 to 2.00, which explains the high values corresponding to fractiles of 10, 1 and 0.5%.

Other studies, e.g. *Partial Factor Design for Reinforced Concrete Buildings during Construction*[13] have suggested higher mean construction variable actions, but with a lower value of the coefficient of variation.

From the above it is suggested that:

- the order of magnitude of the characteristic vertical uniformly distributed load on floors during execution of buildings is close to $1\,kN/m^2$
- the geometrical variability of loads during execution should be covered by an overloading on a limited area.

4.11.1. General

Construction loads (Q_c) may be represented in the appropriate design situations in accordance with EN 1990 and Sections 3.2 and 3.3 of this Part of this Designers' Guide either, as

- one single variable action, or
- where appropriate, different types of construction loads may be grouped and applied as a single variable action.

In the appropriate combination of action expressions (see Sections 3.2 and 3.3 above), single and/or a grouping of construction loads need to be considered to act simultaneously with non-construction loads as appropriate. *cl.4.11.1(1): 1991-1-6*

Advice on simultaneity of non-construction loads, that can also apply when combining non-construction and construction loads, are given in EN 1990 and EN 1991. *Note 1 to cl.4.11.1(1): 1991-1-6*

Six types of construction loads (see Table 4.2) have been introduced in EN 1991-1-6 and the groupings of these loads that are to be taken into account as appropriate during execution are dependent on the individual project. *Note 1 to cl.4.11.1(1): 1991-1-6*

No further guidance is required on Note 3 to Clause 4.11.1(1). *Note 3 to cl.4.11.1(1): 1991-1-6*

The different construction loads Q_{ca}, Q_{cb}, Q_{cc}, Q_{cd}, Q_{ce} and Q_{cf} that have to be taken into account as appropriate are given in Table 4.2. *cl.4.11.1(2): 1991-1-6*

Each will then be described with regard to the representation of the action together with remarks that include the values of loads to be used.

Q_{ca}: Personnel and handtools

Q_{ca}, covering personnel (site staff and visitors) and handtools (see Fig. 4.4) is modelled as a uniformly distributed load q_{ca} and applied as to obtain the most unfavourable effects.

The recommended characteristic value $q_{ca,k}$ of the uniformly distributed load I EN 1991-1-6 which may be defined in the National Annex or for the individual project is $1.0\,kN/m^2$.

Table 4.2. Construction loads (Q_c)

Construction loads (Q_c)		
Type	Symbol	Description
Personnel and handtools	Q_{ca}	Working personnel, staff and visitors, possibly with handtools or other small site equipment
Storage of movable items	Q_{cb}	Storage of moveable items, e.g.: • building and construction materials, precast elements, and • equipment
Non-permanent equipment	Q_{cc}	Non-permanent equipment in position for use during execution, either: • static (e.g. formwork panels, scaffolding, falsework, machinery, containers), or • during movement (e.g. travelling forms, launching girders and nose, counterweights)
Movable heavy machinery and equipment	Q_{cd}	Movable heavy machinery and equipment, usually wheeled or tracked (e.g. cranes, lifts, vehicles, lift-trucks, power installations, jacks, heavy lifting devices)
Accumulation of waste materials	Q_{ce}	Accumulation of waste materials (e.g. surplus construction materials, excavated soil, or demolition materials)
Loads from parts of a structure in temporary states	Q_{cf}	Loads from parts of a structure in temporary states (under execution) before the final design actions take effect, such as loads from lifting operations

Fig. 4.4. Representation of Q_{ca}: personnel and handtools

Table 4.1: 1991-1-6

The UK National Annex states the characteristic value $q_{ca,k}$ of the uniformly distributed load may be defined for the individual project and a *minimum* value of 1.00 is recommended.

Q_{cb}: Storage of movable items

Q_{cb} covering storage of movable items (see Figs 4.5(a) and (b) which shows the storage of movable construction materials) is modelled as free actions and represented as appropriate by:

- a uniformly distributed load q_{cb}
- a concentrated load F_{cb}.

Table 4.1: 1991-1-6

The characteristic values of the uniformly distributed load and the concentrated load may be defined in the National Annex or for the individual project. The UK National Annex does not define any values but leaves it to be defined for the individual project.

Q_{cc}: Non-permanent equipment

Q_{cc} covers non-permanent equipment in position for use during execution, which can either be static equipment (e.g. formwork panels, scaffolding, falsework, machinery, containers) or equipment during movement (e.g. travelling forms, launching girders and nose, counterweights). See Fig. 4.6. Q_{cc} is modelled as free actions and is represented as appropriate by a uniformly distributed load q_{cc}.

Table 4.1: 1991-1-6

These loads Q_{cc} may be defined for the individual project using information given by the supplier of the equipment. Unless more accurate information is available, the loads may be modelled by a uniformly distributed load with a recommended minimum characteristic value of $q_{cc,k} = 0.5\,\text{kN/m}^2$.

For further information and for formwork and falsework design see EN 12812[7] and EN 12811[6].

Q_{cd}: Movable heavy machinery and equipment

Table 4.1: 1991-1-6

This type of load is not generally applicable to buildings and the appropriate information in EN 1991-1-6 is intended for bridges. Further guidance is available in the *Designers' Guide on Actions on Bridges*.

Q_{ce}: Accumulation of waste materials

Q_{ce} covers accumulation of waste materials (e.g. surplus construction materials, excavated soil, or demolition materials). See Fig. 4.7.

Table 4.1: 1991-1-6

Q_{ce} is taken into account by considering possible effects on horizontal, inclined and vertical elements (such as walls), depending on the build-up, and thus mass effects of the accumulation of material.

(a)

(b)

Fig. 4.5(a) and (b). Examples of storage of building and construction materials

These loads may vary significantly, and over short time periods, depending on types of materials, climatic conditions, build-up and clearance rates.

Q_{cf}: Loads from parts of a structure in temporary states

Q_{cf} covers loads from parts of a structure in temporary states (under execution) before the final design actions take effect, e.g. loads from lifting operations. See Fig. 4.8.

Fig. 4.6. Example of non-permanent static equipment (formwork panels, scaffolding, and falsework)

Table 4.1: 1991-1-6

Q_{cf} is taken into account and modelled according to the planned execution sequences, including consequences of those sequences, for example loads and reverse load effects due to particular processes of construction, such as assemblage.

cl.4.11.1(3)P: 1991-1-6

See Section 4.11.2 of this guide for additional loads due to concrete being fresh.

No further guidance is necessary for Clause 4.11.1(3)P.

Note 1 to cl.4.11.1(3)P: 1991-1-6

Recommended values of ψ factors for construction loads which are given in Annex A1 of EN 1991-1-6 for buildings, have been explained in Section 4.1 of this Designers' Guide.

Fig. 4.7. Accumulation of waste materials

Fig. 4.8. Loads from parts of a structure in temporary states

No further guidance is necessary for Note 2 to Clause 4.11.1(3)P. *Note 2 to cl.4.11.1(3)P: 1991-1-6*

In case of heavy equipment or for other specific construction materials, the nominal values supplied by the producers may be assumed as characteristic. *cl.4.11.1(4)P: 1991-1-6*

No further guidance is necessary for Clause 4.11.1(4)P. *cl.4.11.1(5)P: 1991-1-6*

No further guidance necessary for Clause 4.11.1(5)P.

4.11.2. Construction loads during the casting of concrete

Casting of concrete is an extremely common activity and is a good example of grouping of different sources of construction loads in one single action, which will need to be combined with other actions which may arise during this specific execution phase, such as climatic actions (e.g. wind) or other construction loads applied to other parts of the structure.

Actions to be taken into account simultaneously during the casting of concrete may include working personnel with small site equipment (Q_{ca}), formwork and load-bearing members (Q_{cc}) and the weight of fresh concrete (which is one example of Q_{cf}), as appropriate. An example of Q_{ca}, Q_{cb} and Q_{cc} occurring together is shown in Fig. 4.6. *cl.4.11.2(1): 1991-1-6*

Loads according to (a), (b) and (c), as given in Table 4.3, are intended to be positioned to cause the maximum effects. The loads and effects may or may not be symmetrical. *Note 3 to cl.4.11.2(1): 1991-1-6*

The density of fresh concrete can be obtained from EN 1991-1-1 Table A.1. The density of fresh concrete can be taken as $25\,kN/m^2$. *Note 1 to cl.4.11.2(1): 1991-1-6*

Q_{ca}, Q_{cc} and Q_{cf} may be given in the National Annex. During the casting of concrete the following values should be used as recommended values: *Note 2 to cl.4.11.2(1): 1991-1-6*

- For Q_{ca}: $q_{ca,k} = 0.75\,kN/m^2$ as given in Table 4.3.
- For Q_{cc}: $q_{cc,k} = 0.5\,kN/m^2$ given in Table 4.3.
- Alternative values for Q_{ca} and Q_{cc} may be determined for the individual project if a specific assessment is undertaken.
- Values for Q_{cf} should be assessed and determined for the individual project taking account of the information provided in Table 4.3 and this section.

Table 4.3. Recommended characteristic values of actions due to construction loads during casting of concrete

Action	Loaded area	Load in kN/m^2
(a)	Inside the working area 3 m × 3 m (or the span length if less)	10% of the self-weight of the fresh concrete for the design thickness but not less than 0.75 kN/m^2 and not more than 1.5 kN/m^2 Includes Q_{ca} and Q_{cf}
(b)	Outside the working area	0.75 kN/m^2 covering Q_{ca}
(c)	Actual area	Self-weight of the formwork, load-bearing element (Q_{cc}) and the weight of the fresh concrete for the design thickness (Q_{cf})

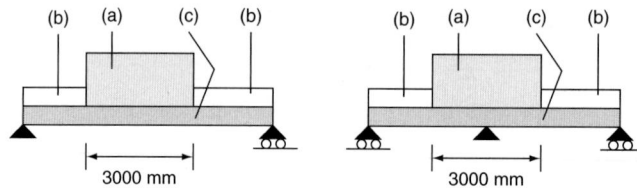

Note: For action (b) a lower value of Q_{ca} than the one in Table 4.1 of EN 1991-1-6. This takes into account the fact that most site personnel will be in area (a) where the main construction activities are taking place.

Note 3 to cl.4.11.2(1): 1991-1-6

Other values may have to be defined, for example when using self-levelling concrete or precast products.

Horizontal actions of fresh concrete may need to be taken into account where appropriate.

4.12. Accidental actions

Accidental actions that may occur on the site, for example

- impacts from construction vehicles, cranes, building equipment or materials in transit (e.g. skip of fresh concrete), and/or
- local failure of final or temporary supports, including dynamic effects, that may result in collapse of load-bearing structural members

cl.4.12(1)P: 1991-1-6

have to be taken into account, where relevant.

Note 1 to cl.4.12(1)P: 1991-1-6

It should be noted that abnormal concentrations of building equipment and building materials on load-bearing structural members should not be taken into account as accidental actions. The transient design situation should be used for this case. However, this situation of abnormal concentrations of load should be avoided.

Note 2 to cl.4.12(1)P: 1991-1-6

When dynamic effects due to accidental actions need to be taken into account a dynamic amplification factor of 2 may be used for the striking action. This can be different subject to assessment (e.g. a dynamic analysis) and specified for the individual project.

Note 3 to cl.4.12(1)P: 1991-1-6

No further guidance is necessary on Note 3 of Clause 4.12(1)P.

cl.4.12(2): 1991-1-6

The action due to falls of equipment either onto or from the structure needs to be defined and taken into account where appropriate. The dynamic effects due to the fall need to be considered and may be given in the National Annex or for the individual project.

cl.4.12(3): 1991-1-6

Where relevant, a human impact load may need to be taken into account as an accidental action, represented by a quasi-static vertical force.

Note to cl.4.12(3): 1991-1-6
NA2.16: 1991-1-6

The recommended design values, confirmed in the UK National Annex, of the human impact force to be used are:

1. 2.5 kN applied over an area 200 mm × 200 mm, to account for stumbling effects
2. 6.0 kN applied over an area 300 mm × 300 mm, to account for falling effects.

The effects of the accidental actions described in this section should be assessed to determine the potential for inducing movement in the structure. The extent and effect of any such movement should be determined, and the potential for progressive collapse assessed. *cl.4.12(4): 1991-1-6*

Additional information may be obtained from EN 1991-1-7 and the chapter in this Designers' Guide relating to accidental actions. *Note to cl.4.12(3): 1991-1-6*

Accidental actions used for design situations should be taken into account for any changes of the execution process. To ensure that the appropriate design criteria are applied at all times, corrective measures need to be taken as work proceeds. *cl.4.12(5): 1991-1-6*

No further guidance required on the statement in Clause 4.12(6) '*Fire actions should be taken into account, where appropriate*'. However, more information may be obtained from EN 1991-1-2 and its Thomas Telford Designers' Guide[1] and EN 1991-1-7 and the chapter in this Designers' Guide relating to accidental actions. *cl.4.12(5): 1991-1-6*

4.13. Seismic actions

No further information is given on this Clause 4.13 Seismic actions, except for stressing the choice of the reference period during execution or repair. *cl.4.13(1): 1991-1-6*

Reference should be made to EN 1998 and in particular the following parts of EN 1998: *cl.4.13(2): 1991-1-6*

- EN 1998-1: 2004: Design for structures for earthquake resistance. Assessment and retrofitting of buildings.[14]
- EN 1998-3: 2004: Design for structures for earthquake resistance. General rules, seismic actions and rules for buildings.[15]

ANNEX A1 (NORMATIVE)

Supplementary rules for buildings

This chapter is concerned with the Annex A1 (normative) Supplementary rules for buildings in EN 1991-1-6: *Eurocode 1 – Actions on Structures: Part 1.6: General Actions – Actions during execution*. The material described in this chapter which concern basis of structural design rules during execution is covered in the following clauses:

- Ultimate limit states *Clause A1.1: 1991-1-6*
- Serviceability limit states *Clause A1.2: 1991-1-6*
- Horizontal actions *Clause A1.3: 1991-1-6*

A1.1. Ultimate limit states

The ultimate limit state verifications during execution should be based on combinations of actions applied with the partial factors for actions γ_F and the appropriate ψ factors for transient (Expression 6.10 or 6.10a/610b from EN 1990), accidental (Expression 6.11b from EN 1990), and seismic (Expression 6.12b from EN 1990) design situations. The values for the actions should consider the length of a particular execution phase in accordance with Section 3.1 of this Part of this Designers' Guide. *cl.A1.1(1): 1991-1-6*

For buildings, the values for γ_G and γ_Q should be obtained from EN 1990, Annex A1 and the appropriate National Annex. ψ factors for loads other than the construction loads should be obtained from EN 1990, Annex A1 Table A.1.1 and the appropriate National Annex. *Note 1 to cl.A1.1(1): 1991-1-6*

The representative values of the variable actions due to construction loads may be set by the National Annex, within a recommended range of $\psi_0 = 0.6$ to 1.0. The recommended value of ψ_0 is 1.0. The UK National Annex specifies the recommended values. *Note 2 to cl.A1.1(1): 1991-1-6* *NA 2.18: BS 1991-1-6*

The minimum recommended value of ψ_2 is 0.2 and EN 1991-1-6 recommends that values below 0.2 are not selected. The UK National Annex specifies the recommended values.

Generally, ψ_1 does not apply to construction loads during execution. *Note 3 to cl.A1.1(1): 1991-1-6*

A1.2. Serviceability limit states

The serviceability limit state verifications during execution should be based on combinations of actions applied with the partial factors for actions γ_F (equal to 1) and the appropriate ψ factors for characteristic (Expression 6.14b from EN 1990) and quasi-permanent (Expression 6.14c from EN 1990) combinations. *cl.A1.2(1): 1991-1-6*

Note to cl.A1.2(1): 1991-1-6

The guidance given in Section A1.1 of this guide for values of ψ factors are also applicable for serviceability limit state verifications.

cl.A1.2(1)P: 1991-1-6

A1.3. Horizontal actions

Note to cl.A1.2(1)P: 1991-1-6

Horizontal actions which result from, for example, wind forces and the effects of sway imperfections and sway deformations need to be taken into account. Further information is given in EN 1990 and EN 1991-1-4.

cl.A1.2(2): 1991-1-6

Note to cl.A1.2(2)2: 1991-1-6

NA 2.19: BS 1991-1-6

Nominal horizontal forces (F_{hn}) may be applied only when such a method can be justified as appropriate and reasonable for a particular case. In such cases, the determined nominal horizontal forces should be applied at locations to give the worst effects, and may not always correspond to those of the vertical loads. The recommended characteristic value of these equivalent horizontal forces is 3% of the vertical loads from the most unfavourable combination of actions. Although the UK National Annex recommends this value it states in addition '*the recommended value may be appropriate; however, values should be determined and defined for the individual project*' (***NA 2.19: BS EN1991-1-6***).

ANNEX A2 (NORMATIVE)

Supplementary rules for bridges

No guidance is given in the Designers' Guide for Annex A2 (normative) Supplementary rules for bridges. Information can be found in the *Designers' Guide for Actions on Bridges*. *Annex A2: 1991-1-6*

ANNEX B (INFORMATIVE)

Actions on structures during alteration, reconstruction or demolition

No additional guidance is considered necessary in this Designers' Guide for Annex B (informative) Actions on structures during alteration, reconstruction or demolition. *Annex B: 1991-1-6*

References for Part 6

1. EN 1991-1-6: 2005. *Eurocode 1 – Actions on Structures. Part 1-6: General Actions – Actions during execution.* European Committee for Standardisation, Brussels, 2005.
2. ISO 12494: 2001. *Atmospheric icing of structures.* International Organisation for Standardisation, Geneva, Switzerland, 2001.
3. British Constructional Steelwork Association. BCSA Publication No. 39/05, *Guide to steel erection in windy conditions.* BCSA, London, 2005.
4. CEB Bulletin 191. *General Principles on Reliability for Structures – A commentary on ISO 2394 approved by the Plenum of the JCSS.* fib, Lausanne, 1988, 62 pp.
5. EN 12810-1: 2003. *Façade scaffolds made of prefabricated components – Part 1: Products specifications.* European Committee for Standardisation, Brussels, 2003.
6. EN 12811-1: 2003. *Temporary works equipment – Part 1: Scaffolds – Performance requirements and general design.* European Committee for Standardisation, Brussels, 2003.
7. EN 12812: 2008. *Falsework – Performance requirements and general design.* European Committee for Standardisation, Brussels, 2008.
8. EN 12813: 2004. *Temporary works equipment – Load bearing towers of prefabricated components – Particular methods of structural design.* European Committee for Standardisation, Brussels, 2004.
9. Gulvanessian, H., Calgaro, J.-A. and Holický, M. *Designers' Guide to EN 1990. Eurocode: Basis for Structural Design.* Thomas Telford, London, 2002, ISBN 0 7277 3011 8.
10. Carper, K. Beware of vulnerabilities during construction. *Construction and equipment,* 3/2004.
11. EN 1993-3: 2006. *Eurocode 3 – Design of steel structures – Part 3-1: Towers, masts and chimneys – Towers and masts.* European Committee for Standardisation, Brussels, 2006.
12. Council on Tall Buildings and Urban Habitat Committee 21 D. *Cast-in-Place Concrete in Tall Building Design and Construction.* McGraw-Hill, Ch. 2: Construction loads.
13. El-Shahhat, A. M., Rosowsky, D. V. and Chen, W. F. Partial Factor Design for Reinforced Concrete Buildings during Construction. *ACI Structural Journal,* July–August 1994.
14. EN 1998-1: 2004. *Design for structures for earthquake resistance. Assessment and retrofitting of buildings.* European Committee for Standardisation, Brussels, 2004.
15. EN 1998-3: 2004. *Design for structures for earthquake resistance. General rules, seismic actions and rules for buildings.* European Committee for Standardisation, Brussels, 2004.

PART 7: EN 1991-1-7: Eurocode 1: Part 1-7: Accidental actions

CHAPTER 1

General

This chapter is concerned with the general aspects of EN 1991-1-7: *Eurocode 1 – Actions on Structures: Part 1.7: General Actions – Accidental actions.*[1] The material described in this chapter is covered in the following clauses:

- Scope — *Clause 1.1: 1991-1-7*
- Normative references — *Clause 1.2: 1991-1-7*
- Assumptions — *Clause 1.3: 1991-1-7*
- Distinction between Principles and Application Rules — *Clause 1.4: 1991-1-7*
- Terms and definitions — *Clause 1.5: 1991-1-7*
- Symbols — *Clause 1.6: 1991-1-7*

1.1. Scope

EN 1991-1-7: *Eurocode 1 – Actions on Structures: Part 1.7: General Actions – Accidental actions* is one of the ten Parts of EN 1991. It gives strategies and rules for safeguarding buildings and other civil engineering works (e.g. bridges) against identifiable and unidentifiable accidental actions. *cl.1.1(1): 1991-1-7*

EN 1991-1-7 defines the strategies based on: *cl.1.1(2): 1991-1-7*

- identified accidental actions, and
- limiting the extent of localised failure (e.g. when the accidental action is unidentified).

Typical examples of accidental actions that can be identified include fire, some explosions, earthquakes, impacts, floods, avalanches and landslides.

Next to these identified accidental actions, structural members may get damaged for a variety of less identifiable reasons such as human errors in design and construction, improper use, exposure to aggressive agencies, failure of equipment, malicious attacks (e.g. terrorist attacks), warfare and so on.

EN 1991-1-7 gives general guidelines on dealing with the effects of identified and unidentified accidental actions. Because of the nature of accidental loads the design approach is different from that for normal loads (e.g. imposed loads and climatic actions). Local damage may be acceptable and non-structural measures (e.g. sprinkler installations or vent openings) may prove to be more cost-effective than structural ones.

EN 1991-1-7 comprehensively deals with many topics including: *cl.1.1(3): 1991-1-7*

- impact (in Section 4 of EN 1991-1-7)
- explosions (in Section 5 of EN 1991-1-7).

The code includes a number of key informative annexes on the following:

- Design for the consequences of localised failure in buildings from an unspecified cause (in Annex A of EN 1991-1-7).
- Information on risk assessment which may have to be used for effects giving rise to high consequences of failure are given in Annex B of EN 1991-1-7.
- Dynamic design for impact. Section 4 gives values of actions due to impact from traffic, trains etc. As an alternative these can be directly obtained using Annex C of EN 1991-1-7.
- Internal explosions (in Annex D of EN 1991-1-7).

cl.1.1(4): 1991-1-7 No additional comment is required for Clauses 1.1(4) and (1.1.5).

cl.1.1(5): 1991-1-7 *cl.1.1(6): 1991-1-7* *Note 3 of cl.3.1(2): 1991-1-7* Although EN 1991-1-7 does not specifically deal with accidental actions caused by external explosions, warfare and terrorist activities, this clause does refer to Clause 3.1(2) of EN 1991-1-7, which in its Note 3 states:

> '*Strategies based on unidentified accidental actions cover a wide range of possible events and are related to strategies based on limiting the extent of localised failure. The adoption of strategies for limiting the extent of localised failure may provide adequate robustness against those accidental actions identified in 1.1(6), ...*'

cl.1.1(6): 1991-1-7 In addition EN 1991-1-7 does not deal with **the residual stability of buildings or other civil engineering works damaged by seismic action or fire, etc**. Fire and earthquakes are dealt with in specific parts in the Eurocode system.

Events, which are generally denoted as accidents, such as persons falling through windows or roofs, are outside the scope of EN 1991-1-7. The reason is this type of accident has no potential for damaging the structural system.

This Designers' Guide will only give guidance on the parts of EN 1991-1-7 dealing with buildings. Bridges are covered by the Thomas Telford *Designers' Guide on Actions on Bridges*.[2]

EN 1990 requirements affecting EN 1991-1-7

Very relevant to the logic used in developing EN 1991-1-7 and the determination of Accidental actions, are the following clauses from EN 1990:

- 2.1 Basic Requirements EN 1990 Clauses 2.1(4)P and 2.1(5)P, and
- 2.2 Reliability differentiation EN 1990 Clause 2.2(3)

which are given below.

2.1(4)P A structure shall be designed and executed in such a way that it will not be damaged by events such as:

- *explosion*
- *impact, and*
- *the consequences of human errors,*

to an extent disproportionate to the original cause.

Note 1. The events to be taken into account are those agreed for an individual project with the client and the relevant authority.

Note 2. Further information is given in EN 1991-1-7.

2.1(5)P Potential damage shall be avoided or limited by appropriate choice of one or more of the following:

- *avoiding, eliminating or reducing the hazards to which the structure can be subjected*
- *selecting a structural form which has low sensitivity to the hazards considered*
- *selecting a structural form and design that can survive adequately the accidental removal of an individual member or a limited part of the structure, or the occurrence of acceptable localised damage*

- *avoiding as far as possible structural systems that can collapse without warning*
- *tying the structural members together.*

2.2(3) The choice of the levels of reliability for a particular structure should take account of the relevant factors, including:

- *the possible cause and/or mode of attaining a limit state*
- *the possible consequences of failure in terms of risk of life, injury, potential economical losses*
- *public aversion to failure*
- *the expense and procedures necessary to reduce the risk of failure.*

1.2. Normative references

No comment is necessary on the quoted references, all of which now have the status of EN. *cls.1.2: 1991-1-7*

1.3. Assumptions

EN 1991-1-7 (***Clause 1.3: 1991-1-7***) directly references EN 1990 Clause 1.3. *cl.1.3: 1991-1-7*

The statements and assumptions given in EN 1990 *Clause 1.3* apply to all the Eurocode parts, and designers guidance for these are given in the Thomas Telford *Designers' Guide to EN 1990.*[3]

1.4. Distinction between Principles and Application Rules

EN 1991-1-7 directly references EN 1990 Clause 1.4. *cl.1.4: 1991-1-7*

The statements and assumptions given in EN 1990 *Clause 1.4* apply to all the Eurocode parts and designers guidance for these are given in the Thomas Telford *Designers' Guide to EN 1990*.

1.5. Terms and definitions

Most of the definitions given in EN 1991-1-7 derive from ISO 2394[4], ISO 3898[5], and ISO 8930[6] and ISO 4355. In addition reference should be made to EN 1990 which provides a basic list of terms and definitions which are applicable to EN 1990 to EN 1999, thus ensuring a common basis for the Eurocode suite. *cl.1.5: 1991-1-7*

For the structural Eurocode suite, attention is drawn to the following key definitions, which may be different from current national practices:

- '*Action*' means a load, or an imposed deformation (e.g. temperature effects or settlement).
- '*Effects of Actions*' or '*Action effects*' are internal moments and forces, bending moments, shear forces and deformations caused by actions.

From the many definitions provided in EN 1990, those that apply for use with EN 1991 are described in Chapter 1 *Clause 1.4(a), (b), (c) and (d): 1991-1-1.* *cl.1.4(a), (b), (c) and (d): 1991-1-1*

The following comments are made with regard to particular definitions in order to help the understanding of EN 1991-1-7.

(a) **Accidental action** is not defined in EN 1991-1-7. Accidental actions are actions with low probability, severe consequences of failure and usually of short duration. The EN 1990 definition is reproduced below.

> '*An action, usually of short duration, but of significant magnitude, that is unlikely to occur on a given structure during the design working life.*'

Note: An accidental action can be expected in many cases to cause severe consequences unless appropriate measures are taken.

cl.1.5.2: 1991-1-7 (b) A definition is given for **consequence class** as follows.

'*Classification of the consequences of failure of the structure or part of it.*'

Note that the concept of consequence classes is introduced in Annex B of EN 1990, and is discussed in the Thomas Telford *Designers' Guide to EN 1990*.

cl.1.5.14: 1991-1-7 (c) A definition is given for **robustness** as follows.

'*The ability of a structure to withstand events like fire, explosions, impact or the consequences of human error, without being damaged to an extent disproportionate to the original cause.*'

To help with the better appreciation of this definition for robustness, the definition for progressive collapse is given below.

'*A chain reaction of failures following damage to a relatively small portion of a structure. The damage resulting from progressive collapse is disproportionate to the damage that initiated the collapse.*'

Note that in extreme circumstances if the accidental event is massive then progressive collapse may be considered proportionate.

An example of a disproportionate collapse is the collapse of Ronan Point shown in Fig. 1.1.

Fig. 1.1. The disproportionate collapse of Ronan Point

The concrete panel building was subjected to an accidental event on 16 May 1968, caused by a gas explosion which occurred at 5.45 a.m. in the kitchen of a flat located at the 18th floor. The explosion caused a progressive collapse with a 'domino' effect, involving the south-east corner of the building.

(d) The definitions for **substructure** and **superstructure** can be the cause of confusion, because the definitions are different for buildings and bridges. The definitions are reproduced below but with the definition for buildings given in ***bold italics***.

substructure

'***That part of a building structure that supports the superstructure. In the case of buildings this usually relates to the foundations and other construction work below ground level.*** *In the case of bridges this usually relates to foundations, abutments, piers and columns etc.*' *cl.1.5.15: 1991-1-7*

superstructure

'***That part of a building structure that is supported by the substructure. In the case of buildings this usually relates to the above ground construction.*** *In the case of bridges this usually relates to the bridge deck.*' *cl.1.5.16: 1991-1-7*

1.6. Symbols

The notation in *Clause 1.6: EN 1991-1-7* is based on ISO 3898. *cl.1.6(1): 1991-1-7*

EN 1990 *Clause 1.6* provides a comprehensive list of symbols, some of which may be appropriate for use with EN 1991-1-7. The symbols given in *Clause 1.6(1): 1991-1-7* are additional notations specific to this part of EN 1991-1-7.

CHAPTER 2

Classification of actions

This chapter is concerned with the rules relating to classification of the actions in EN 1991-1-7: *Eurocode 1 – Actions on Structures: Part 1.7: General Actions – Accidental actions.* The material described in this chapter is covered in the following clause:

- Classification of actions *Clause 2.1: 1991-1-7*

Clause 2.1(1)P gives a table which specifies the relevant clauses and sub-clauses in EN 1990, which apply to the design of a structure subjected to Accidental actions, including the following: *cl.2.1(1)P: 1991-1-7*

(a) *EN 1990 Clause 4.1.2(8)* states that '*for accidental actions the design value A_d should be specified for individual projects*'.

- where A is an accidental action, and
- A_d is the design value of an accidental action.

(b) The combination of actions for the accidental design situation is given by *Clause 6.4.3.3 of EN 1990* as:

$$\sum_{j\geq 1} G_{k,j}\,''+''\,P''+''\,A_d\,''+''\,(\psi_{1,1}\text{ or }\psi_{2,1})Q_{k,1}\,''+''\,\sum_{i>1}\psi_{2,i}Q_{k,i} \qquad (6.11b/EN\,1990)$$

The choice between $\psi_{1,1}Q_{k,1}$ or $\psi_{2,1}Q_{k,1}$ should be related to the relevant accidental design situation (impact, fire or survival after an accidental event or situation). In the UK National Annex for EN 1990, $\psi_{1,1}Q_{k,1}$ should only be used.

(c) In accordance with *EN 1990 Clause A1.3.2*. The partial factors for actions for the ultimate limit states in the accidental design situations should be 1.0.

EN 1991-1-7 classifies Accidental actions due to impact as free actions unless otherwise specified. *cl.2.1(1)P: 1991-1-7*

It does not classify accidental actions due to explosions which may in some cases be treated as fixed action at the point of consideration (e.g. on a key element defined in EN 1991-1-7 as '*a structural member upon which the stability of the remainder of the structure depends*'. *cl.1.5.10: 1991-1-7*

Differences between an Accidental and Variable actions

Figure 2.1 (obtained from the background document for EN 1991-1-7, which can be obtained from the CEN/TC250 Secretariat, BSI) shows the typical difference between a variable and an accidental action as far as the time characteristics are concerned.

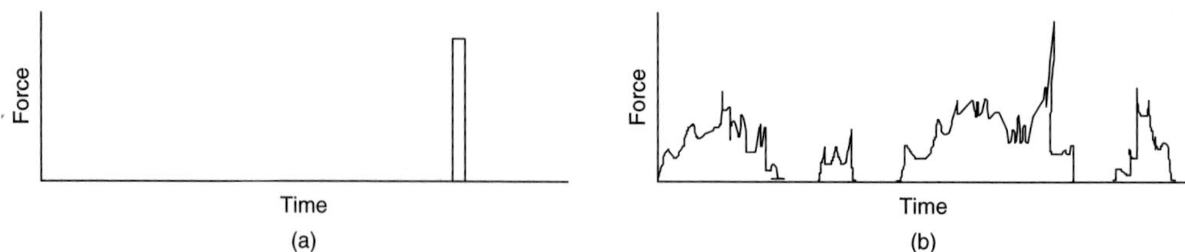

Fig. 2.1. Typical time characteristics of (a) accidental and (b) variable load

A typical accidental action (Fig. 2.1(a)) will most probably not occur during the design working life of the structure. If it does occur, it will normally last only a short time, varying from a few seconds (for explosions) to some days (for floods).

A variable action (Fig. 2.1(b)) is nearly always present, although its value may be small for a substantial part of the time. However, significant high values will in most cases (e.g. wind, snow) occur many times during the design life of the structure.

Another difference between a variable and an accidental action is the action intensity. The accidental action is so intense that it is considered an extremely rare event, unlikely to occur during normal use of the structure.

CHAPTER 3

Design situations

This chapter is concerned with the general concepts of design situations relating to EN 1991-1-7: *Eurocode 1 – Actions on Structures: Part 1.7: General Actions – Accidental actions*. The material described in this chapter is covered in the following clauses:

- General — *Clause 3.1: 1991-1-7*
- Accidental design situations – strategies for identified accidental actions — *Clause 3.2: 1991-1-7*
- Accidental design situations – strategies for limiting the extent of localised failure — *Clause 3.3: 1991-1-7*
- Accidental design situations – use of consequence classes — *Clause 3.4: 1991-1-7*

3.1. General

In accordance with *Clause 3.2 of EN 1990*, the term 'design situation' is defined to mean the circumstances in which the structure may be required to fulfil its function. The selected design situations have to be sufficiently severe and varied enough to encompass all conditions that can reasonably be foreseen to occur during the execution and use of the structure.

The phrase 'which can reasonably be foreseen' is somewhat ambiguous in the case of accidental situations, the characteristics of which are that they cannot easily be foreseen in detail or maybe not at all. In EN 1991-1-7 therefore a distinction is made between so-called identified and unidentified actions. See Section 3.2 of this Part of this Designers' Guide.

EN 1990 Clause 3.2 identifies the following design situations for the verification of ultimate limit states: *cl.3.1(1)P: 1991-1-7*

- persistent design situations, which refer to the conditions of normal use
- transient design situations, which refer to temporary conditions applicable to the structure, e.g. during execution or repair
- accidental design situations, which refer to exceptional conditions applicable to the structure or to its exposure, e.g. explosions
- seismic design situations.

Each of these design situations is linked to a particular expression for the combination of action effects as follows:

- persistent and transient design situations, which refer to *expressions (6.10), or (6.10a) and (6.10b) in EN 1990*
- accidental design situations, which refers to *expression (6.11b) in EN 1990*
- seismic design situations, which refers to *expression (6.12b) in EN 1990.*

In addition, thermal actions need to be determined for the verification of serviceability limit states and the following expressions for the combination of action effects given in EN 1990:

- the characteristic combination which refers to *expression (6.14b) of EN 1990*
- the frequent combination which refers to *expression (6.15b) of EN 1990*
- the quasi-permanent combination which refers to *expression (6.16b) of EN 1990.*

Accidental actions are used only with the accidental design situation which refers to *expression (6.11b) in EN 1990.*

Design for accidental actions and the acceptance of localised damage
The objective of design is to reduce risks at an economical acceptable price. Risk may be defined as the danger that undesired events represent. Risk is expressed in terms of the probability and consequences of undesired events. Risk-reducing measures consist in reducing the probability of an event occurring, and the consequences of the occurring event. Therefore risk-reducing measures should be given high priority in design for accidental actions, and also be taken into account in design.

No structure can be expected to resist all actions that could arise due to an extreme cause, but there should be a reasonable probability that a structure will not be damaged to an extent disproportionate to the original cause, which is an EN 1990 requirement.

As a result of this requirement, local failure (which in most cases may be identified as a component failure) may be accepted in accidental design situations, provided that it does not lead to a system failure. The consequence is that redundancy and non-linear effects both regarding material behaviour and geometry have a greater influence in design to mitigate accidental actions than in the case of variable actions.

EN 1991-1-7 makes a distinction between identified and unidentified actions.

The identified actions may be analysed using classical (advanced) structural analysis.

For the unidentified actions more general robustness requirements (e.g. prescribed tying forces) have been introduced. These are described in Chapter 5 and 6 of this Part of this Designers' Guide.

cl.3.1(2): 1991-1-7 The strategies to be considered for accidental design situations are illustrated in Fig. 3.1, which is reproduced from EN 1991-1-7.

Note 1 to cl.3.1(2): 1991-1-7 The strategies and rules to be taken into account for the individual project should be agreed with the client and the relevant authority.

Note 2 to cl.3.1(2): 1991-1-7 The two strategies shown in Fig. 3.1 recognise that an accidental action can be an *identified* (e.g. a delivery lorry impacting a large supermarket) or an *unidentified* (e.g. an internal gas explosion in a block of flats) accidental action. For a design based on the strategies for *identified accidental actions* see Section 3.2 below. Notional values for *limiting the extent of local failure* are given in Section 3.3 below.

For a design based on the strategies for *unidentified accidental actions* see Section 3.3 below. This strategy covers a wide range of possible events and is related to strategies based on *limiting the extent of localised failure.* Guidance for buildings is given in Annex A in this Part and in Chapter 6 of this Designers' Guide.

This strategy, based on limiting the extent of localised failure, can also be used for identified accidental actions. Figure 3.2 shows various hazards which may either be variable or accidental actions that can occur on a multi-storey building. Concerning the hazards caused by the crash of a helicopter, or an impact from a vehicle on the ground floor of a building, although the accidental action for impact may be determinable, a design based on limiting the extent of localised failure will also prove satisfactory.

Note 3 to cl.3.1(2): 1991-1-7 The adoption of strategies for limiting the extent of localised failure may provide adequate robustness against those accidental actions caused by external explosions, warfare and some malicious (e.g. terrorist) activities, or any other action resulting from an unspecified cause.

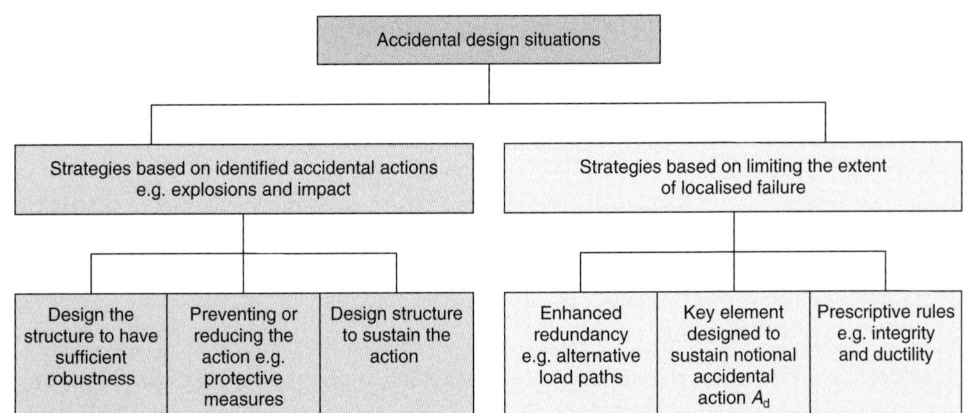

Fig. 3.1. Strategies for accidental design situations

Most of the rules in EN 1991-1-7 for limiting the extent of localised failure and thus helping ensure against disproportionate collapse are based on the UK Codes of Practice BS 8110,[7] BS 5950[8] and BS 5628[9] and other official guidance on meeting the Building Regulations in the UK. These rules have proved satisfactory over the past three decades. Their efficiency was dramatically demonstrated during the IRA bomb attacks that occurred in the City of London in 1992 and 1993. Although the rules were not intended to safeguard buildings against terrorist attack, the damage sustained by those buildings close to the seat of the explosions that were designed to meet the regulatory requirement relating to disproportionate collapse was found to be far less compared with other buildings that were subjected to a similar level of abuse.

For the strategy based on identified accidental actions, notional values for identified accidental actions (e.g. for certain impacts) are given in EN 1991-1-7. These values which are discussed in Chapter 4 of this Part of this Designers' Guide may be altered in the National Annex or for an individual project and agreed for the design by the client and the relevant authority. *Note 4 to cl.3.1(2): 1991-1-7*

For some structures subjected to accidental actions from which there is no risk to human life, and where economic, social or environmental consequences are negligible (e.g. a rural agricultural building where people do not normally enter), the complete collapse of the structure caused by an extreme event may be acceptable. The circumstances when such a collapse is acceptable may be agreed for the individual project with the client and the relevant authority. *Note 5 to cl.3.1(2): 1991-1-7*

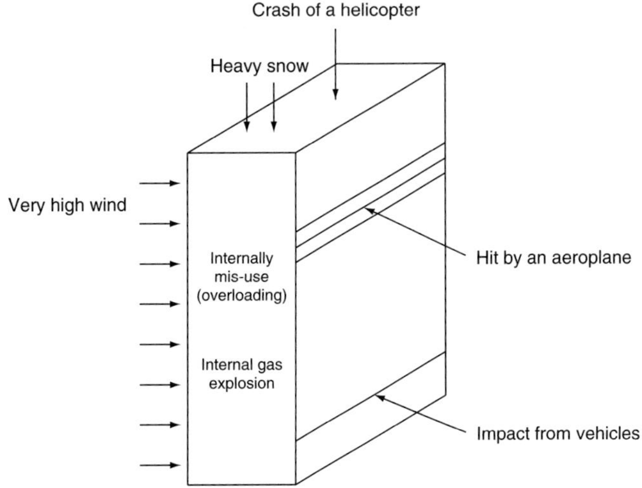

Fig. 3.2. Hazards on a building

3.2. Accidental design situations – strategies for identified accidental actions

Note 1 to cl.3.2(1): 1991-1-7

The accidental actions that need to be taken into account are those agreed for the individual project with the client and the relevant competent authority. This is in accordance with Note 1 of Clause 2.4(1) of EN 1990.

cl.3.2(1): 1991-1-7

In theory a majority of structures are susceptible to identified accidental actions, e.g. impact from heavy vehicles, aeroplanes etc. However, to design a structure to every accidental action to which it may be susceptible is not viable economically and therefore the accidental actions that should be taken into account in accordance with *Clause 3.2(1): 1991-1-7* depend upon:

- the measures taken for preventing or reducing the severity of an accidental action as described in items (a) and (b) below
- the probability of occurrence of the identified accidental action which can be obtained from statistical or historic information
- the consequences of failure due to the identified accidental action; as an example Table 3.1 below is given which is reproduced from Annex B of EN 1990.

Public perception does not accept fatalities and injuries due to structural failure (at home, at the workplace, during recreational and other activities etc.), for the design working life of a structure compared to fatalities arising from other hazards and events. The concept of 'acceptable' risk of human death resulting from structural failure raises very sensitive questions, concerned with public perception. Fatalities due to accidents, taken from recent figures[5] given in Table 3.2 are considered to reflect the public perception of acceptability of fatalities for different types of accidents and exposure to serious hazards.

Note 2 to cl.3.2(1): 1991-1-7

The level of acceptable risk, depends mainly on the consequences of failure, public perception and political and economic matters. In practice, the occurrence and consequences of accidental actions can be associated with a certain risk level as described in Annex B of EN 1990 and explained in the Thomas Telford EN 1990 Designers' Guide. If this level cannot be accepted, additional measures are necessary. A zero risk level, however, is impracticable and in most cases it is necessary to accept a certain level of risk. Such a risk level can be determined by various factors, such as the potential number of casualties, the economic consequences and the cost of safety measures, etc.

No further comment is required on Note 3 to Clause 3.2(1): 1991-1-7.

cl.3.2(2): 1991-1-7

For a majority of situations a localised failure due to an accidental action is acceptable and possibly unavoidable, provided the localised failure will not endanger the stability of the whole structure and that the overall load-bearing capacity of the structure is maintained and allows necessary emergency measures to be taken.

Table 3.1. Definition of consequences classes

Consequences class	Description	Examples of buildings and civil engineering works
CC3	**High** consequence for loss of human life, *or* economic, social or environmental consequences **very great**	Grandstands, public buildings where consequences of failure are high (e.g. a concert hall)
CC2	**Medium** consequence for loss of human life, economic, social or environmental consequences **considerable**	Residential and office buildings, public buildings where consequences of failure are medium (e.g. an office building)
CC1	**Low** consequence for loss of human life, *and* economic, social or environmental consequences **small or negligible**	Agricultural buildings where people do not normally enter (e.g. storage buildings, greenhouses)

Table 3.2. 'Accepted' risks of death due to exposure to various hazards

Hazard	Risk ($\times 10^{-6}$ p.a.)*	Hazard	Risk ($\times 10^{-6}$ p.a.)*
Building hazards		***Occupations (UK)***	
Structural failure (*UK*)	0.14	Chemical and allied industries	85
Building fires (*Australia*)	4	Shipbuilding and marine engineering	105
		Agriculture	110
		Construction industries	150
		Railways	180
		Coalmining	210
		Quarrying	295
		Mining (non-coal)	750
		Offshore oil and gas (*1967–1976*)	1650
		Deep sea fishing (*1959–1978*)	2800
Natural hazards (US)		***Sports (US)***	
Hurricanes (1901–1972)	0.4	Cave exploration (*1970–1978*)	45
Tornadoes (1953–1971)	0.4	Glider flying (*1970–1978*)	400
Lightning (1969)	0.5	Scuba diving (*1970–1978*)	420
Earthquakes (California)	2	Hang gliding (*1977–1979*)	1500
		Parachuting (*1978*)	1900
General accidents (US 1969)		***All causes (UK 1977)***	
Poisoning	20	Whole population	12 000
Drowning	30	Woman aged 30	600
Fires and burns	40	Man aged 30	1000
Falls	90	Woman aged 60	10 000
Road accidents	300	Man aged 60	20 000

* Risk expressed as probability of death for typical exposed person per calendar year.

For building structures, such emergency measures can involve the safe evacuation of persons from the premises and its surroundings immediately after the accidental action has occurred. *Note 1 of cl.3.2(2): 1991-1-7*

No comment is necessary on *Note 2 of Clause 3.2(2): 1991-1-7*, as this has been described in the Thomas Telford Guide for Actions on Bridges. *Note 2 of cl.3.2(2): 1991-1-7*

Within the strategy based on identified accidental actions, measures should be taken to mitigate the risk of accidental actions. In accordance with EN 1991-1-7 these measures can include, as appropriate, one or more of the following: *cl.3.2(3): 1991-1-7*

(a) By preventing the action from occurring, e.g. in the case of building over a road by providing adequate clearances between the trafficked lanes and the structure. Or reducing the magnitude of the action to an acceptable level through the structural design process, e.g. in the case of buildings by providing sacrificial venting components with a low mass and strength to reduce the effect of explosions;
(b) Protecting the structure against the effects of an accidental action by reducing the effects of the action on the structure, e.g. by providing protective bollards, safety barriers or continuous protection in the form of heavy elements built into the ground.
(c) Providing sufficient robustness to the structure by adopting one or more of the following approaches:
   - By designing certain components of the structure upon which stability depends as key elements (see Sections A.4 to A.8 of this Part of this Designers' Guide) to increase the likelihood of the structure's survival following an accidental event.
   - Designing structural members, and selecting materials, to have sufficient ductility capable of absorbing significant strain energy without rupture.
   - Incorporating sufficient redundancy in the structure to facilitate the transfer of actions to alternative load paths following an accidental event.

Note 1 of cl.3.2(3): 1991-1-7

It may not always be possible to protect the structure by reducing the effects of an accidental action, or preventing an action from occurring. This is because an action is dependent upon factors which, over the design working life of the structure, may not necessarily be part of the design assumptions. Preventative measures may involve periodic inspection and maintenance during the design working life of the structure.

No further comment is necessary on *Note 2 of Clause 3.2(3): 1991-1-7*.

The design philosophy necessitates that accidental actions are treated in a special manner with respect to partial safety factors for actions and load combinations. Partial load factors to be applied are defined in EN 1990 to be 1.0 for all actions (permanent, variable and accidental).

cl.3.2(4)P: 1991-1-7

Hence, accidental actions need to, where appropriate, be applied simultaneously in combination with permanent and other variable actions in accordance with EN 1990, 6.4.3.3, using the following expression:

$$\sum_{j\geq 1} G_{k,j} \text{"+"} P \text{"+"} A_d \text{"+"} (\psi_{1,1} \text{ or } \psi_{2,1}) Q_{k,1} \text{"+"} \sum_{i>1} \psi_{2,i} Q_{k,i}$$

Note to cl.3.2(4)P: 1991-1-7

For ψ, values for variable actions may be obtained from Annex A of EN 1990 or the appropriate national Annex.

Clause 3.2(5)P: 1991-1-7

EN 1991-1-7 requires that the safety of the structure immediately following the occurrence of the accidental action needs to be taken into account.

Note to cl.3.2(5)P: 1991-1-7

The expression given above applies with the following qualification in: 'Combinations for accidental design situations either involve an explicit accidental action A_d (e.g. an impact) or refer to a situation after an accidental event with $A_d = 0$.'

After an accidental event the structure will normally not have the required strength in persistent and transient design situations and will have to be strengthened for a possible continued application. In temporary phases there may be reasons for a relaxation of the requirements, e.g. by allowing wind or wave loads for shorter return periods to be applied in the analysis after an accidental event.

3.3. Accidental design situations – strategies for limiting the extent of localised failure

The design for unidentified accidental load is presented in Annex A of EN 1991-1-7. Design rules for providing robustness by limiting the extent of localised failure were developed and made a regulatory requirement in the early 1970s following the partial collapse of Ronan Point (see Fig. 1.1), a block of flats in east London caused by a gas explosion. In addition, largely prescriptive design rules were introduced in UK Codes of Practice. The rules have changed little over the intervening years. They aim to provide a level of building robustness, thus safeguarding buildings against a disproportionate extent of collapse following local damage being sustained from an accidental event.

The rules have proved satisfactory over the past three decades. Their efficiency was dramatically demonstrated during the bomb attacks that occurred in the City of London in 1992 and 1993. Although the rules were not intended to safeguard buildings against malicious attack, the damage sustained by those buildings close to the seat of the explosions that were designed to meet the regulatory requirement relating to disproportionate collapse was found to be far less compared with other buildings (e.g. the Muhra building in Oklahoma) that were subjected to a similar level of abuse.

cl.3.3(1)P: 1991-1-7
cl.3.3(2): 1991-1-7

In the design, the potential failure of the structure arising from an unspecified cause is required to be mitigated as follows by adopting one or more of the following approaches:

(a) Designing key elements, which is defined in (*Clause 1.5(10): 1991-1-7*) as a structural member upon which the stability of the remainder of the structure (after an accidental

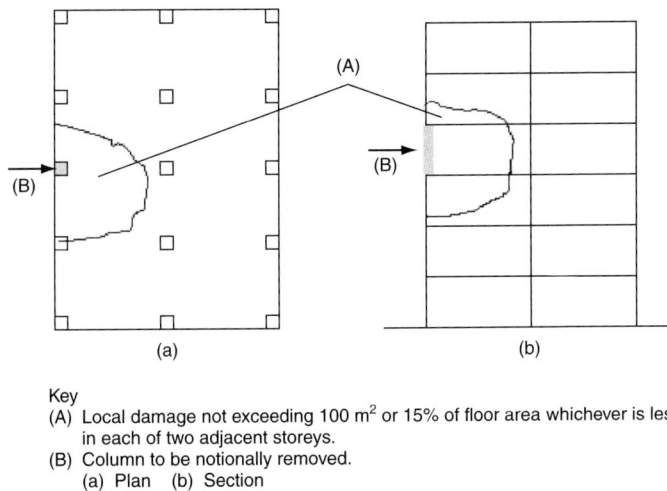

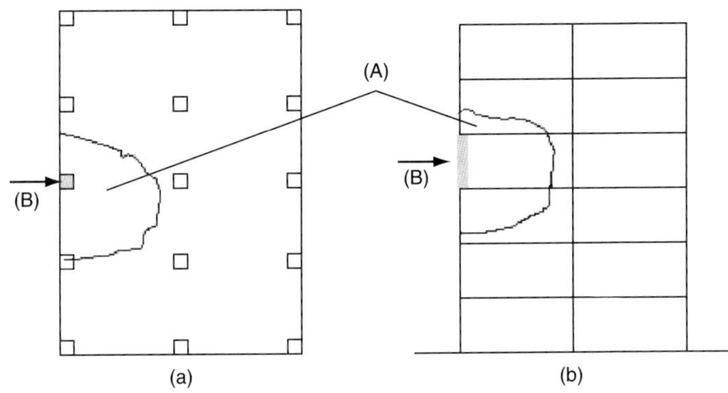

Key
(A) Local damage not exceeding 100 m^2 or 15% of floor area whichever is less in each of two adjacent storeys.
(B) Column to be notionally removed.
(a) Plan (b) Section

Fig. 3.3. Recommended limit of admissible damage

cl.3.3(2)a: 1991-1-7

action has caused local damage) depends. The key element needs to sustain the effects of a model of accidental action A_d.

Note 1 to cl.3.3(2)a: 1991-1-7

UK NA 2.4: 1991-1-7

UK NA 2.5: 1991-1-7

The definition of the model and the design value of A_d is allowed in the National Annex. For buildings, EN 1991-1-7 recommends a model of a uniformly distributed notional load applicable in any direction to the key element and any attached components (e.g. claddings, etc.). The recommended value for the uniformly distributed load given by EN 1991-1-7 is 34 kN/m^2 for building structures. The UK National Annex specifies the use of the recommended model and 34 kN/m^2 for the uniformly distributed load (***UK NA 2.4: 1991-1-7***) (***UK NA 2.5: 1991-1-7***).

cl.3.3(2)b: 1991-1-7

(b) Designing the structure so that in the event of a localised failure (e.g. failure of a single member) the stability of the remaining whole structure or of a significant part of it is not endangered.

Note 2 to cl.3.3(2)b: 1991-1-7

The choice of the acceptable limit of 'localised failure' is allowed in the National Annex. The indicative limit for building structures is 100 m^2 or 15% of the floor area, whichever is less, on two adjacent floors caused by the removal of any supporting column, pier or wall. See Fig. 3.3.

UK NA 2.5: 1991-1-7

Although the UK National Annex specifies 100 m^2 or 15% (***UK NA 2.5: 1991-1-7***) it should be noted that the official guidance in the Building Regulations in the UK at the present time (2009) recommends 70 m^2.

The above limits are likely to provide the structure with sufficient robustness regardless of whether an identified accidental action has been taken into account.

cl.3.3(2)c: 1991-1-7

(c) Designing the structure by applying prescriptive design/detailing rules that provide acceptable robustness for the structure (e.g. three-dimensional tying for additional integrity, or a minimum level of ductility of structural members subject to impact) (***Clause 3.3(2)c: 1991-1-7***).

Example 3.1

Considering *Clauses 3.3(2)a, b and c: 1991-1-7* a design may provide effective horizontal and vertical ties in accordance with the BSI structural codes of practice.

Where such provisions cannot be made, it is recommended that the structure be designed to bridge over the limited, localised area of collapse caused by the notional loss of member.

If this is not possible, such a member should be designed as a protected (or key) element capable of sustaining additional loads related to a pressure of 34 kN/m^2 as described in *Clause 3.3(2)a: 1991-1-7*.

Fig. 3.4. Example of an unidentified accidental action

Note 3 to cl.3.3(2)b: 1991-1-7

The National Annex may state which of the approaches given in Section 3.3 are to be considered for various structures. Examples for buildings are discussed in Chapter 6 in this Part of this Designers' Guide.

Figure 3.4 shows an example of a collapse, caused by an accidental action due to a collapsed crane. This action would not normally have been identified at the time of the design for the damaged older buildings.

3.4. Accidental design situations – use of consequence classes

cl.3.4(1): 1991-1-7
Note 1 to cl.3.4(1): 1991-1-7

Selecting strategies and the appropriate measures for designing to accidental design situations can be based on the consideration of the following consequence classes as set out in EN 1990 Annex B where further information is given.

- CC1 Low consequences of failure
- CC2 Medium consequences of failure
- CC3 High consequences of failure

See also Section 3.2 of this Part of this Designers' Guide.

Note 2 to cl.3.4(1): 1991-1-7

The measures selected in the design can be different in some parts of the structure which may belong to a different consequence class. An example is a structurally separate low-rise wing of a building that is serving a less critical function than the main building.

Note 3 to cl.3.4(1): 1991-1-7

Preventative and/or protective measures can be employed and these are intended to remove or to reduce the probability of damage to the structure. The design may take this into consideration by assigning the structure to a lower consequence class or more appropriately a consideration of a reduction of forces on the structure.

The National Annex can provide a categorisation of structures according to the consequence classes mentioned above. A suggested classification of consequence classes relating to buildings is provided in Chapter 6. *Note 4 to cl.3.4(1): 1991-1-7*

The accidental design situations for the different consequence classes given above may be considered in the design as follows: *cl.3.4(2): 1991-1-7*

- For structures under CC1 (e.g. greenhouses, single-occupancy dwellings): no specific consideration is necessary for accidental actions provided that the robustness and stability rules given in EN 1990 to EN 1999, as applicable to CC1 structures, are met (see Chapter 6 of this Part).
- For structures under CC2 (e.g. offices), depending upon the specific circumstances of the structure, a simplified analysis by static equivalent action models may be adopted or prescriptive design/detailing rules may be applied (see Chapter 6 of this Part).
- For structures under CC3 (e.g. major grandstands, concert halls) an examination of the specific case will need to be carried out to determine the level of reliability and the depth of structural analyses required. This may require a risk analysis to be carried out and the use of refined methods such as dynamic analyses, non-linear models and interaction between the load and the structure (see Chapter 6 of this Part).

Examples have been given above for CC1, CC2 and CC3. However, the National Annex can give reference to, as non-conflicting, complementary information, appropriate design approaches for higher and lower consequence classes. *Note to cl.3.4(2): 1991-1-7*

CHAPTER 4

Impact

This chapter is concerned with the rules relating to accidental actions due to impact in EN 1991-1-7: *Eurocode 1 – Actions on Structures: Part 1.7: General Actions – Accidental actions*. The material described in this chapter is covered in the following clauses:

- Field of application — *Clause 4.1: 1991-1-7*
- Representation of actions — *Clause 4.2: 1991-1-7*
- Accidental actions caused by road vehicles — *Clause 4.3: 1991-1-7*
- Accidental actions caused by forklift trucks — *Clause 4.4: 1991-1-7*
- Accidental actions caused by derailed rail traffic under or adjacent to structures — *Clause 4.5: 1991-1-7*
- Accidental actions caused by ship traffic — *Clause 4.6: 1991-1-7*
- Accidental actions caused by helicopters — *Clause 4.7: 1991-1-7*

4.1. Field of application

This section of EN 1991-1-1 is concerned with accidental actions due to the following events: *cl.4.1(1): 1991-1-7*

- Impact from road vehicles. EN 1991-1-6 excludes guidance on collisions on lightweight structures and gives the following examples of lightweight structures: footbridges, lighting columns. See Section 4.3 of this Part of this Designers' Guide. *Note 1 to cl.4.1(1): 1991-1-7*
- Impact from forklift trucks which is described in Section 4.4 of this Part of this Designers' Guide.
- Impact from trains with exclusions on lightweight structures which is described in Section 4.5 of this Part of this Designers' Guide.
- Impact from ships.
- The hard landing of helicopters on roofs of buildings which is described in Section 4.7 of this Part of this Designers' Guide.

The guidance in this Designers' Guide is limited to accidental actions due to impact on buildings. The Thomas Telford Designers' Guide relating to Actions on Bridges covers accidental actions due to impact on bridges. *Note 2 to cl.4.1(1): 1991-1-7* *Note 3 to cl.4.1(1): 1991-1-7*

No further guidance is required on Notes 2 and 3 of Clause 4.1.

For buildings, actions due to impact need to be taken into account for the following: *cl.4.1(2)P: 1991-1-7*

- buildings used for car parking, where information is given on impact due to smaller collision forces (i.e. private vehicles or light vans) in EN 1991-1-1 Annex B and described in Part 1, Chapter 8 of this Designers' Guide

- buildings in which vehicles or forklift trucks are permitted (see Section 4.4)
- buildings that are located adjacent to either road or railway traffic (see Sections 4.3 and 4.5), and
- *cl.4.1(4)P: 1991-1-7* buildings where the roof contains a designated helicopter landing pad (see Section 4.7).

cl.4.1(3): 1991-1-7 Clause 4.1(3) is meant for bridges, but the part of the guidance given in this clause is also applicable for buildings. In particular the actions due to impact should consider any mitigating measures provided, the type of traffic (e.g. delivery lorries near a hypermarket) and the consequences of the impact.

4.2. Representation of actions

cl.4.2(1): 1991-1-7 In principle, the mechanical effects of an impact should be determined by a dynamic analysis, taking into account the effects of time and the real behaviour of materials. But, in common cases, actions due to impact are represented by an equivalent static force, i.e. an alternative representation for the dynamic force intended to cover the dynamic response of the structure without refined calculations.

cl.1.5.5: 1991-1-7 *Note 1 to cl.4.2(1): 1991-1-7* The collision force is a dynamic force, i.e. a force, with an associated contact area at the point of impact, that varies in time and which may cause significant dynamic effects on the structure. It depends on the interaction between the impacting object and the structure.

cl.1.5.5: 1991-1-7 This simplified representation, as shown in Fig. 4.1 (reproduced from EN 1991-1-7 Fig. 1.1), gives a simplified representation of a dynamic force, the structural response and the static equivalent force.

Note 2 to cl.4.2(1): 1991-1-7 Impact loading is the result of a collision between two objects. The basic variables for impact analysis are the impact velocity of the impacting object and the mass distribution, deformation behaviour and damping characteristics of both the impacting object and the structure. In the case of buildings, the most common colliding objects are vehicles, and exceptionally railways, helicopters and aeroplanes, that have an intended course. But the occurrence of a human or mechanical failure may lead to a deviation of the intended course and hence the angle of impact will be relevant. After the initial impact, the course of the impacting object will depend on its properties and movement.

Note 3 to cl.4.2(1): 1991-1-7 No further guidance is required on Note 3 of Clause 4.2(1).

cl.4.2(2): 1991-1-7 As a simplification it may be assumed that the impacting body absorbs all the energy. Generally, this assumption gives conservative results.

Note to cl.4.2(2): 1991-1-7 No further guidance is required on Note 3 of Clause 4.2(3).

cl.4.2(3): 1991-1-7 Within Section 4.2 and Annex C, EN 1991-1-7 defines the concepts of **hard and soft impact**.

cl.4.2(4): 1991-1-7 **Soft impact** corresponds to collision effects in the case of structures which are designed to absorb impact energy by elastic-plastic deformations of members. For this case the equivalent static loads may be determined by taking into account both plastic strength and the deformation capacity of such members.

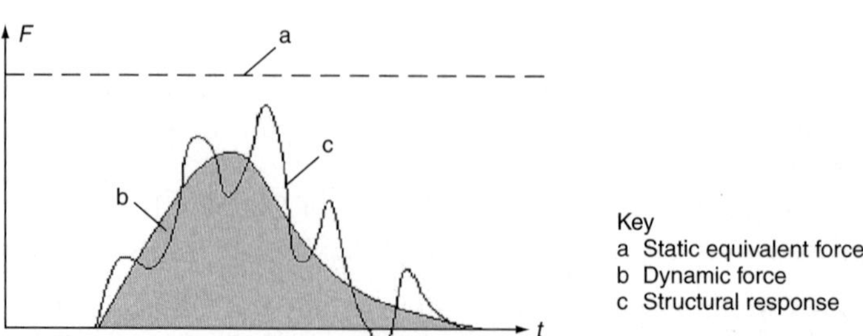

Fig. 4.1. Definitions related to actions due to impact (EN 1991-1-7, Fig. 1.1)

Annex C of EN 1991-1-7 gives detailed procedures for considering soft impacts (***Annex C: 1991-1-7***). *Annex C: 1991-1-7*

Hard impact corresponds to collision effects in the case of structures for which the energy is mainly dissipated by the impacting body. The impact forces given in Section 4 of EN 1991-1-7 assume the phenomona of hard impact and hence the dynamic or equivalent static forces may be determined from the guidance given in the rest of this chapter of this Designers' Guide. *cl.4.2(5): 1991-1-7*

In actual practice, collision effects are intermediate between hard and soft impact as shown in the example of Fig. 8.1 in Part 1 of this Designers' Guide. However, for simplicity, the impact load is determined using the 'rigid structure' assumption, i.e. using a 'hard impact' model. The impacting force may be represented by an equivalent static force. This simplified model may be used for the verification of static equilibrium, for strength verifications and for the determination of deformations of the impacted structure. *cl.4.2(3): 1991-1-7*

Annex C of EN 1991-1-7 gives information on design values for masses and velocities of colliding objects as a basis for a dynamic analysis. *Annex C: 1991-1-7*

4.3. Accidental actions caused by road vehicles

4.3.1. Impact on columns and walls of buildings

Design values for actions due to impact on columns and walls of buildings, see Fig. 4.2, adjacent to various types of roads need to be considered. *cl.4.3.1(1): 1991-1-7*

For hard impact (see 4.2 of this designer's guide) from road traffic, indicative equivalent static design forces are given in EN 1991-1-7 and reproduced below in Table 4.1.

The National Annex may confirm the table or select alternative values which should take account of the consequences of the impact, the expected volume and type of traffic, and any mitigating measures provided. *Note 1 to cl.4.3.1(1): 1991-1-7*

Fig. 4.2. Impact on building

Table 4.1. Indicative equivalent static design forces due to vehicular impact on members supporting structures over or adjacent to roadways

Category of traffic	Force $F_{dx}{}^{*}$ (kN)	Force $F_{dy}{}^{*}$ (kN)
Motorways and country national and main roads	1000	500
Country roads in rural area	750	375
Roads in urban area	500	250
Courtyards and parking garages with access to:		
• Cars	50	25
• Lorries**	150	75

* x = direction of normal travel, y = perpendicular to the direction of normal travel.
** The term 'lorry' refers to vehicles with maximum gross weight greater than 3.5 tonnes.

Note 2 to cl.4.3.1(1): 1991-1-7

The force given in Table 4.1 may be given in the National Annex as a function of the distance s of the centreline of the trafficked lanes nearest to the structural member.

cl.NA 12: UK NA 1991-1-7

In accordance with the UK National Annex for buildings where the distance s of the centreline of the trafficked lane nearest to the structural member is greater than 10 m, the equivalent static design force due to vehicle impact need not be considered.

Note 3 to cl.4.3.1(1): 1991-1-7

The National Annex may define types or elements of the structure that may not need to be considered for vehicular collision. If a risk analysis is needed to define these elements, guidance on risk analysis is given in Annex B of EN 1991-1-7.

cl.NA 13: UK NA 1991-1-7

The UK National Annex also states that vehicle collision need not be considered for buildings in CC1 and CC2 (see Tables 3.1 and A1) provided that the Annex A of EN 1991-1-7 is used. However, where required by the client or the competent authority, consideration of impacts on particular buildings (e.g. a large supermarket or a high-profile office building) caused by road vehicles, Table 4.1 of EN 1991-1-7 should be used.

cl.NA 2-11-1: UK NA 1991-1-7

In accordance with the UK National Annex for buildings, the equivalent static design force due to vehicular impact given in Table NA1 of the National Annex to EN 1991-1-7: 2006 should be used in category CC3 (see Tables 3.1 and A1) in the absence of mitigating measures. The values in Table NA1 are about 50% greater than those given in Table 4.1 of EN 1991-1-7 and this Designers' Guide.

Note 4 to cl.4.3.1(1): 1991-1-7

No further guidance is required for Notes 4 and 5 of Clause 4.3.1(1) as these concern road and rail bridge traffic matters respectively and are covered in the Thomas Telford Designers' Guide relating to Actions on Bridges.

Note 5 to cl.4.3.1(1): 1991-1-7

Rules for the application of F_{dx} and F_{dy} may be defined in the National Annex or for the individual project. It is generally recommended that F_{dx} does not act simultaneously with F_{dy}.

cl.4.3.1(2): 1991-1-7

The National Annex may define the conditions of impact from road vehicles. The recommended conditions in EN 1991-1-7 are as follows (see Fig. 4.3):

cl.4.3.1(3): 1991-1-7

- For impact from lorries the collision force F may be applied at any height h between 0.5 m to 1.5 m above the level of the carriageway. The recommended application area is $a = 0.5$ m (height) by 1.50 m (width) or the member width, whichever is the smaller.
- For impact from cars the collision force F may be applied at $h = 0.50$ m above the level of the carriageway. The recommended application area is $a = 0.25$ m (height) by 1.50 m (width) or the member width, whichever is the smaller.

4.3.2. Impact on superstructures

cl.4.3.2(1): 1991-1-7

Design values for actions due to impact from lorries and/or loads carried by the lorries on members of the superstructure (e.g. the underside or the side of a building over a roadway) should be taken into account unless adequate clearances or suitable protection measures to avoid impact are provided. See Fig. 4.4.

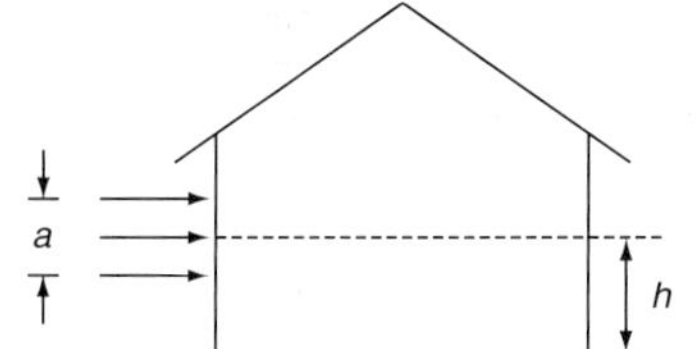

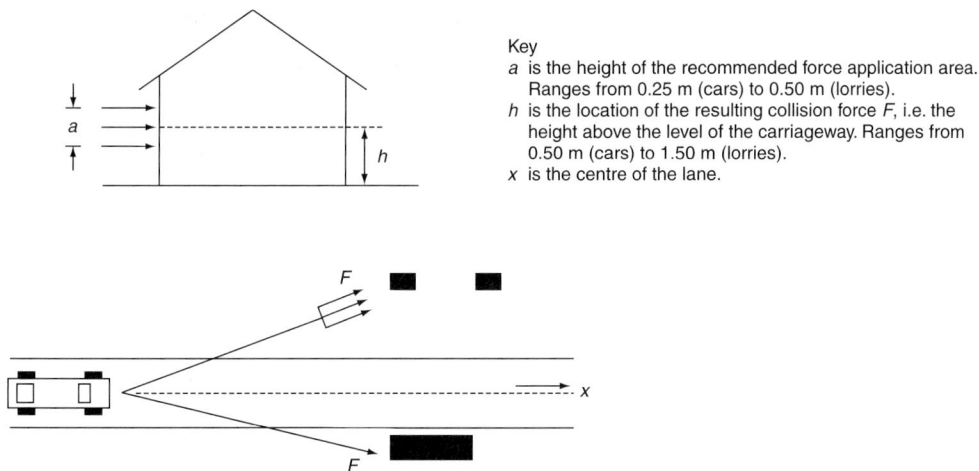

Key
a is the height of the recommended force application area. Ranges from 0.25 m (cars) to 0.50 m (lorries).
h is the location of the resulting collision force *F*, i.e. the height above the level of the carriageway. Ranges from 0.50 m (cars) to 1.50 m (lorries).
x is the centre of the lane.

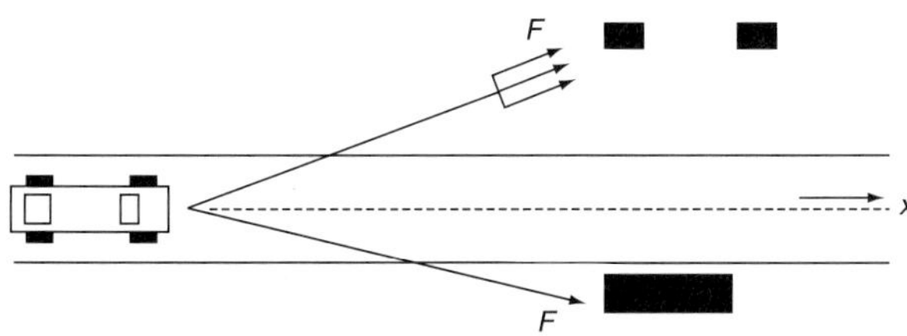

Fig. 4.3. Collision force for columns and walls of buildings

Note 1 to cl.4.3.1(2): 1991-1-7

The design values for actions due to impact together with the values for adequate clearances may be defined in the National Annex. The indicative equivalent static design forces are given in Table 4.2. The recommended value for adequate clearance, excluding future re-surfacing of the roadway under the bridge, to avoid impact given in EN 1991-1-7 is $h_0 = 5.0\,\text{m}$. No impact need be considered for a vertical clearance beyond an upper limit equal to $h_0 + b$, b being defined at the national level. The recommended value is $b = 1\,\text{m}$. See Fig. 4.5.

Note 2 to cl.4.3.1(2): 1991-1-7

The choice of the values may take account of the consequences of the impact, the expected volume and type of traffic, and any mitigating (protective and preventative) measures provided.

For $h_0 \leq h \leq h_1 = h_0 + b$ (see Fig. 4.5) the magnitude of the impact force given in Table 4.2 may be reduced linearly. Figure 4.5, reproduced from Fig. 4.2 of EN 1991-1-7,

Fig. 4.4. Clearance under a building above a carriageway

Table 4.2. Indicative equivalent static design forces due to impact on superstructures

Category of traffic	Equivalent static design force F_{dx} (kN)
Motorways and country national and main roads	500
Country roads in rural area	375
Roads in urban area	250
Courtyards and parking garages	75

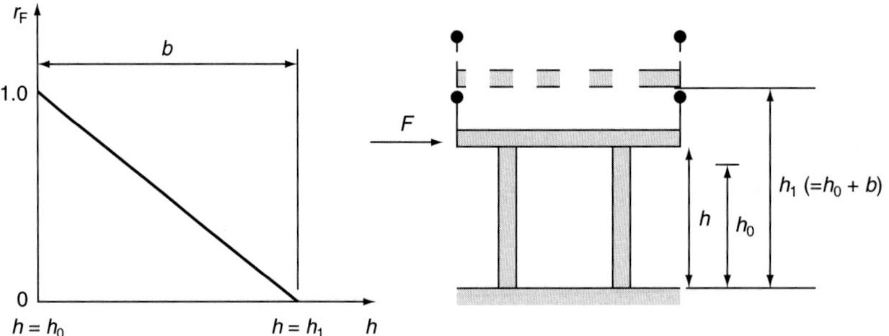

h is the physical clearance between the road surface and the underside of the bridge deck
h_0 is the minimum height of clearance between the road surface and the underside of the bridge deck below which an impact on the superstructure need to be taken into account. The recommended value of h_0 in EN 1991-1-6 is 5.0 m
h_1 is the value of the clearance between the road surface and the underside of the bridge deck. For values of h_1 and above, the impact force F need not be considered. The recommended value of h_1 is 6.0 m (+ allowances for future re-surfacing, vertical sag curve and deflection of bridge)
b is the difference in height between h_1 and h_0, i.e. $b = h_1 - h_0$. The recommended value for b in EN 1991-1-6 is 1.0 m. A reduction factor for F is allowed for values of b between 0 and 1 m, i.e. between h_0 and h_1

Fig. 4.5. Recommended value of the factor r_F for vehicular collision forces on horizontal structural members above roadways, depending on the clearance height h

Note 3 to cl.4.3.1(2): 1991-1-7

shows the conditions of the recommended reduction factor r_F applicable to F_{dx} between h_0 and h_1.

Figure 4.6 gives a representation of the impact force based on the recommended values of the Eurocode.

cl.NA 2-17: UK NA 1991-1-7

The UK National Annex recommends that $h_0 = h = 5.7$ m.

Note 4 cl.4.3.2(1): 1991-1-7

EN 1991-1-7 recommends that account be taken of the impact loads F_{dx} given in Table 4.2 acting with an upward inclination, the recommended value of upward inclination being 10°, on the underside surfaces of the structural member as shown in Fig. 4.7. This rule is intended to cover the risk of lifting of a crane under a structural member. See Fig. 4.3.

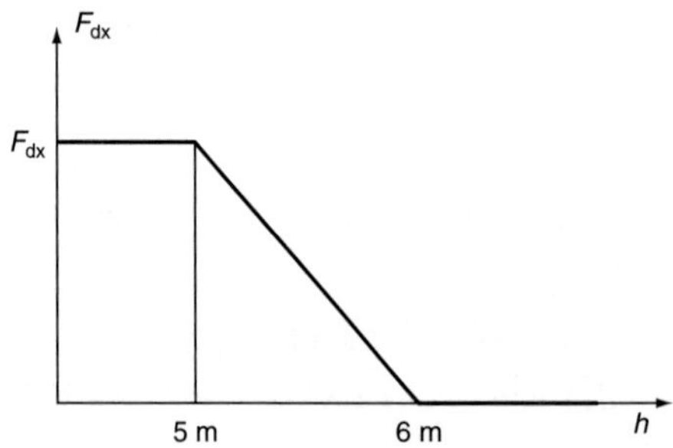

Fig. 4.6. Recommended values in EN 1991-1-7 for F_{dx} for clearance of underside of a structure over a road surface

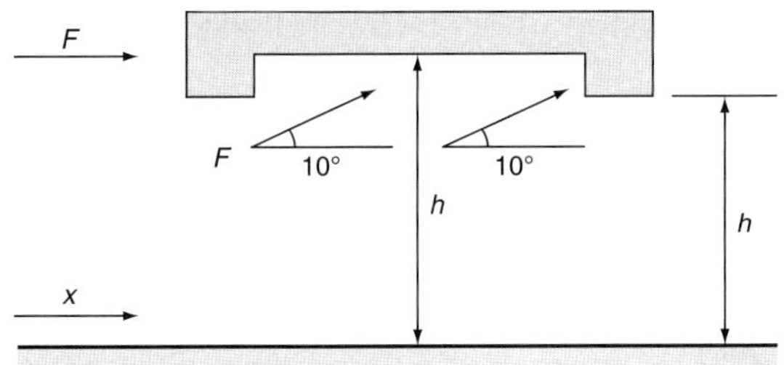

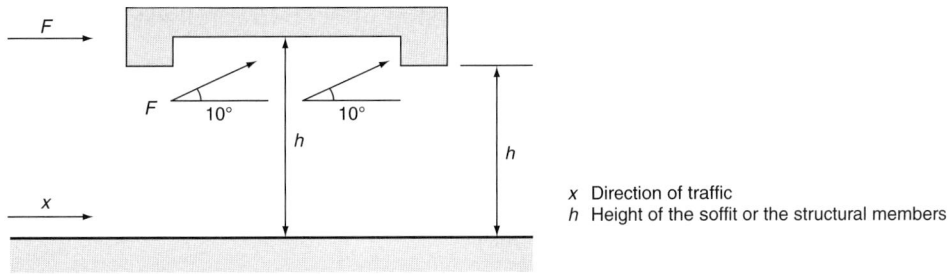

Fig. 4.7. Impact force on members of the superstructure

No further guidance is necessary for *Note 5 Clause 4.3.2(1): EN 1991-1-7.*

EN 1991-1-7 defines only an impact force in the direction of normal travel, F_{dx}. EN 1991-1-7 recommends that, where appropriate, forces perpendicular to the direction of normal travel, F_{dy}, should also be taken into account. In these cases, EN 1991-1-7 recommends that F_{dy} does not act simultaneously with F_{dx}. *cl.4.3.2(2): 1991-1-7* *Note to cl.4.3.2(2): 1991-1-7*

The UK National Annex recommends that F_{dy} may be taken as 50% of F_{dx}. *cl.NA 2-16: UK NA 1991-1-7*

Regarding the area of application of the impact force(s) on the members of the superstructure, EN 1991-1-7 recommends a rectangular area of impact; a square with sides of 25 cm length. *cl.4.3.2(3): 1991-1-7* *Note to cl.4.3.2(3): 1991-1-7*

4.4. Accidental actions caused by forklift trucks

EN 1991-1-7 specifies that, where applicable, design values for accidental actions due to impact from forklift trucks need to be determined. The design value may be determined with a dynamic analysis taking account of the dynamic behaviour of the forklift truck and the structure. The structural response may allow for non-linear deformation.

As an alternative to a dynamic analysis an equivalent static design force F may be applied as described below. *cl.4.4(1): 1991-1-7*

For the design of buildings EN 1991-1-7 recommends an equivalent static design force F due to accidental impact from forklift trucks may be taken as follows:

$$F = 5W$$

where W is the sum of the net weight and hoisting load of a loaded forklift truck (given in Part 3 Chapter 6 of this Designers' Guide) applied at a height of 0.75 m above floor level, although higher or lower values may be more appropriate in some cases.

The National Annex may give alternative values of the equivalent static design force F. It is recommended that the value of F is determined according to advanced impact design for soft impact in accordance with C.2.2. Alternatively, it is recommended that F may be taken as $5W$, where W is the sum of the net weight and hoisting load of a loaded truck (see EN 1991-1-1, Table 6.5), applied at a height of 0.75 m above floor level. However, higher or lower values may be more appropriate in some cases. *Note to cl.4.4(1): 1991-1-7*

The UK National Annex recommends the procedure in EN 1991-1-7. *cl.NA 2-21: UK NA 1991-1-7*

4.5. Accidental actions caused by derailed rail traffic under or adjacent to structures

Accidental actions due to rail traffic need to be taken into account where appropriate. *cl.4.5(1): 1991-1-7*

The National Annex may give the types of rail traffic for which the rules in this clause are applicable. *Note to cl.4.4(1): 1991-1-7*

The types of rail traffic to be considered for design of structure supports in the UK National Annex are those described in UIC 777-2R[6]. *cl.NA 2-22: UK NA 1991-1-7*

4.5.1. Structures spanning across or alongside operational railway lines

4.5.1.1. General

EN 1991-1-7 gives guidance to determine the design values for actions due to impact on supporting members (e.g. columns and walls) caused by derailed trains passing under or adjacent to building structures.

When designing structures that are built over tracks, the reasonably foreseeable development of railway infrastructure, particularly the track layout and the structural clearances, should be taken into consideration.

The strategy for design should consider other appropriate measures (both preventative and protective) to reduce as far as is reasonably practicable the effects of an accidental impact from a derailed train against structural members located above or adjacent to the tracks.

Examples of preventative and protective measures are given below.

- Increasing the lateral distance between the structural members being considered and the centreline of the track.
- Increasing the longitudinal distance between the structural members being considered and any switch or crossing on the approach to the structure.
- For structures over a railway line, the provision of a robust superstructure by having a continuous superstructure, so that the superstructure is not disproportionately damaged if one of the columns is removed due to an impact.
- Avoidance of supports located on a line that is crossed by a line extended in the direction of the turnout of a switch. If this is not reasonably practicable, the provision of dwarf walls should be considered, taking into account their effect on other adjacent infrastructure.
- For structures adjacent to the railway line the provision of continuous walls or wall-type supports instead of columns.
- Provision of deflecting devices or absorbing devices.

cl.4.5.1.1(1): 1991-1-7

Note 1 and 2 to cl.4.5.1.1(1): 1991-1-7

No additional guidance is necessary for Note 1 and Note 2 of Clause 4.5.1.1(1): EN 1990.

Structures that may be subject to impact from derailed railway traffic are classified in the next section of this Designers' Guide.

4.5.1.2. Classification of structures

cl.4.5.1.2(1): 1991-1-7

The Eurocode distinguishes two classes of permanent structures that may be subject to impact from derailed railway traffic (rules concerning temporary structures may be given at the national levels). These classes are defined in Table 4.3, which derives from Table 4.3 of the Eurocode.

Note 1 to cl.4.5.1.2(1): 1991-1-7

The structures to be included in either Classes A or B may be defined in the National Annex or for the individual project.

Note 2 to cl.4.5.1.2(1): 1991-1-7

The National Annex may give reference to the classification of temporary structures used by the public as well as auxiliary construction works as non-contradictory, complementary information.

1991-1-7, Table 4.3

Table 4.3. Classes of structures subject to impact from derailed railway traffic (from ***EN 1991-1-7, Table 4.3***)

Class A	Structures that span across or near to the operational railway that are either permanently occupied or serve as a temporary gathering place for people (such as waiting rooms) or consist of more than one storey (such as car parks and warehouses).
Class B	Massive structures that span across the operational railway such as bridges carrying vehicular traffic or single-storey buildings that are not permanently occupied or do not serve as a temporary gathering place for people.

Further information and background on this classification system given in Table 4.3 is given in relevant International Union of Railways (UIC)-documents. *Note 3 to cl.4.5.1.2(1): 1991-1-7*

The UK National Annex gives broadly the above advice (i.e. to be decided by the individual project) for the above in *Clauses NA 2-23: EN 1991-1-7* and *NA 2-24: EN 1991-1-7*. *cl. UK NA 2.23 and 2.24: UK NA 1991-1-7*

4.5.1.3. Accidental design situations in relation to the classes of structure

The consideration of situations involving the derailment of rail traffic under or on the approach to a structure classified as Class A or B should be taken into account as an accidental design situation, in accordance with EN 1990. *cl.4.5.1.3(1): 1991-1-7*

No additional guidance is required for Clause 4.5.1.3(2). *cl.4.5.1.3(2): 1991-1-7*

4.5.1.4. Class A structures

For class A structures, where the maximum speed of rail traffic at the location is less than or equal to 120 km/h, the Eurocode gives indicative design values for the static equivalent forces due to impact on supporting structural members. *cl.4.5.1.4(1): 1991-1-7*

'Indicative' values are given in EN 1991-1-7 (see Table 4.4 below). Table 7.5, deriving from Table 4.4 of the Eurocode, gives the indicative values. *Note to cl.4.5.1.4(1): 1991-1-7*

The UK National Annex to EN 1991-1-7 specifies that the indicative values in the above table should be used in the UK. *UK NA 2-24: 1991-1-7*

These values may be reduced where supporting structural members are protected, for example by solid plinths or platforms with a minimum height of 38 cm above the top of the rail. *cl.4.5.1.4(2): 1991-1-7*

EN 1991-1-7 does not recommend the scale of the reductions. These reductions may be given in the National Annex. The UK National Annex advises that the reduction of impact forces should be agreed for the individual project, and also refers the designer to the Guidance for reduction of impact forces for supports on platforms in UIC 777-2R[6]. *Note to cl.4.5.1.4(2): 1991-1-7* *UK NA 2-26: 1991-1-7* *UK NA 2-27: 1991-1-7)*

EN 1991-1-7 Clause 4.5.1.4(3) and its *Note* recommends that the forces F_{dx} and F_{dy} (see Table 4.4) should be applied at a height of 1.8 m above track level. This value is a Nationally Determined Parameter. This value is specified by the UK National Annex. *Note to cl.4.5.1.4(3): 1991-1-7*

The design should take into account F_{dx} and F_{dy} separately as appropriate (***Clause 4.5.1.4(3): EN 1991-1-7***). *cl.4.5.1.4(3): 1991-1-7*

EN 1991-1-7 does not give guidance for the determination of forces for rail speeds greater than 120 km/hour. However, if the maximum speed of rail traffic at the location is lower or equal to 50 km/h, the values of the forces given in Table 4.4 may be reduced. *cl.4.5.1.4(5): 1991-1-7*

This recommended value for the reduction, which may be given in the National Annex, is 50% with a recommendation, which is also specified in the UK National Annex, that further information may be found in UIC 777-2R[6]. *Note to cl.4.5.1.4(3): 1991-1-7* *UK NA 2-26: 1991-1-7*

Table 4.4. Indicative horizontal static equivalent design forces due to impact for class A structures over or alongside railways (from ***EN 1991-1-7, Table 4.4***) *1991-1-7, Table 4.4*

Distance d from structural elements to the centreline of the nearest track (m)	Force F_{dx}* (kN)	Force F_{dy}* (kN)
Structural elements: $d < 3$ m	To be specified for the individual project. Further information is set out in Annex B (of EN 1991-1-7)	To be specified for the individual project. Further information is set out in Annex B (of EN 1991-1-7)
For continuous walls and wall type structures: $3\text{ m} \leq d \leq 5\text{ m}$	4000	1500
$D > 5$ m	0	0

* x = track direction; y = perpendicular to track direction.

cl.4.5.1.4(5): 1991-1-7

The values for F_{dx} and F_{dy}, given in Table 4.4, correspond to impact due to derailment at lower speeds. They do not cover a direct impact by a high-speed train derailing at full velocity. Where the maximum permitted speed of rail traffic at the location is greater than 120 km/h, EN 1991-1-7 Clause 4.5.1.4(5) recommends the provision of preventative and/or protective measures and to determine equivalent static forces assuming that consequence class CC3 applies.

Note to cl.4.5.1.4(1): 1991-1-7
UK NA 2-29: 1991-1-7

The values for F_{dx} and F_{dy}, which may take into account additional preventative and/or protective measures, may be given in the National Annex or for the individual project. This recommended guidance is specified in the UK National Annex.

4.5.1.5. Class B structures

cl.4.5.1.5(1): 1991-1-7

For class B structures, particular requirements need to be specified in the National Annex or for the individual project. These particular requirements may be based on a risk assessment.

Note to Clause 4.5.1.5(1): 1991-1-7
UK NA 2-30: 1991-1-7

The UK National Annex states that the requirements for the factors and measures to be considered within the risk assessment should be agreed for the individual project.

4.5.2. Structures located in areas beyond track ends

cl.4.5.2(1): 1991-1-7

Actions caused by overrunning of rail traffic beyond the end of a track or tracks (for example at a terminal station) should be taken into account as an accidental design situation in accordance with EN 1990 when the structure or its supports are located in the area immediately beyond the track ends.

The area immediately beyond the track ends may be specified either in the National Annex or for the individual project (Note to Clause 4.5.2(1): EN 1991-1-7).

As an example in the UK National Annex, permanent new structures, including buildings and columns supporting structures, should not be located within a zone extending 20 m behind the face of the buffer stop and 5 m either side of the projected centreline of the track approaching the buffer stop (NA 2-31: EN 1991-1-7).

cl.4.5.2(3): 1991-1-7

No additional guidance is required for Clause 4.5.2(3).

cl.4.5.2(4): 1991-1-7

No additional guidance is required for Clause 4.5.2(4).

cl.4.5.2(5): 1991-1-7

Where supporting structural members are required to be located near to track ends, an end impact wall should be provided in the area immediately beyond the track ends in addition to any buffer stop. Values of static equivalent forces due to impact onto an end impact wall should be specified.

Note to cl.4.5.2(5): 1991-1-7

Particular measures and alternative design values for the static equivalent force due to impact may be specified in the National Annex or for the individual project. The Note to Clause 4.5.2(5) recommends design values for the static equivalent force due to impact on the end impact wall of $F_{dx} = 5000$ kN for passenger trains and $F_{dx} = 10\,000$ kN for shunting and marshalling trains. The Note to Clause 4.5.2(5) further recommends that these forces are applied horizontally and at a level of 1.0 m above track level.

UK NA 2-31: 1991-1-7

The UK National Annex specifies the above recommended values.

4.6. Accidental actions caused by ship traffic

cl.4.6: 1991-1-7

Clause 4.6 of EN 1991-1-7 gives guidance on accidental actions caused by ship (both river and canal, and sea-going) ships. It gives indicative values for F_{dx} and F_{dy} for a variety of ship types, areas and heights of impact etc. These clauses are in the main intended for supporting structures of bridges. If guidance is required for buildings (e.g. a building at risk from a ship impact), then the reader is referred to the Thomas Telford Designers' Guide on Actions of Bridges.

4.7. Accidental actions caused by helicopters

cl.4.7(1): 1991-1-7

For buildings with roofs designated as a landing pad for helicopters, an emergency landing force should be taken into account.

Clause 4.7(1) gives the following expression (4.3) for determining the vertical equivalent static design force F_d:

$$F_d = C\sqrt{m} \qquad \text{(4.3 of EN 1991-1-7)}$$

where:

C is $3\,\text{kN}\,\text{kg}^{-0.5}$
m is the mass of the helicopter (kg).

The force due to impact should be considered as acting on any part of the landing pad as well as on the roof structure within a maximum distance of 7 m from the edge of the landing pad. The area of impact should be taken as 2 m × 2 m. *cl.4.7(2): 1991-1-7*

CHAPTER 5

Internal explosions

This chapter is concerned with the rules relating to accidental actions due to internal explosions in EN 1991-1-7: *Eurocode 1 – Actions on structures: Part 1.7: General Actions – Accidental actions*. The material described in this chapter is covered in the following clauses:

- Field of application *Clause 5.1: 1991-1-7*
- Representation of actions *Clause 2.2: 1991-1-7*
- Principles of design *Clause 4.3: 1991-1-7*

5.1. Field of application

The effects of explosions need to be taken into account in the design of all parts of a building where gas is burned or regulated; or where explosive material such as dust, explosive gases, or liquids forming explosive vapour or gas is stored or transported (e.g. chemical facilities, vessels, bunkers, sewage constructions, dwellings with gas installations, energy ducts, road and rail tunnels). *cl.5.1(1)P: 1991-1-7*

No further information is required for Clause 5.1(2). *cl.5.1(2): 1991-1-7*

No further information is required for Clause 5.1(3). *cl.5.1(3): 1991-1-7*

EN 1991-1-7 defines various types of internal explosions which are listed above (***Clause 5.1(4): EN 1991-1-7*** in Clause 5.1(4)). *cl.5.1(4): 1991-1-7*

Gas explosions

Gas explosions account for the majority of accidental explosions in buildings. Gas is widely used and, excluding vehicular impact, the incidence of occurrence of gas explosions in buildings is an order of magnitude higher than other accidental loads causing medium or severe damage that may lead to progressive collapse.

Many gas explosions within buildings occur from leakage into the building from external mains. According to Mainstone *et al.*, in their 1978 BRE Report *Structural Damage in Buildings caused by Gaseous Explosions and Other Accidental Loadings,*[12] '*There should be no relaxation . . . for buildings without a piped gas supply, since a risk would usually remain of gas leaking into the building from outside*'. It would be impractical in most circumstances to ensure gas will not be a hazard to any particular building. Therefore a gas explosion is taken as the principal design accidental action, together with impact.

In this context an explosion is defined as rapid chemical reaction of dust or gas in air. It results in high temperatures and high overpressure. Explosion pressures propagate as pressure waves.

The following are necessary for an explosion to occur:

- fuel, in the proper concentration
- an oxidant, in sufficient quantity to support the combustion
- an ignition source strong enough to initiate combustion.

The fuel involved in an explosion may be a combustible gas (or vapour), a mist of combustible liquid, a combustible dust, or some combination of these. The most common combination of two fuels is that of a combustible gas and a combustible dust, called a 'hybrid mixture'.

The pressure generated by an internal explosion depends primarily on the type of gas or dust, the percentage of gas or dust in the air and the uniformity of gas or dust–air mixture, the size and shape of the enclosure in which the explosion occurs, and the amount of venting of pressure release that may be available.

In completely closed rooms with infinitely strong walls, gas explosions may lead to pressures up to 1500 kN/m^2 and dust explosions up to 1000 kN/m^2, depending on the type of gas or dust. In practice, pressures generated are much lower due to imperfect mixing and the venting that occurs due to failure of doors, windows and other openings. Windows make good explosion vents and venting is also afforded by failure of non-structural relatively weak wall panels.

The response in real structures to explosions is highly complex and this is discussed by Mainstone in his 1976 BRE Report *The Response of Buildings to Accidental Explosions*.[13]

5.2. Representation of actions

cl.5.2(1): 1991-1-7 Where appropriate, the determination of explosion pressures on structural members should take account of reactions transmitted to the structural members by non-structural members.

Note 1 to cl.5.2(1): 1991-1-7 Note 1 of Clause 5.2(1) gives the following definition of an explosion: '*a rapid chemical reaction of dust, gas or vapour in air. It results in high temperatures and high overpressures. Explosion pressures propagate as pressure waves*'.

Note 2 to cl.5.2(1): 1991-1-7 The pressure produced by an internal explosion depends on a number of issues of which the following are predominant:

- the type of dust, gas or vapour
- the percentage of dust, gas or vapour in the air
- the uniformity of the dust, gas or vapour air mixture
- the ignition source
- the presence of obstacles in the enclosurc
- the size, the shape and the strength of the enclosure in which the explosion occurs
- the amount of venting or pressure release that may be available.

cl.5: EN.2(2): 1991-1-7 The probable presence of the items listed above needs to be taken into account as appropriate for the layout of the room or group of rooms under consideration, etc.

For construction works classified as CC1 (see Chapter 3 and Table A1 of this Part of *cl.5.2(3): 1991-1-7* this Designer's Guide), provided the design complies with the rules for connections and interaction between components provided in EN 1992 to EN 1999, no specific consideration of the effects of an explosion will normally be necessary.

cl.5.2(3): 1991-1-7 For construction works classified as CC2 or CC3 (see Chapter 3 and Table A1 of this Part of this Designer's Guide), the design will need to incorporate key elements of the structure. The key elements have to be designed to resist actions by either using an analysis based upon equivalent static load models, or by applying prescriptive design/detailing rules. Additionally, for structures classified as CC3 more advanced designs will be necessary including a dynamic analysis.

The methods given in Annex A gives information for the design against explosions for construction works classified as CC1, CC2 and CC3. The reader may also find additional information in Annex D of EN 1991-1-6 Chapter 9. *Note 1 to cl.5.2(3): 1991-1-7*

Advanced design for explosions which will normally be needed for Construction Works classified as CC3 will include one or more of the following aspects as appropriate: *Note 2 to cl.5.2(3): 1991-1-7*

- explosion pressure calculations, including the effects of confinements and venting panels
- dynamic non-linear structural calculations
- probabilistic aspects and analysis of consequences
- economic optimisation of mitigating measures.

5.3. Principles of design

It is a requirement of EN 1990, Clause 2.1(4) (reproduced in Chapter 1 of this Part of this Designers' Guide) that structures shall be designed to resist progressive collapse resulting from an internal explosion. *cl.5.3(1)P: 1991-1-7*

No additional information is required for the Note to Clause 5.3(1)P. *Note to cl.5.3(1)P: 1991-1-7*

In accordance with the philosophy of design for achieving robustness and the strategy based on limiting the extent of localised failure (see Fig. 3.1), the failure of a limited part of the structure due to an internal explosion may be permitted provided this does not include key elements upon which the stability of the whole structure depends and lead to disproportionate failure. *cl.5.3(2): 1991-1-7*

Within the strategy based on limiting the extent of localised failure, the consequences of explosions may be limited by applying one or more of the following measures: *cl.5.3(3): 1991-1-7*

- designing the structure to resist the explosion peak pressure
- using venting panels with defined venting pressures
- separating adjacent sections of the structure that contain explosive materials
- limiting the area of structures that are exposed to explosion risks
- providing specific protective measures between adjacent structures exposed to explosion risks to avoid propagation of pressures.

Annex D of EN 1991-1-7 provides information on obtaining peak and maximum pressures due to a variety of type of internal explosions and venting.

In accordance with EN 1991-1-7, '*while the peak pressures may be higher than the values determined by the methods given in Annex D, such peak pressures have to be considered in the context of a maximum load duration of 0.2 s and assume plastic ductile material behaviour*'. The point is that in reality the peak will generally be larger, but the duration is shorter and the duration of 0.2 s seems to be a reasonable approximation. *Note to cl.5.3(3): 1991-1-7*

Advice in EN 1991-1-7 on explosive pressures, designing and locating venting panels, etc.

The explosive pressure should be assumed to act effectively simultaneously on all of the bounding surfaces of the enclosure in which the explosion occurs. *cl.5.3(4): 1991-1-7*

If the possible ignition source is known and/or where pressures may be higher, the venting panels should be placed close to these positions. However it needs to be ensured that:

- the location of venting panels is such that their failure will not endanger personnel or ignite other material, and
- a venting panel is restrained so that it does not become a missile in the event of an explosion.

In addition, the design should limit the possibilities that the effects of the fire cause any impairment of the surroundings or initiates an explosion in an adjacent room. *cl.5.3(5): 1991-1-7*

Specifically for the design of venting panels, they should be opened at a low pressure and should be as light as possible. In determining the capacity of the venting panel, account needs to be taken of the dimensioning and construction of the supporting frame of the panel. Where windows are used as venting panels the risk of injury to persons from glass fragments *cl.5.3(6): 1991-1-7* *cl.5.3(7)P: 1991-1-7*

Note to cl.5.3(6): 1991-1-7

or other structural members needs to be taken into account. Windows respond in a brittle manner because the thinness of the glass makes very little deformation possible before there is complete disintegration. For this reason, coupled with their relatively light weight and low static strengths, they make good explosion vents.

cl.5.3(6): 1991-1-7

Note to cl.5.3(6): 1991-1-7

In the phenomena for internal explosions, after the first positive phase of the explosion with an overpressure, a second phase follows with an underpressure. No further information is given in this Designers' Guide but this effect should be considered in the design where relevant. Assistance by specialists is recommended.

CHAPTER 6

Annex A (informative): Design of consequences of localised failures in buildings from an unspecified cause

This chapter is concerned with the rules relating to the accidental actions due to design of consequences of localised failures in buildings from an unspecified cause in Annex A of EN 1991-1-7: *Eurocode 1 – Actions on Structures: Part 1.7: General Actions – Accidental actions*. The material described in this chapter is covered in the following clauses:

- Scope — *Clause A.1: 1991-1-7*
- Introduction — *Clause A.2: 1991-1-7*
- Consequence classes of buildings — *Clause A.3: 1991-1-7*
- Recommended strategies — *Clause A.4: 1991-1-7*
- Horizontal ties — *Clause A.5: 1991-1-7*
- Vertical ties — *Clause A.6: 1991-1-7*
- Nominal section of load-bearing wall — *Clause A.7: 1991-1-7*
- Key elements — *Clause A.8: 1991-1-7*

Introduction

Annex A1 of EN 1991-1-7 has been written to accord with the Requirements in *EN 1990: 2002 Eurocode – Basis of Structural Design* to provide robustness, and reliability differentiation.

Annex A of EN 1991-1-7 provides informative rules and methods for designing buildings to sustain an extent of localised failure from an undefined cause without disproportionate collapse.

The Annex also provides an example of how the consequences of building failure may be classified into 'Consequence Classes' in accordance with EN 1990 corresponding to reliability levels of robustness.

This classification of consequences has been partly derived from the outcome of a research study commissioned by the Department for Communities and Local Government (CLG) to Allott & Lomax[14], whose report *Proposals for Amending Part A – Structure. A Consultation Package Available to the ODPM*[9] recommended the use of a risk-based analysis consistent with the philosophy adopted by EN 1990 and EN 1991-1-7. The resulting recommendations produced from the study were subsequently calibrated against current UK practice by the Building Research Establishment and refined accordingly. The rules were further refined following the proposal being subjected to public consultation in the autumn of 2001.

A.1. Scope

cl.A1: 1991-1-7 Annex A of EN 1991-1-7 gives rules and methods for designing buildings, to sustain an extent of localised failure from an unspecified cause without disproportionate collapse. Other approaches may be equally valid, but adoption of this strategy is likely to ensure that a building, depending upon the consequence class (see Section 3.4 of this Part of this Designers' Guide), is sufficiently robust to sustain a limited extent of damage or failure without collapse.

A.2. Introduction

cl.1.5.12: 1991-1-7 *cl.A2(1): 1991-1-7* In accordance with strategies based on limiting the extent of localised failure, see Section 3.1 and Fig. 3.1 of this Part of this Designers' Guide, a building can be designed so that neither the whole building nor a significant part of it will collapse if localised failure were sustained. A localised failure is defined as '*that part of a structure that is assumed to have collapsed, or been severely disabled, by an accidental event*'. Adopting this strategy should provide a building with sufficient robustness to survive a reasonable range of undefined accidental actions.

cl.A2(2): 1991-1-7 The design should take into account the requirement that a building needs to survive a minimum period following an accident in order to facilitate the safe evacuation and rescue of personnel from the building and its surroundings. For certain types of occupancy, e.g. for buildings used for handling hazardous materials, provision of essential services, or for national security reasons, longer periods of survival will probably be required, and this should be agreed with the competent authority.

The rules in Annex A of EN 1991-1-7 are based on UK Codes of Practice (e.g. BS 8110, BS 5950 and BS 5628) and Regulatory requirements introduced in the early 1970s following the collapse of Ronan Point caused by a gas explosion.

Background to the UK requirements for achieving robustness in buildings

The UK 'robustness' regulations were implemented following the progressive collapse of one corner of Ronan Point in 1968 which resulted from a gas explosion on the 18th floor of the 22-storey building. The event had a significant effect on the engineering community in the UK and this led to changes in the Building Regulations to deal specifically with damage due to an accident such as a gas explosion for buildings of five storeys or more. The 1976 Building Regulations required that the building needs to be so constructed that structural failure consequent on removal of any one member in a storey should be localised and limited in its extent to a certain area of that storey. If this was not possible, that member was required to sustain, without structural failure, an additional load corresponding to a pressure of 34 kN/m^2 (5 psi). This value was chosen with reference to evidence recovered at Ronan Point.

Subsequent revisions were aimed at reducing the sensitivity of a building to disproportionate collapse. One design option was to provide effective horizontal and vertical ties in accordance with the structural codes of practice. Where such provisions could not be made, it was recommended that the structure should be able to bridge over a limited area of collapse resulting from the loss of an untied member. Finally, if this was not possible, such a member was required to be designed as a protected (or key) element capable of sustaining additional loads related to a pressure of 34 kN/m^2 acting in any direction. In practice, the 34 kN/m^2 was used to determine a notional load which was applied sequentially to key elements in each possible direction and not as a specific overpressure which would result from a gas explosion. These requirements were considered to provide more robust structures which would generally be more resistant to disproportionate failure due to various causes, such as impact, and were not solely related to gas explosions.

The regulations relating to disproportionate collapse were translated into functional requirements in 1985, together with official guidance being published in Approved Document A of the the Building Regulations. More detailed guidance was subsequently incorporated into the British Standards Institution (BSI) structural design Codes of Practice for common structural materials. The accidental loading requirement was also included in the 1996 BS Loading Code BS 6399-1[20] with the following text. '*When an accidental load is required for a key or protected element approach to design (see appropriate material design code) that load shall be taken as 34 kN/m^2.*'

The current BSI structural material codes dealing with robust design are:

- BS 5628: Code of practice for use of masonry. Part 1 – 1992: Structural use of unreinforced masonry. Clause 37.[15]
- BS 5950: Structural use of steelwork in building. Part 1 – 2000: Code of practice for design – Rolled and welded sections. Clause 2.4.5.[8]
- BS 8110: Structural use of concrete. Part 1 – 1997: Structural use of plain, reinforced and prestressed concrete. Clause 2.2.2.2.[17]
- BS 8110: Structural use of concrete. Part 2 – 1985: Code of practice for special circumstances. Clause 2.6.[18]

A.3. Consequence classes of buildings

One of the most innovative aspects of EN 1990 is the introduction of consequence classes. These classes specify how different categories of building should be designed to deal with accidental actions.

The concept of consequence classes was introduced in Annex B of EN 1990 as a way to meet the Requirement in Section 2.2 of EN 1990 on reliability differentiation. The criterion for classification of consequences is the importance, in terms of consequences of failure, of the structure or structural member concerned. The consequence classes given in EN 1990 are reproduced in Table 6.1.

Table A.1 of EN 1991-1-7 (reproduced in this Designers' Guide as Table A.1) gives a recommended categorisation of building types/occupancies to consequence classes. This categorisation relates to the low (CC1), medium (CC2) and high (CC3) consequence classes given in Table 6.1. *cl.A3: 1991-1-7*

Table 6.1. Definition of consequence classes in EN 1990

Consequences class	Description	Examples of buildings and civil engineering works
CC1	Low consequences for loss of human life; economic, social or environmental consequences small or negligible	Agricultural buildings where people do not normally enter (e.g. storage buildings, greenhouses)
CC2	Medium consequences for loss of human life; economic, social or environmental consequences considerable	Residential and office buildings, public buildings where consequences of failure are medium
CC3	High consequences for loss of human life; or economic, social or environmental consequences very great	Grandstands, public buildings where consequences of failure are high

Table 6.2. Differences in consequence class definitions between EN 1991-1-7 and the UK

Class	EN 1991-1-7 (items not in UK Building Regulations)	UK Building Regulations (items not in EN 1991-1-7)
1	Single occupancy houses not exceeding 4 storeys	Houses not exceeding 4 storeys
2A	Retailing premises not exceeding 3 storeys of less than 1000 m^2 floor area at each storey	Retailing premises not exceeding 3 storeys of less than 2000 m^2 floor area at each storey
3	Stadia accommodating more than 5000 spectators	Grandstands accommodating more than 5000 spectators Buildings containing hazardous substances and/or processes

EN 1991-1-7 allows the classification to be specified in National Annexes. The categorisation given in the Approved Document A of the Building Regulations of England and Wales[11] is closely aligned, with minor differences, with that in EN 1991-1-7. There are minor differences that are outlined in Table 6.2.

Note 1 to Table A1: 1991-1-7 For buildings intended for mixed use (e.g. domestic and office), the most onerous 'consequences class' should be used.

Note 2 to Table A1: 1991-1-7 When determining the number of storeys, basement storeys may be excluded provided the basement storeys fulfil the requirements of 'Consequences Class 2b Upper Risk Group'.

Note 3 to Table A1: 1991-1-7 Table A.1 is not exhaustive and can be adjusted.

Table A.1. Categorisation of consequence classes

Consequence class	Example of categorisation of building type and occupancy
1	• Single-occupancy houses not exceeding 4 storeys • Agricultural buildings • Buildings into which people rarely go, provided no part of the building is closer to another building, or area where people do go, than a distance of $1\frac{1}{2}$ times the building height
2a Lower Risk Group	• 5-storey single-occupancy houses • Hotels not exceeding 4 storeys • Flats, apartments and other residential buildings not exceeding 4 storeys • Offices not exceeding 4 storeys • Industrial buildings not exceeding 3 storeys • Retailing premises not exceeding 3 storeys of less than 1000 m^2 floor area in each storey • Single-storey educational buildings • All buildings not exceeding 2 storeys to which the public are admitted and which contain floor areas not exceeding 2000 m^2 at each storey
2b Upper Risk Group	• Hotels, flats, apartments and other residential buildings greater than 4 storeys but not exceeding 15 storeys • Educational buildings greater than single storey but not exceeding 15 storeys • Retailing premises greater than 3 storeys but not exceeding 15 storeys • Hospitals not exceeding 3 storeys • Offices greater than 4 storeys but not exceeding 15 storeys • All buildings to which the public are admitted and which contain floor areas exceeding 2000 m^2 but not exceeding 5000 m^2 at each storey • Car parking not exceeding 6 storeys
3	• All buildings defined above as Class 2 Lower and Upper Consequences Class that exceed the limits on area and number of storeys • All buildings to which members of the public are admitted in significant numbers • Stadia accommodating more than 5000 spectators • Buildings containing hazardous substances and/or processes

A.4. Recommended strategies

Adopting the following recommended strategies given in Annex A of EN 1991-1-7 and shown in Table 6.3 of this Designers' Guide should provide a building with an acceptable level of robustness to sustain localised failure without a disproportionate level of collapse and satisfying Clause 2.1(4)P of EN 1990, the requirement for robustness.

Table 6.3. Recommended strategies for Consequence Classes CC1, CC2 and CC3

Consequence Class of Building as Classified in Table A.1	Recommended design strategy*
1	Provided a building has been designed and constructed in accordance with the rules given in EN 1990 to EN 1999 for satisfying stability in normal use, no further specific consideration is necessary with regard to accidental actions from unidentified causes.
2a Lower Risk Group	In addition to the recommended strategies for Consequences Class 1, the provision of effective horizontal ties, or effective anchorage of suspended floors to walls, as defined in A.5.1 and A.5.2 respectively for framed and load-bearing wall construction should be provided.
2b Upper Risk Group	In addition to the recommended strategies for Consequences Class 1, the provision of: • Horizontal ties, as defined in A.5.1 and A.5.2 respectively for framed and load-bearing wall construction (see 1.5.11), together with vertical ties, as defined in A.6, in all supporting columns and walls should be provided. • Where it is inappropriate or impossible to provide vertical and horizontal ties, the building should be checked to ensure that upon the notional removal of each supporting column and each beam supporting a column, or any nominal section of load-bearing wall as defined in A.7 (one at a time in each storey of the building), the building remains stable and that any local damage does not exceed a certain limit (see Fig. 6.1). • Where the notional removal of such columns and sections of walls would result in an extent of damage in excess of the agreed limit, or other such limit specified, then such elements should be designed as a 'key element' (see A.8).
3	A systematic risk assessment of the building should be undertaken taking into account both foreseeable and unforeseeable hazards.

*Clause references are to both Annex A of EN 1991-1-7, and this Designers' Guide.

For buildings in Consequence Class 2a, details of effective anchorage may be given in the National Annex. *Note 1 to cl.A(4): 1991-1-7*

For buildings in Consequence Class 3, some guidance on risk analysis is included in Annex B of this Designers' Guide. For more comprehensive guidance see Annex B of EN 1991-1-7 and Chapter 7. *Note 2 to cl.A(4): 1991-1-7*

For buildings the limit of admissible local failure may be different for each type of building. The recommended value is 15% of the floor, or 100 m^2, whichever is smaller, in each of two adjacent storeys. See Fig. A.1 (reproduced below as Fig. 6.1). In the UK, the Approved Document A of the Building Regulations specifies 70 m^2 instead of 100 m^2. *Note 3 to cl.A(4): 1991-1-7*

A.5. Horizontal ties

N.B. This Section A.5 of this Part of this Designers' Guide gives information on horizontal ties only. For vertical ties see Section A.6.

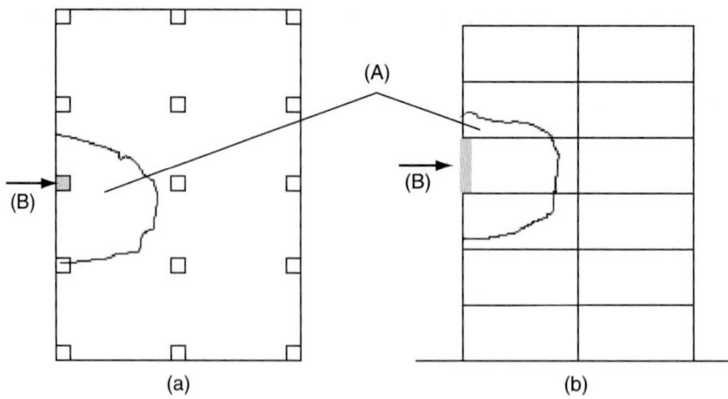

Key
(A) Local damage not exceeding 15% of floor area or 100 m^2, whichever is the smaller, in each of two adjacent storeys
(B) Column to be notionally removed
(a) Plan (b) Section

Fig. 6.1. Recommended limit of admissible damage

A.5.1. Framed structures

cl.A5.1(1): 1991-1-7

Horizontal ties should be provided around the perimeter of each floor and roof level and internally in two orthogonal directions to tie the column and wall elements securely to the structure of the building. The ties should be continuous and be arranged as closely as practicable to the edges of floors and lines of columns and walls. At least 30% of the ties should be located within the close vicinity of the centre lines of the columns and the walls.

Note to cl.A5.1(1): 1991-1-7

See Fig. A.2 (reproduced here as Fig. 6.2).

Horizontal ties for example, may consist of depending on the form of the structure:

- rolled steel sections
- steel bar reinforcement in concrete slabs, or
- steel mesh reinforcement and profiled steel sheeting in composite steel–concrete floors (if directly connected to the steel beams with shear connectors).

cl.A5.1(2): 1991-1-7

The ties can consist of a combination of the above types.

In accordance with EN 1991-1-7, each continuous tie, including its end connections, should be capable of sustaining:

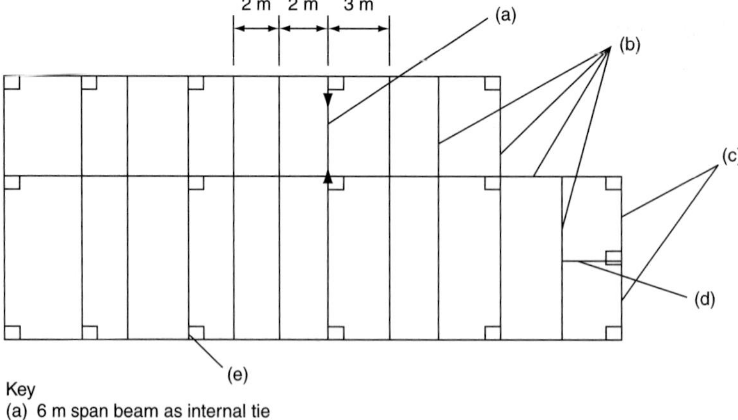

Key
(a) 6 m span beam as internal tie
(b) All beams designed to act as ties
(c) Perimeter ties
(d) Tie anchored to a column
(e) Edge column

Fig. 6.2. Example of horizontal ties

- a design tensile load of T_i for the accidental limit state in the case of internal ties, and
- T_p, in the case of perimeter ties (see Fig. 6.2)

equal to the following values.

For internal ties:

$$T_i = 0.8(g_k + \psi q_k)sL \text{ or } 75\,\text{kN, whichever is the greater} \quad \text{(A.1)}$$

For perimeter ties:

$$T_p = 0.4(g_k + \psi q_k)sL \text{ or } 75\,\text{kN, whichever is the greater} \quad \text{(A.2)}$$

where:

s is the spacing of ties
L is the span of the tie (e.g. see Fig. A.2)
ψ is the relevant factor in the expression for combination of action effects for the accidental design situation (i.e. ψ_1 in accordance with expression (6.11b) of EN 1990 given in EN 1990, and is normally taken as 0.5). *cl.A5.1(3): 1991-1-7*

EN 1991-1-7 does not make any differentiation for lightweight building structures (e.g. those whose primary structure is timber or cold-formed thin gauge steel). This is made in the UK NA, where in the case of lightweight building structures the values for minimum horizontal tie forces in expressions A.1 and A.2 should be taken as 15 kN and 7.5 kN respectively. *cl.NA3.1: UK NA to BS 1991-1-7*

Example 6.1. Horizontal ties for framed structures

Consider a *framed structure*, **5 storeys** with story height structures $\boldsymbol{h} = \mathbf{3.6\,m}$, *consequence class 2, upper group*. The span is $\boldsymbol{L} = \mathbf{7.2\,m}$ and the span distance $\boldsymbol{s} = \mathbf{6\,m}$. See Fig. 6.3.

Assume $\boldsymbol{g_k} = \boldsymbol{q_k} = \mathbf{4\,kN/m^2}$ the combination factor = **0.5**.

In that case, in accordance with expressions (A.1) and (A.2), the required internal ($\boldsymbol{T_i}$) and perimeter ($\boldsymbol{T_p}$) tie forces may be calculated from:

$$\boldsymbol{T_i = 0.8(g_k + \psi q_k)_{sL}} \text{ or } 75\,\text{kN, whichever is greater (A.1), and}$$

$$\boldsymbol{T_p = 0.4(g_k + \psi q_k)_{sL}} \text{ or } 75\,\text{kN, whichever is greater (A.2)}$$

leading to $\boldsymbol{T_i} = 0.8\{4 + 0.5 \times 4\}(6 \times 7.2) = \mathbf{207.36\,kN}$ and $\boldsymbol{T_p} = \mathbf{103.68\,kN}$

In the design of ties any members that are being used for sustaining actions other than accidental actions, may be utilised. *cl.A5.1(4): 1991-1-7*

A.5.2. Load-bearing wall construction

For Class 2 buildings (Lower Risk Group) (see Table A.1), in load-bearing (e.g. masonry) construction, a cellular form of construction designed to facilitate interaction of all components including an appropriate means of anchoring the floor to the walls should be adopted to provide the appropriate robustness. *cl.A5.2(1): 1991-1-7*

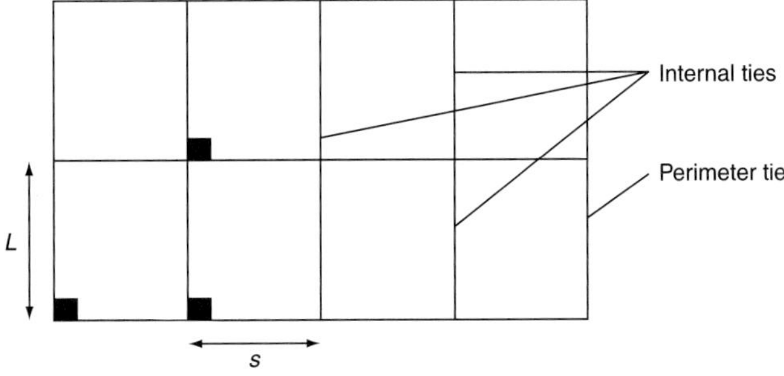

Fig. 6.3. (Reference Example 1)

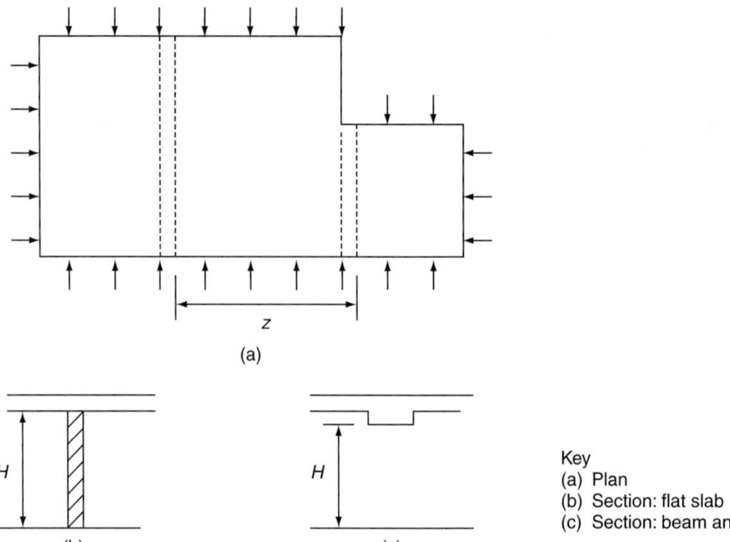

Fig. 6.4. illustration of factors *H* and *z*

cl.A5.2(2): 1991-1-7

For Class 2 buildings (Upper Risk Group) (see Table A.1), in load-bearing construction, continuous horizontal ties are required in the floors to achieve the appropriate robustness.

These should be internal ties which need to be uniformly distributed throughout the floors in both orthogonal directions and peripheral ties extending around the perimeter of the floor slabs within a 1.2 m width of the slab. The design tensile load in the ties should be determined as follows.

For internal ties:

$$T_{\text{I}} = \text{the greater of } F_{\text{t}}\,\text{kN/m or } \frac{F_{\text{t}}(g_{\text{k}} + \psi q_{\text{k}})}{7.5}\frac{z}{5}\,\text{kN/m} \tag{A.3}$$

For peripheral ties:

$$T_{\text{p}} = F_{\text{t}} \tag{A.4}$$

where:

F_{t} is 60 kN/m or $20 + 4n_{\text{s}}$ kN/m, whichever is less
n_{s} is the number of storeys
z is the lesser of:

- 5 times the clear storey height H, or
- the greatest distance in metres in the direction of the tie, between the centres of the columns or other vertical load-bearing members whether this distance is spanned by:
  - a single slab, or
  - a system of beams and slabs.

Factors H (in metres) and z are illustrated in Fig. 6.4.

N.B. The units for T_{p} is in kN/m to give it consistency with the units for F_{c} from which it is derived. It is advised that the numerical value for T_{p} is taken in kN acting over a 1.2 m width.

Example 6.2. Horizontal ties for load-bearing structures

Consider a *load-bearing structure*, 5 storeys with story height structures $h = 3.6$ m, consequences class 2, upper group. The span is $L = 7.2$ m and the span distance $s = 6$ m. (*Load-bearing wall type of structure, consequences class 2, Upper Group.*)

$g_{\text{k}} = q_{\text{k}} = 4\,\text{kN/m}^2$ the combination factor $\psi = 0.5$.

$F_t = \min(60, 20 + 5 \times 4) = 40\,\text{kN/m}$ and

z is the lesser of $5h = 5 \times 3.6 = 18\,\text{m}$ or the greatest distance in metres in the direction of the tie, between the centres of the columns or other vertical load-bearing members whether this distance is spanned by a single slab or by a system of beams and slabs (i.e. 7.2 m or 6 m depending on the direction of the tie considered).

From expression (A.3) for *internal ties*:

$$T_I = \text{the greater of } F_t(\text{kN/m}) \text{ or } \frac{F_t(g_k + \psi q_k)}{7.5}\frac{z}{5}\ \text{kN/m}$$

Hence $\boldsymbol{T_i} = \max\{40; [40(4 + 0.5 \times 4)/7.5] \times 7.2/5\} = \max\{40; 46.1\} = 46.1\,\textbf{kN/m}$ and $\boldsymbol{T_i} = \max\{40; [40(4 + 0.5 \times 4)/7.5] \times 6.0/5\} = \max\{40; 38.4\} = 40\,\textbf{kN/m}$ in the orthogonal direction.

From expression (A.4) for *peripheral ties*:

$$T_p = F_t$$

where:

F_t is 60 kN/m or $20 + 4n_s$ kN/m, whichever is less
n_s is the number of storeys.

$F_t = 20 + 4 \times 5 = 40\,\text{kN/m}$ which is less than 60 kN/m hence $T_p = 40\,\text{kN/m}$

A.6. Vertical ties

In the design each column and wall should be tied continuously from the foundations to the roof level (***Clause A.6(1): EN 1991-1-7***). *cl.A.6(1): 1991-1-7*

In the case of ***framed buildings*** (e.g. steel or reinforced concrete framed structures) the columns and walls carrying vertical actions should have the capacity to resist an accidental design tensile force whose magnitude is equal to the largest design vertical permanent and variable load reaction applied to the column from any one storey. Such accidental design loading should not be assumed to act simultaneously with permanent and variable actions that may be acting on the structure (***Clause A.6(2): EN 1991-1-7***). *cl.A.6(2): 1991-1-7*

Example 6.3. Vertical ties for framed structures

Consider the ***framed structure*** of Example 1, 5 storeys with storey height structures $h = 3.6\,\text{m}$, consequence class 2, upper group. The span is $\boldsymbol{L = 7.2\,m}$ and the span distance $\boldsymbol{s = 6\,m}$ **and** $\boldsymbol{g_k = q_k = 4\,kN/m^2}$. For determining the vertical tensile force T_v the combination factor $\boldsymbol{\psi = 1.0}$ is taken.

In accordance with Clause A6(2):

$$T_v = (g_k + \psi q_k)(L \times s)$$

Thus $\boldsymbol{T_v} = (4 + 4)(7.2 \times 6) = \textbf{345.6 kN/column}$

For ***load-bearing wall construction*** (defined in Clause 1.5.11 of EN 1991-7 as '*non-framed masonry cross-wall construction mainly supporting vertical loading. Also includes lightweight panel construction comprising timber or steel vertical studs at close centres with particle board, expanded metal or alternative sheathing*' the vertical ties may be considered effective provided:

i. for masonry walls their thickness is at least 150 mm thick and if they have a minimum compressive strength of $5\,\text{N/mm}^2$ in accordance with EN 1996-1-1
ii. the clear height of the wall, H, measured in metres between faces of floors or roof does not exceed $20t$, where t is the thickness of the wall in metres

iii. if they are designed to sustain the following vertical tie force T:

$$T = \frac{34A}{8000}\left(\frac{H}{t}\right)^2 \text{ N, or } 100\,\text{kN/m of wall, whichever is the greater} \qquad \text{(A.5)}$$

where:

A is the cross-sectional area in mm^2 of the wall measured on plan, excluding the non-load-bearing leaf of a cavity wall.

cl.A.6(3): 1991-1-7 iv. the vertical ties are grouped at 5 m maximum centres along the wall and occur no further than 2.5 m from an unrestrained end of the wall.

Example 6.4. Vertical ties for load-bearing construction

Consider a load-bearing structure of Example 6.2 of 5 storeys with clear height of the wall $\boldsymbol{H} = \mathbf{3.6}$ m. The thickness of the walls $\boldsymbol{t} = \mathbf{0.2}$ m.

Using Expression (A.5) and adjusting the expression to give the results in kN/m:

$$T_v = 34 \times 0.2/8 \times (3.6/2)^2 = \mathbf{275\,kN/m}$$

A.7. Nominal section of load-bearing wall

(1) The nominal length of load-bearing wall construction referred to in Table A2 for Consequence Class **2b Upper Risk Group** should be taken as follows:

- for a reinforced concrete wall, a length not exceeding $2.25H$
- for an external masonry, or timber or steel stud wall, the length measured between lateral supports provided by other vertical building components (e.g. columns or transverse partition walls)
- for an internal masonry, or timber or steel stud wall, a length not exceeding $2.25H$

where:

cl.A.7(1): 1991-1-7 H is the storey height in metres.

A.8. Key elements

cl.A.8(1): 1991-1-7 In accordance with 3.3(1)P, for building structures a 'key element', as referred to in Table A2 for Consequence Class **2b Upper Risk Group**, should be capable of sustaining an accidental design action of A_d applied in the horizontal and vertical directions (in one direction at a time) to the member and any attached components having regard to the ultimate strength of such components and their connections. This accidental design loading is applied using expression (6.11b) of EN 1990, and may be applied either as a concentrated or distributed load.

Note to cl.A.7(1): 1991-1-7 The recommended value of A_d for building structures is $34\,\text{kN/m}^2$. This value is used in the UK National Annex.

CHAPTER 7

Annex B (informative): Information on Risk Analysis

This chapter is concerned with the rules relating to information on risk analysis in Annex B of EN 1991-1-7: *Eurocode 1 – Actions on Structures: Part 1.7: General Actions – Accidental actions*. The material described in this chapter is covered in the following clauses:

- Introduction — *Clause B.1: 1991-1-7*
- Definitions — *Clause B.2: 1991-1-7*
- Description of the scope of risk analysis — *Clause B.3: 1991-1-7*
- Methods of risk analysis — *Clause B.4: 1991-1-7*
- Risk acceptance and mitigating measures — *Clause B.5: 1991-1-7*
- Risk mitigating measures — *Clause B.6: 1991-1-7*
- Reconsideration — *Clause B.7: 1991-1-7*
- Communication of results and conclusions — *Clause B.8: 1991-1-7*
- Application to buildings and civil engineering structures — *Clause B.9: 1991-1-7*

Application of risk assessment to Class 3 buildings

With reference to the range of Class 3 buildings (see Table A.2 in Annex A of this Designers' Guide) and corresponding strategies in this guidance, it is apparent that much is left to the discretion of the designer. The requirements for Class 3 buildings state:

> '*A systematic risk assessment of the building should be undertaken taking into account both foreseeable and unforeseeable hazards.*'

The design for Class 3 structures should take account of the design rules for Class 2b Upper Risk Group structures as a minimum. The design should also include considerations of specific defined incidents or the provision of general robustness through the provision of horizontal and vertical ties, alternate load paths or the provision of key elements as described in Annex A1 of this Designers' Guide. A systematic risk analysis may result in enhancements to the Class 2b Upper Risk Group requirements and/or the provision of further hazard mitigating measures.

B.1. Introduction

Risk analysis is the topic discussed in Annex B of EN 1991-1-7 which gives guidance for the planning and execution of risk assessment in the field of buildings and civil engineering

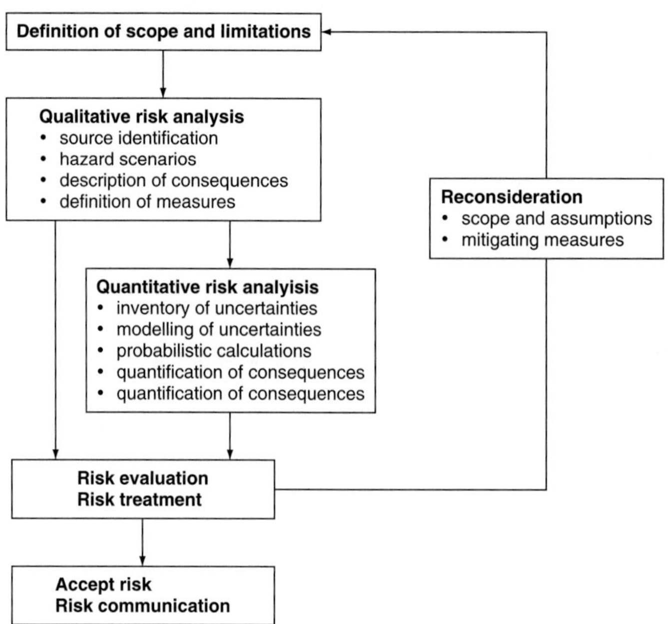

Fig. 7.1. Overview of risk analysis

cl.B.1(1): 1991-1-7 structures. In Clause B.1 a pictorial overview of the envisaged risk analysis is given. This is reproduced below as Fig. 7.1.

cl.B.2.1 to B2.8: 1991-1-7

B.2. Definitions

The following definitions are given:

- ***consequence*** in Clause B.2.1
- ***hazard scenario*** in Clause B.2.2
- ***risk*** in Clause in B.2.3
- ***risk acceptance criteria*** in Clause B.2.4
- ***risk analysis*** in Clause B.2.5
- ***risk evaluation*** in Clause B.2.6
- ***risk management*** in Clause B.2.7
- ***undesired event*** in Clause B.2.8.

It is recommended that the reader fully appreciates and understands these definitions before using Annex B.

Two of the definitions are repeated below as they are very important for this chapter.

cl.1.15.13: 1991-1-7 *cl.B.2.3: 1991-1-7* **Risk** which is '*a measure of the combination (usually the product) of the probability or frequency of occurrence of a defined hazard and the magnitude of the consequences of the occurrence*'.

cl.B.2.5: 1991-1-7 **Risk analysis** which is '*a systematic approach for describing and/or calculating risk. Risk analysis involves the identification of undesired events, and the causes, likelihoods and consequences of these events*' (see Fig. B.1 of 1991-1-7, reproduced here as Fig. 7.1).

B.3. Description of the scope of risk analysis

An effective risk analysis needs to fully describe:

- the subject
- background, and

- objectives.

All:

- technical
- environmental
- organisational, and
- human circumstances that are relevant to the activity and the problem being analysed

need to be stated in sufficient detail.

All:

- presuppositions
- assumptions, and
- simplifications

made in connection with the risk analysis should be stated. *cl.B.3: 1991-1-7*

B.4. Methods of risk analysis

As shown in Fig. 7.1, there are two types of risk analysis. First, a risk analysis has a descriptive (qualitative) analysis looking at the various hazards and their consequences, and second, a numerical (quantitative) analysis that can be more rigorous, but which may only be relevant and practicable in certain circumstances. *cl.B.4(1): 1991-1-7*

It should be appreciated that risk analysis is not an exact science and the main gain is obtained through a systematic examination and recording of the potential hazards, the structure's vulnerability to these hazards, and if necessary the measures that can be taken to minimise any undesirable consequences. Therefore it is not appropriate to provide a set procedure for any specific type of building.

B.4.1. Qualitative risk analysis

A qualitative risk analysis is a systematic consideration of the various hazard scenarios, recording what has been considered and the actions that are to be taken. Further information on risk analyses can be found in References 12–14. Identification of hazards and hazard scenarios is a crucial task to a risk analysis. It requires a detailed examination and understanding of the system. Various techniques have been developed to assist the engineer in performing this part of the analysis (e.g. PHA, HAZOP, fault tree, event tree, decision tree, causal networks, etc.). *cl.B.4.1(1): 1991-1-7*

The risk analysis should initially identify the potential hazards to which a building may be exposed.

In structural risk analysis the following conditions can, for example, present hazards to the structure:

- high values of ordinary actions
- low values of resistances, possibly due to errors or unforeseen deterioration
- ground and other environmental conditions different from those assumed in the design
- accidental actions such as fire, explosion, flood (including scour), impact or earthquake
- unspecified accidental actions
- possible malicious attack on a building.

When defining the various hazard scenarios the following should be taken into account:

- the anticipated or known variable actions on the structure
- the environment surrounding the structure
- the proposed or known inspection regime of the structure
- the concept of the structure, its detailed design, materials of construction and possible points of vulnerability to damage or deterioration
- the consequences of type and degree of damage due to the identified hazard scenario.

For each hazard a profile should be developed including a determination of various parameters including the magnitude, duration, frequency, probability and extent of a hazard. The stakeholders (i.e. controlling authority, the owner) should be consulted concerning the hazards to be included in the design, and it may be reasonably decided to exclude some hazards. Next, for each hazard it is necessary to understand how it will affect the structure and what impact this would have on the building contents, especially the building occupiers (immediate and longer term) and possibly the local community. This requires an assessment of the vulnerability of the construction, contents and economic value of its functions.

The systematic evaluation of the vulnerability of the assets to the various hazards should provide the basis to develop a design strategy, perhaps eliminating some hazards, reducing others, and designing to resist specific hazards. For particular buildings the robustness measures for class 2B buildings should be adopted for unidentified hazards (design or construction errors, unexpected deterioration, malicious attacks etc.).

Once a risk has been determined it has to be decided whether it will be accepted or whether mitigating measures will be specified. To mitigate the risk, the following measures may be taken:

- structural measures – that is, by designing over-strong structural elements or by designing for second load paths in case of local failures
- non-structural measures – that is, by a reduction of the event probability, the action intensity or the consequences.

In some situations there may be conflicting requirements for different hazards and these should be considered carefully. For example, the use of laminated windows may reduce the damage to external explosions (usually detonations) but serve to increase the pressures in internal gas explosions (deflagrations).

The results of the risk analysis should be presented as a list of consequences and probabilities and their degree of acceptance should be discussed. The probabilities may need to be expressed in relative terms, i.e. high, moderate, low, unless more detailed information is available. Recommendations for measures to mitigate risk that naturally arise from the risk analysis should be stated.

It is necessary to compile the results of the work and reasons for the various decisions, in a written report which may need to be reviewed by another expert. The report may need to be re-evaluated periodically or when there is a change in circumstances which may affect specific hazards and thereby require different mitigation measures.

B.4.2. Quantitative risk analysis

cl.B.4.1(1): 1991-1-7

In the *quantitative* part of the risk analysis the probabilities for all undesired events and their subsequent consequences should be estimated. The probability estimations are usually at least partly based on judgement and may for that reason differ substantially from actual failure frequencies. If failure can be expressed numerically the risk may be presented as the mathematical expectation of the consequences of an undesired event. A possible way of presenting risks is indicated in Fig. 7.2.

In a rigorous quantitative assessment, mathematical calculations are used to determine the expected consequences from the whole range of hazards. This provides a potentially more accurate picture of the risk (and the benefits of avoiding it) than the qualitative method, provided the basic data related to the risk are sufficient to warrant a rigorous mathematical treatment.

It is important therefore to consider whether the available information is sufficient to make such a quantitative risk analysis worthwhile. The statistical data available concerning internal gas explosions in housing in the UK are probably as detailed as any statistical data on a specific accidental action. Nevertheless, it is considered by Ellis[13] that it is still not possible to determine the intensity of a gas explosion with any confidence, except for single-room explosions which may be less severe than multi-room explosions. Therefore

Severe	×				
High	×				
Medium		×			
Low			×		
Very low				×	
consequence / probability	0.00001	0.001	0.001	0.01	>0.1
× represents examples of maximum acceptable risk levels					

Classification: The severity of potential failure is identified for each hazard scenario and classified as Severe, High, Medium, Low or Very low. They may be defined as follows:

- *Severe*: Sudden collapse of structure occurs with high potential for loss of life and injury.
- *High*: Failure of part(s) of the structure with high potential for partial collapse and some potential for injury and disruption to users and public.
- *Medium*: Failure of part of the structure. Total or partial collapse of structure unlikely. Small potential for injury and disruption to users and public.
- *Low*: Local damage.
- *Very low*: Local damage of small importance.

Fig. 7.2. Possible presentation diagram for the outcome of a quantitative risk analysis

for this important accidental action hazard, it is not yet sensible to place any confidence in a quantitative analysis. The quantity and quality of data for use in such analyses are a prime consideration.

The assumptions upon which the quantitative risk analysis is based should be reconsidered when the results of the analysis are available. Sensitivities of factors used in the analysis should be quantified.

B.5. Risk acceptance and mitigating measures

Once the level of risk is identified, decisions may be taken as to whether mitigating (structural or non-structural) measures should be specified. *cl.B.5.1(1): 1991-1-7*

The ALARP (as low as reasonably practicable) principle is the technique normally used to consider risk acceptance. According to the ALARP principle, two risk levels are specified:

1. If the risk is below the lower bound of the broadly tolerable (i.e. ALARP) region, no measures need to be taken.
2. If the risk is above the upper bound of the broadly tolerable region, the risk is considered as unacceptable.

If the risk is between the upper and lower bound, an economical optimal solution should be sought. *cl.B.5.1(2): 1991-1-7*

When evaluating the risk of a certain period of time related to the failure event on the basis of the consequences, an appropriate reduction factor may be applied. *cl.B.5.1(3): 1991-1-7*

The risk acceptance levels will need to be specified and be formulated on the basis of the following two acceptance criteria concerning the individual acceptable level of risk and the socially acceptable level of risk: *cl.B.5.1(4): 1991-1-7*

1. *The individual acceptable level of risk*: individual risks are usually expressed as fatal accident rates. They can be expressed as an annual fatality probability or as the probability per time unit of a single fatality when actually being involved in a specific activity.

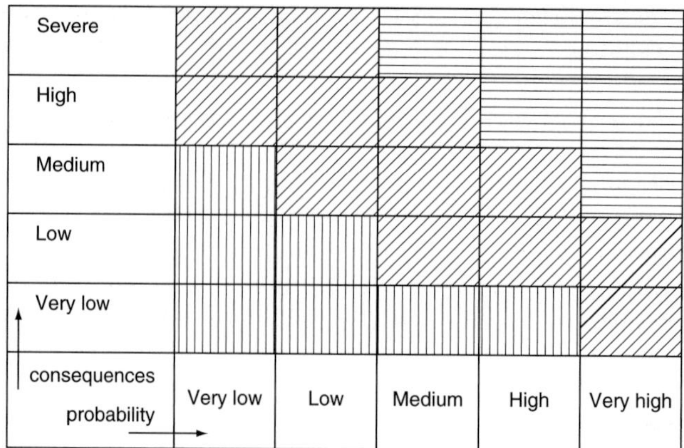

Fig. 7.3. Risk matrix for the outcome of a qualitative risk analysis. (Note: The severity of potential failure is identical to the one given in Fig. 7.2. The probabilities could be the same (on a yearly or event basis).)

2. *The socially acceptable level of risk*: the social acceptance of risk to human life, which may vary with time, is often presented as an F–N curve, indicating a maximum yearly probability F of having an accident with more than N casualties.

   Alternatively, concepts such as:

- *value for prevented fatality (VPF)*, or
- *quality index of life*

may be used.

Risk acceptance levels will normally have to be specified for the appropriate individual project and may be determined from:

- particular national regulations and requirements
- particular codes and standards, or
- from experience and/or theoretical knowledge that may be used as a basis for decisions on acceptable risk.

Note to cl.B.5.1(4): 1991-1-7
cl.B.5.1(5): 1991-1-7

The risk acceptance criteria may be expressed qualitatively or numerically (quantitatively). In the case of a qualitative risk analysis the following criteria may be used:

- The general aim should be to minimise the risk without incurring a substantial cost penalty.
- For the consequences within the vertically hatched area of Fig. 7.3, the risks associated with the scenario can normally be accepted.
- For the consequences within the diagonally hatched area of Fig. 7.3, a decision on whether the risk of the scenario can be accepted and whether risk mitigation measures can be adopted at an acceptable cost will need to be made.
- For the consequences considered to be unacceptable (those falling within the horizontally hatched area of Fig. 7.3 are likely to be unacceptable), appropriate risk mitigation measures (see Section B.6) should be taken.

B.6. Risk mitigating measures

cl.B.6.1(1): 1991-1-7

Where the risk is unacceptable the risk mitigation measures may be selected from one or more of the following:

- ***Elimination or reduction of the hazard.*** This may be achieved by making an adequate design, modifying the design concept, and providing the countermeasures to combat the hazard, etc.

- ***By-passing the hazard.*** This may be achieved by changing the design concepts or occupancy, for example through the protection of the structure, provision of sprinkler system, etc.
- ***Controlling the hazard.*** This may be achieved by controlled checks, warning systems or monitoring.
- ***Overcoming the hazard.*** This may be achieved by providing increased reserves of strength or robustness, availability of alternative load paths through structural redundancy, or resistance to degradation, etc.
- ***Permitting controlled collapse of a structure*** where the probability of injury or fatality may be reduced, for example for impact on lighting columns or signal posts.

B.7. Reconsideration

As indicated by Fig. B.1 (see Fig. 7.1) for the results of the risk analysis to become acceptable, the scope, design and assumptions may have to be revised and re-evaluated against the scenarios until it is possible to accept the structure with the selected mitigation measures. *cl.B.7.1(1): 1991-1-7*

B.8. Communication of results and conclusions

Communicating the results of the risk analysis to the stakeholders without ambiguity is essential, as is keeping a thorough record of the work.

The results of the qualitative and (if available) the quantitative analysis should be presented as a list of consequences and probabilities, and their degree of acceptance should be communicated with all stakeholders. *cl.B.8.1(1): 1991-1-7*

All the data that have been used to carry out a risk analysis and their sources should be recorded. *cl.B.8.1(2): 1991-1-7*

All the essential assumptions, presuppositions and simplifications that have been made should be summarised so that the validity and limitations of the risk analysis are made clear and recorded. *cl.B.8.1(3): 1991-1-7*

The recommendations for the measures to mitigate risk should be stated and be based on conclusions from the risk analysis. *cl.B.8.1(4): 1991-1-7*

B.9. Application to buildings and civil engineering structures

Introduction to B.9

This section concerns the practical application of Sections B.1 to B.8 to buildings and civil engineering structures. This Designers' Guide will only cover the building examples given in Annex B of EN 1990.

B.9.1. General

To mitigate the risk in relation to extreme events in buildings (and civil engineering structures) the design should consider one or more of the following measures: *cl.B.9.1(1): 1991-1-7*

(a) ***Structural measures***, where the structure and the structural members have been designed to have reserves of strength or alternative load paths in case of local failures (i.e. the strategy based on limiting the extent of localised failure as described in Sections 3.1 of both this Designers' Guide and EN 1991-1-7).

(b) ***Non-structural measures***, which can include the reduction of:
- the probability of the event occurring
- the action intensity, or
- the consequences of failure.

cl.B.9.1(2): 1991-1-7 The probability of occurrence of all accidental and extreme actions (e.g. actions due to fire, earthquake, impact, explosion, extreme climatic actions) should be considered for a suitable set of possible hazard scenarios. The consequences for each should then be estimated in terms of the number of casualties and economic losses. Detailed information is presented in Sections B.9.2 and B.9.3 below.

cl.B.9.1(3): 1991-1-7 The above approach (Section B9.2) is generally more suitable for those accidental and extreme events that constitute foreseeable hazards. The approach is less suitable for unforeseeable hazards arising from, for example, design or construction errors and unexpected deterioration. It is for this reason that more global damage tolerance design strategies have been developed (see Annex A), e.g. the requirements on sufficient ductility and tying of elements.

cl.B.9.1(3): 1991-1-7 In this respect consider the situation whereby a structural member (e.g. a beam, a column) has been damaged by an event, to such an extent that the member has lost its normal load-bearing capacity. By applying the strategy based on limiting the extent of localised failure as described in Section 3.1, the remaining part of the structure is (for a relatively short period of time defined as the repair period T) to withstand the 'normal' loads with some prescribed reliability as shown probabilistically in expression (B.1):

$$P(R < E \text{ in } T \mid \text{one element removed}) < p_{\text{target}} \tag{B.1}$$

where R is the resistance of the member and E is the effects of the Actions.

The target reliability p_{target} depends upon:

- the normal safety target for the building
- the period under consideration (hours, days or months), and
- the probability that the element under consideration is removed (by causes other than those already considered in design).

cl.B.9.1(4): 1991-1-7 For conventional structures (e.g. using traditional materials and the types of buildings generally in Classifications 1 and 2a and 2b in Table A1 of Annex A) all relevant collapse possibilities should be included in the design. Where this can be justified, and for some cases in agreement with the client and the controlling authority, failure causes that have only a remote likelihood of occurring may be disregarded. The approach given in B.9.1(2) should be taken into account. In many cases, and in order to avoid complicated analyses, the strategy given in B.9.1(3) may be investigated.

cl.B.9.1(5): 1991-1-7 For unconventional structures (e.g. very large structures, those with new design concepts, those using new materials) the probability of having some unspecified cause of failure should be considered as substantial. A combined approach of the methods described in B.9.1(2) and B.9.1(3) should be taken into account.

B.9.2. Structural risk analysis

cl.B.9.2(1): 1991-1-7 Risk analysis of structures subject to accidental actions may be approached using the three steps explained below, taking Fig. 7.4 as an example:

1. Step 1 where assessment of the probability of occurrence of different hazards with their intensities (Fig. 7.4(a)) is made.
2. Step 2 where the assessment of the probability of different states of damage and the corresponding consequences for given hazards (Fig. 7.4(b)) is made.
3. Step 3 where the assessment of the probability of inadequate performance(s) of the damaged structure together with the corresponding consequence(s) (Fig. 7.4(c)) is made.

cl.B.9.2(2): 1991-1-7 In accordance with Annex B of EN 1991-1-7, the total risk R is assessed by

$$R = \sum_{i=1}^{N_H} p(H_i) \sum_{j}^{N_D} \sum_{k=1}^{N_S} p(D_j|H_i)p(S_k|D_j)C(S_k) \tag{B.2}$$

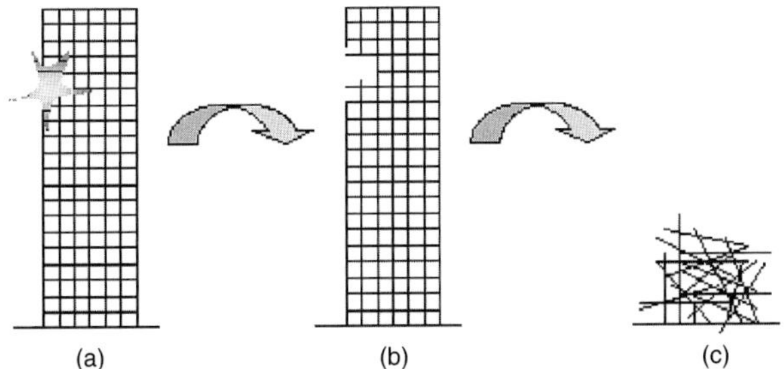

Step 1: Identification and modelling of relevant accidental hazards and the assessment of the probability of occurrence of different hazards with different intensities.
Step 2: Assessment of damage states to structure from different hazards and the assessment of the probability of different states of damage and corresponding consequences for given hazards.
Step 3: Assessment of the performance of the damaged structure and the assessment of the probability of inadequate performance(s) of the damaged structure together with the corresponding consequence(s).

Fig. 7.4. Illustration of steps in risk analysis of structures subject to accidental actions

where:

- it is assumed that the structure is subjected to N_H different hazards, and
- that the hazards may damage the structure in N_D different ways (can be dependent on the considered hazards), and
- that the performance of the damaged structure can be discretised into N_S adverse states S_K with corresponding consequences $C(S_k)$
- $P(H_i)$ is the probability of occurrence (within a reference time interval) of the *i*th hazard
- $P(D_j|H_i)$ is the conditional probability of the *j*th damage state of the structure given the *j*th hazard, and
- $PS_k|D_j|$ is the conditional probability of the *k*th adverse overall structural performance S given the *i*th damage state.

$P(S_k|D_j)$ and $C(S_k)$ can be highly dependent on time (e.g. in case of fire and evacuation, respectively) and the overall risk needs to be assessed and compared to acceptable risks accordingly. *Note 1 to cl.B.9.2(2): 1991-1-7*

Expression (B.2), although it is primarily the basis for the risk assessment of structures subject to rare and accidental loads, but it can be applied to structures subjected to ordinary loads. *Note 2 to cl.B.9.2(2): 1991-1-7*

The economic feasibility of a risk assessment can be taken into account by having different strategies for risk control and risk reduction. For example:

- The risk may be reduced by the reduction of the probability that the hazards will occur, i.e. by reducing $P(H)$. The risk of explosions in buildings might be reduced by removing explosive materials from the building.
- The risk may be reduced by reducing the probability of significant damages for given hazards, i.e. $P(D|H)$. For example, damage which might follow as a consequence of the initiation of fires can be mitigated by passive and active fire control measures (e.g. foam protection of steel members and sprinkler systems).
- The risk may be reduced by reducing the probability of adverse structural performance given structural damage, i.e. $P(S|D)$. This might be undertaken by designing the structures with a sufficient degree of redundancy, thus allowing for alternative load transfer should the static system change due to damage. *cl.B.9.2(3): 1991-1-7*

B.9.3. Modelling of risks from extreme events

B.9.3.1. General format

As part of a risk analysis, potential extreme hazards such as earthquakes, explosions, collisions, etc. will need to be investigated. The general model for such an event may consist

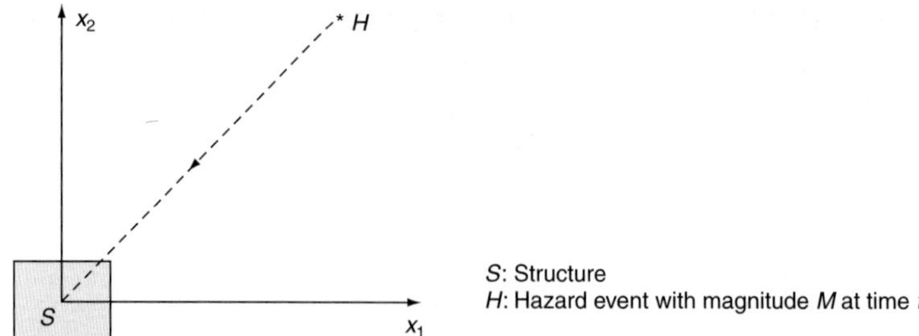

Fig. 7.5. Components for the extreme event modelling

of the following components (see Fig. 7.5):

- a triggering event at some place and at some point in time
- the magnitude M of the energy involved in the event and possibly some other parameters
- the physical interactions between the event, the environment and the structure, leading to the exceedance of some limit state in the structure.

cl.B.9.3.1(1): 1991-1-7

cl.B.9.3.1(2): 1991-1-7

The occurrence of the triggering event for hazard H in B.9.3.1(1) may often be modelled as events in a Poisson process of intensity $\lambda(t, x)$ per unit volume and time unit, t representing the point in time and x the location in space (x_1, x_2, x_3). The probability of occurrence of failure during the time period up to time T is then (for constant λ and small probabilities) given by expression (B.3):

$$P_f(T) \approx N \int_0^{\infty} P(F|M = m) f_M(m)\,\mathrm{d}m \qquad \text{(B.3)}$$

where:

$N = \lambda T$ is the total number of relevant initiating events in the considered period of time
$f_M(m)$ is the probability density function of the random magnitude M of the hazard.

The probability of failure can depend on the distance between the structure and the location of the event. In that case an explicit integration over the area or volume of interest is necessary.

B.9.3.2. Application to impact from vehicles

Guidance is given in this clause for the determination of the probability of failure in the event of an impact that will occur if a vehicle, travelling along the roadway, leaves its intended course at a critical place with sufficient speed. No further guidance is given in this Designers' Guide and if necessary for the design the reader should consult Clause B.9.3.2 of Annex B of EN 1991-1-7.

cl.B.9.3.2: 1991-1-7

B.9.3.3. Application to impact from ships

Guidance is given in this clause for the determination of the probability of failure in the event of an impact that will occur if a ship impacts a structure. This clause is more relevant for bridge supporting structures and no further guidance is given on this clause in this Designers' Guide and if necessary for the design the reader should consult Clause B.9.3.3 of Annex B of EN 1991-1-7.

cl.B.9.3.3: 1991-1-7

B.9.4. Guidance for application of risk analysis related to impact from rail traffic

Clause B.9.4(1) gives a very comprehensive list of issues which should be taken into account when assessing the risk to people from derailed trains on the approach to Class A structures, where the maximum permitted line speed is over 120 km/h, and Class B structures. Class A and Class B structures are defined in Table 4.3 of EN 1991-1-7 and this Designers' Guide. No further guidance is necessary on the listed factors.

cl.B.9.4(1): 1991-1-7

Clause B.9.4(2) gives a very comprehensive list of issues which should be taken into account **for Class B structures either singly or in combination in determining the appropriate measures to reduce the risk to people from a derailed train on the approach to a structure**. Class B structures are defined in Table 4.3 of EN 1991-1-7 and this Designers' Guide. No further guidance is necessary on the listed factors. *cl.B.9.4(2): 1991-1-7*

CHAPTER 8

Annex C (informative): Dynamic design for impact

This chapter is concerned with the rules relating to information on the dynamic design for impact in Annex B of EN 1991-1-7: *Eurocode 1 – Actions on Structures: Part 1.7: General Actions – Accidental actions*. The material described in this Chapter is covered in the following clauses:

- General *Clause C.1: 1991-1-7*
- Impact dynamics *Clause C.2: 1991-1-7*
- Impact from aberrant road vehicles *Clause C.3: 1991-1-7*
- Impact from ships *Clause C.4: 1991-1-7*

Introduction to this chapter
Annex C is primarily applicable to bridge supporting structures and superstructures. In this chapter of the Designers' Guide the information given on impact dynamics in EN 1991-1-7 will be discussed. If the reader requires information on the application of impact dynamics for impacts from aberrant road vehicles and ships, these are covered in the Thomas Telford Designers' Guide on Actions on Bridges.

C.1. General

Impact, in the context of EN 1991-1-7, is an interaction phenomenon between a moving object and a structure, in which the kinetic energy of the object is suddenly transformed into energy of deformation. To determine the dynamic interaction forces, the mechanical properties of both the object and the structure should be determined.

In many cases static equivalent forces (e.g. those in Chapter 4 of EN 1991-1-7 and this Designers' Guide) are commonly used in design. *cl.C.1(1): 1991-1-7*

Advanced design of structures to sustain actions due to impact may include explicitly one or both of the following aspects:

- dynamic effects
- non-linear material behaviour.

In this Annex guidance is only given for dynamic effects. *cl.C.1(2): 1991-1-7*

Information on the probabilistic aspects and the analysis of consequences can be obtained from Annex C. *Note to cl.C.1(2): 1991-1-7 cl.C.1(3): 1991-1-7* No further information is given in this Designers' Guide on Clause C.1(3), as this clause is applicable primarily for bridges and guidance may be obtained from the Thomas Telford Designers' Guide on Actions on Bridges.

C.2. Impact dynamics

Impact is characterised as either:

cl.C.2(1): 1991-1-7

- *hard impact*, where the energy is mainly dissipated by the impacting body, or
- *soft impact*, where the structure is designed to deform in order to absorb the impact energy.

C.2.1. Hard impact

cl.C.2.1(1): 1991-1-7 For hard impact, the equivalent static forces may be obtained from Chapter 4 Clauses 4.3 to 4.7 of EN 1991-1-7 or this Designers' Guide. As an alternative, an approximate dynamic analysis may be performed following the simplified approximations given below in Clauses C.2.1(2) and (3).

cl.C.2.1(2): 1991-1-7 For a hard impact analysis, the following assumptions are made:

- the structure is rigid and immovable, and
- the colliding object deforms linearly during the impact phase.

The maximum resulting dynamic interaction force is given by expression (C.1)

$$F = v_r\sqrt{km} \tag{C.1}$$

where:

v_r is the object velocity at impact
k is the equivalent elastic stiffness of the object (i.e. the ratio between force F and total deformation)
m is the mass of the colliding object.

The force due to impact may be considered as a rectangular pulse on the surface of the structure. In that case the duration of the pulse follows from:

$$F\Delta t = mv \quad \text{or} \quad \Delta t = \sqrt{m/k} \tag{C.2}$$

If relevant, a non-zero rise time can be applied (see Fig. C.1).

When the colliding object is modelled as an equivalent impacting object of uniform cross-section (see Fig. 8.1), expressions (C.3) and (C.4) should be used to determine k and m:

$$k = EA/L \tag{C.3}$$

$$m = \rho AL \tag{C.4}$$

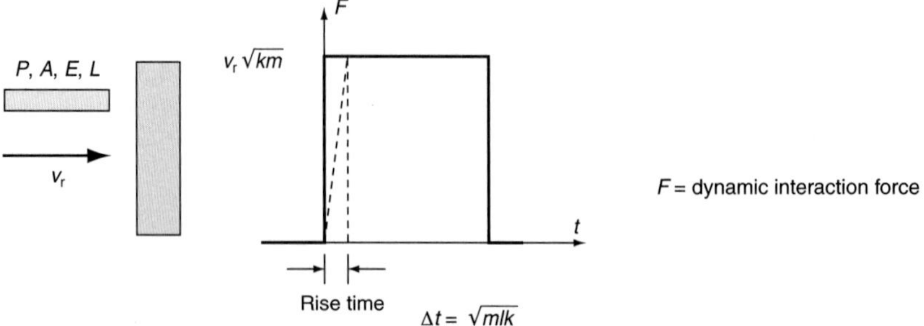

Fig. 8.1. Impact model

where:

L is the length of the impacting object
A is the cross-sectional area
E is the modulus of elasticity
ρ is the mass density of the impacting object.

No additional information is required for Clause C2.1 (3). *cl.C.2.1(3): 1991-1-7*

C.2.2. Soft impact

For the case where the structure is assumed elastic and the colliding object rigid, the expressions given in Section C.2.1 apply, with k being the stiffness of the structure. *cl.C.2.2(1): 1991-1-7*

For the case where the structure is designed to absorb the impact energy by plastic deformations, the design will need to be such that the structure's ductility is sufficient to absorb the total kinetic energy of the colliding object given below:

$$\frac{1}{2}mv_r^2$$

For the case of rigid-plastic response of the structure, the above requirement is satisfied by the condition of expression (C.5):

$$\frac{1}{2}mv_r^2 \leq F_o y_o \qquad \text{(C.5)}$$

where:

F_o is the plastic strength of the structure, i.e. the limit value of the static force F
y_o is its deformation capacity, i.e. the displacement of the point of impact that the structure can undergo. *cl.C.2.2(2): 1991-1-7*

No additional information is required for the note to Clause C2.2 (2). *Note to cl.C.2.2(2): 1991-1-7*

C.3. Impact from aberrant road vehicles

This Section C.3 covers impact from aberrant road vehicles for which values for the impacting force to be used in the design are given in Section 4.3 of this Part of this Designers' Guide.

An alternative method to determine the impact force is given in C.3 of EN 1991-1-7, and the reader may use this if required. However, this Section is more applicable for bridge supporting structures and superstructures for which information is given in the Thomas Telford Designers' Guide on Actions on Bridges. *cl.C.3: 1991-1-7*

C.4. Impact by ships

This Section C.4 covers ship impact in inland waterways and sea waterways.

This Section is more applicable for bridge supporting structures and superstructures for which information is given in the Thomas Telford Designers' Guide on Actions on Bridges. *cl.C.4: 1991-1-7*

CHAPTER 9

Annex D (informative): Internal explosions

This chapter is concerned with the rules relating to internal explosions in Annex D of EN 1991-1-7: *Eurocode 1 – Actions on Structures: Part 1.7: General Actions – Accidental actions*. The material described in this chapter is covered in the following clauses:

- Dust explosions in rooms, vessels and bunkers *Clause D.1: 1991-1-7*
- Natural gas explosions *Clause D.2: 1991-1-7*
- Explosions in road and rail tunnels *Clause D.3: 1991-1-7*

Introduction to Annex D

The following definitions are used in this chapter on Annex D.

- ***Deflagration*** which is defined as the propagation of a combustion zone at a velocity that is less than the speed of sound in the unreacted medium. *cl.1.5.3: 1991-1-7*
- ***Detonation*** which is defined as the propagation of a combustion zone at a velocity that is greater than the speed of sound in the unreacted medium. *cl.1.5.4: 1991-1-7*
- ***Venting panel*** which is defined as a non-structural part of the enclosure (wall, floor, ceiling) with limited resistance that is intended to relieve the developing pressure from deflagration in order to reduce pressure on structural parts of the building. *cl.1.5.17: 1991-1-7*

D.1. Dust explosions in rooms, vessels and bunkers

When the design concerns dust explosions in rooms, vessels and bunkers, the type of dust is represented by a material parameter K_{St}, which characterises the confined explosion behaviour of the appropriate dust type. K_{St} may be experimentally determined by standard methods for each type of dust. *cl.D.1(1): 1991-1-7*

A higher value for K_{St} leads to higher pressures and shorter rise times for internal explosion pressures. The value of K_{St} depends on factors such as changes in the chemical composition, particle size and moisture content. A sample number of Indicative values for K_{St} are given in Table 9.1 of this Designers' Guide. The table does list all the types of dust covered by EN 1991-1-7 but the values for K_{St} for all the types of dust should be obtained from Annex D of EN 1991-1-7. *Note 1 to cl.D.1(1): 1991-1-7*

No further guidance is required for Note 2 of Clause D.4(1). *Note 2 to cl.D.1(1): 1991-1-7*

Table 9.1. K_{St} values for dusts

Type of dust	K_{St} (kN/m^2 × m/s)
Brown coal	18 000
Cellulose	27 000
Coffee	9 000
Corn, corn crush	
Corn starch	
Grain	13 000
Milk powder	16 000
Mineral coal	
Mixed provender	
Paper	
Pea flour	
Pigment	
Rubber	
Rye flour, wheat flour	
Soya meal	
Sugar	15 000
Washing powder	27 000
Wood, wood flour	22 000

Note 3 to cl.D.1(1): 1991-1-7

No further guidance is required for Note 3 of Clause D.4(1).

Clause D.1(2) gives an expression (D.1), which is not reproduced here, to determine the venting area (A) of cubic and elongated rooms, vessels, and bunkers for dust explosions within a single room. The venting area (A) is dependent upon the parameters below when it is determined using expression (D.1):

- p_{max} is the maximum pressure of the dust
- K_{St} is the deflagration index of a dust cloud, see Table D.1
- $p_{red.max}$ is the anticipated maximum reduced pressure in the vented vessel
- p_{stat} is the static activation pressure with the size of existing venting areas
- V is the volume of room, vessel, bunker.

When using expression (D.1) it should be remembered that it is only valid with restrictions which are comprehensively listed in Clause D.1(2) of Annex D of EN 1991-1-7.

cl.D.1(2): 1991-1-7

Furthermore, when determining the venting area A using expression (D.1), care needs to be observed regarding the units for the various parameters that are used.

Clause D.1(3) gives an expression (D.2), which is not reproduced here, to determine the venting area (A) of rectangular enclosures. The venting area (A) is dependent upon the parameters below when it is determined using expression (D.2):

- p_{max} is the maximum pressure of the dust
- K_{St} is the deflagration index of a dust cloud, see Table D.1
- p_{Bem} is the design strength of the structure
- p_{stat} is the static activation pressure with the size of existing venting areas
- V is the volume of the rectangular enclosure.

When using expression (D.2) it should be remembered that it is only valid with restrictions which are comprehensively listed in Clause D.1(3) of Annex D of EN 1991-1-7.

cl.D.1(3): 1991-1-7

Furthermore when determining the venting area A using expression (D.2), care needs to be observed regarding the units for the various parameters that are used.

cl.D.1(4): 1991-1-7

Clause D.1(4) should be used for elongated rooms to determine the increase in the venting area for the geometric parameters given in the Clause.

D.2. Natural gas explosions

This clause gives guidance for buildings which have natural gas installed. In such situations the structure may be designed to withstand the effects of an internal natural gas explosion using a nominal equivalent static pressure determined by expressions (D.4) and (D.5):

$$p_d = 3 + p_{stat} \qquad \text{(D.4)}$$

or

$$p_d = 3 + p_{stat}/2 + 0.04/(A_v/V)^2 \qquad \text{(D.5)}$$

whichever is the greater,

where:

p_{stat} is the uniformly distributed static pressure at which venting components will fail, in (kN/m)
A_v is the area of venting components, in m^2
V is the volume of rectangular enclosure (m^3).

Expressions (D.4) and (D.5) are valid for a room up to 1000 m^3 total volume. *cl.D.2(1): 1991-1-7*

The pressure due to deflagration can be taken to act effectively simultaneously on all of the bounding surfaces of the room. *Note to cl.D.2(1): 1991-1-7*

For rooms in buildings which have different venting components with different p_{stat} values, the largest value of p_{stat} should be used in the design.

If the value for p_d is determined to be greater than 50 kN/m^2 it does not need to be taken into account, and thus the value of 50 kN/m^2 can be used. *cl.D.2(2): 1991-1-7*

When using expressions (D.4) and (D.5) the ratio of the area of venting components (A_v) and the volume (V) should comply with expression (D.6) below: *cl.D.2(3): 1991-1-7*

$$0.05(1/m) \leq A_v/V \leq 0.15 \qquad \text{(D.6)}$$

D.3. Explosions in road and rail tunnels

In this Clause D3, expressions are given to determine the pressure time function in road and rail tunnels in the case of a detonation and deflagration. These clauses do not apply for buildings and if guidance is required the reader may refer to the Thomas Telford Designers' Guide to Actions on Bridges. *cl.D.3(1): 1991-1-7*

References for Part 7

1. EN 1991-1-7: 2006. *Eurocode 1 – Actions on Structures – Part 1-7: General Actions – Accidental actions.*
2. Calgaro, J.-A., Tschumi, M. and Gulvanessian, H., *Designers' guide to EN 1991 for bridges. Actions on bridges.* Thomas Telford, London, 2002.
3. Gulvanessian, H., Calgaro, J.-A. and Holický, M. *Designers' guide to EN 1990 Eurocode: Basis of design.* Thomas Telford, London, 2002.
4. ISO. *ISO 2394:1998. General principles on reliability.* ISO, Genéve, 1998.
5. ISO. *ISO 3898:1997. Basis for the design of structures – notation – general symbols.* ISO, Genéve, 1997.
6. ISO. *ISO 8930:1987. General principles on reliability for structures – list of equivalent terms.* ISO, Genéve, 1987.
7. BS 8110: *Structural use of concrete.* Part 1 – 1997 *Structural use of plain, reinforced and prestressed concrete.* Part 2 – 1985 *Code of practice for special circumstances.* BSI, London.
8. BS 5950: *Structural use of steelwork in building.* Part 1 – *Code of practice for design – Rolled and welded sections.* BSI, London, 2000.
9. BS 5628: *Code of practice for use of masonry.* Part 1 – *Structural use of unreinforced masonry.* BSI, London, 1992.
10. Reid, Stuart, G. *Perception and communication of risk, and the importance of dependability. Structural Safety 21.* Elsevier Science, 1999, pp. 373–384.
11. UIC 777-2R. *Structures built over railway lines – Construction requirements in the track zone – Edition: 2.* Union Internationale de Chemins de Fer, Railway Technical Publications (ETF), 2002.
12. Mainstone, R. J., Nicholson, H. G. and Alexander, S. J. *Structural Damage in Buildings Caused by Gaseous Explosions and Other Accidental Loadings.* BRE, Watford, 1978.
13. Mainstone, R. J. *The Response of Buildings to Accidental Explosions.* BRE, Garston, Watford, 1976.
14. Allott and Lomax. *Guidance on robustness and provision against accidental actions.* CLG internal report, ref: CI/21/2/66, London, May, 1999.
15. BSI. BS 5628: *Code of practice for use of masonry. Part 1 – 1992. Structural use of unreinforced masonry.* BSI, London, 1992.
16. BSI. BS 5850: *Structural use of steelwork in building. Part 1 – 2000. Code of practice for design – rolled and welded sections.* BSI, London, 2000.
17. BSI. BS 8110: *Structural use of concrete. Part 1 – 1992. Structural use of plain, reinforced and prestressed concrete.* BSI, London, 1992.
18. BSI. BS 8110: *Structural use of concrete. Part 2 – 1985. Code of practice for special circumstances.* BSI, London, 1985.
19. *Proposals for Amending Part A – Structure. A Consultation Package Available to the ODPM.*
20. BS 6399: 1996. *Loading for buildings.* Part 1. *Code of practice for dead and imposed loads.* BSI, London, 1996.
21. *Approved document A (2004) of the Building Regulations for England and Wales.* HMSO, London, 2004.
22. CIB WG 32. *A framework for risk management and risk communications.* CIB, CIB Report 259, Rotterdam.
23. Ellis, B. R. and Currie, D. M. Gas explosions in buildings in the UK – regulation and risk. *The Structural Engineer,* **76**, No. 19, 1998.
24. FEMA 386-2. *Understanding your risks – Identifying hazards and estimating losses.* US Federal Emergency Management Agency, August 2001, USA.

Index

Page numbers in *italics* refer to illustrations or tables.

accepted risks, 232, *233*, 269–270
accidental actions, 221–284
  actions during execution, 210–211
  EN 1991-1-7, 221–284
    accepted risks, 232, *233*, 269–270
    action classification, 5, 227–228
    action representation, 240–241, 252–253
    bomb attacks, 234
    buildings, 222, 241–245, 255–264, 265–275
    classification of actions, 5, 227–228
    collapse, 224–225, 231, 234–236, 257
    collisions, 221–222, 240–249
    consequence classes, 224, 232, 236–237, 255–264
    design principles, 253–254
    design situations, 229–237
    explosions, 221–222, 251–254, 256–257, 281–283
    fire, 222, 254
    forklift trucks, 239, 245
    helicopters, 239, 249
    horizontal ties, 259–263
    identified actions, 230–231, 232–234
    impacts, 221–222, 240–249, 274–275, 277–279
    internal explosions, 251–254, 281–283
    localised damage/failure, 230–231, 232, 234–236, 255–264
    rail traffic, 239, 245–248, 274–275, 283
    representation of actions, 240–241, 252–253
    risk analysis, 222, 265–275
    road vehicles, 239, 241–245, 274, 279, 283
    robustness, 224–225, 230, 233–234, 256–257
    seismic action, 222, 229
    ship traffic, 248–249, 274, 279
    ties, 235–236, 259–264
    trains, 239, 245–248, 274–275
    unidentified actions, 230–231, 255–264
    variable actions, 227–228, 234
    vertical ties, 263–264
    walls, 241–242, 261–264
  snow loads, 71, 74–75, 80, 85–86, 93, 99
action classification, 13–14, 16–18
  actions during execution, 181–182, 185–187
  snow loads, 71
  thermal actions, 163–164
action representation, 13–14
  accidental actions, 240–241, 252–253
  actions during execution, 182, 197–211
  imposed loads, 35–37
  self-weight, 31–32
actions during execution
  EN 1991-1-6, 181–218
    accidental actions, 210–211
    action classification, 181–182, 185–187
    action representation, 182, 197–211
    alterations, 182, 217
    atmospheric icing, 204
    auxiliary construction works, 182
    bridges, 182, 215
    buildings, 181–218
    characteristic value, 190–194, 197–199
    classification of actions, 181–182, 185–187
    climatic action, 191–194
    combination value, 197, 198
    concrete construction work, 193, 200, 209–210
    construction, 181–218
    construction loads, 185–187, 198–199, 204–210
    demolition, 182, 217
    design situations, 182, 189–195
    geotechnical actions, 200–201
    handling factors, 199–200
    horizontal actions, 214
    hydration effects, 202
    impacts, 193
    limit states, 182, 194–195, 213–214
    non-structural members, 199–200
    predeformations, 201–202

actions during execution (*continued*)
EN 1991-1-6, 181–218
prestressing, 201, *202*
quasi-permanent value, 197, 198
reconstructions, 182, 217
representation of actions, 182, 197–211
return periods, 190–191, 192–193
seismic actions, 211
self-weight, 199–200
serviceability limit states, 182, 194–195, 213–214
shrinkage effects, 202
snow loads, 192, 193, 203
supplementary rules, 182, 213–215
temperature, 202
thermal actions, 191–192, 202
ultimate limit states, 194, 213
variable actions, 197–199
water, 203
wind actions, 192–194, 202–203
administration area loads, 38–44
aerodynamic coefficients, 153–154
aerodynamic shapes, 83, *84*
aeroelasticity, 155
alteration actions, 182, 217
angle of repose, 14
architectural massing junctions, 160
arrangement of imposed loads, 37–38
arrangement of snow loads, 83–89, 107
atmospheric icing, 204
auxiliary construction works, 182

background response factors, 127
beams, imposed loads, 23–24, 37–38
bomb attacks, 234
bracing systems, 160, 235–236, 259–264
bridges, 33, 175, 182, 215
buffeting, 151
buildings
accidental actions, 222, 241–245, 255–264, 265–275
actions during execution, 181–218
construction loads, 185–187, 198–199, 204–210
imposed loads, 11–12, 22–27, 31–33, 35–53
material storage, 205, 206, *207*
self weight, 31–33
thermal actions, 167–173, 177–178
bulk weight density, 14, 68, 121–122

cantilever beams, 23–24
car parks, 49–50, 57–58
casting concrete, 209–210
characteristic value of actions, 5, 6–7
accidental actions, 227
actions during execution, 190–194, 197–199
imposed loads, 20, 23, 38–53
self-weight, 20, 32–33
snow loads, 68, 73–74, 77–81, 111–115
thermal actions, 165–166
characteristic value coefficient *k*, 5–6, *7*
classification of actions, 16–18
accidental actions, 227–228
actions during execution, 181–182, 185–187
snow loads, 71
thermal actions, 163–164
clearance factors, 243–244
climatic actions, 191–194
coefficients
characteristic value, 5–6, *7*
linear expansion, 160, 167–168, 175
snow loads, 87–99, 117–119
of variations, 117–119
wind actions, 128, 146–148, 153–154
collapse
accidental actions, 224–225, 231, 234–236, 257
concrete casting, 200
roof collapse, 86–87, *88*
collisions, 221–222, 240–249
columns, 38, 241–242
combination value, 5, *6*
accidental actions, 227, 234
actions during execution, 197, 198
self-weight, 23
snow loads, 73
thermal actions, 165–166
commercial area loads, 38–44
communication, risk analysis, 271
concrete construction work, 193, 200, 209–210
concrete structure self weight, 32
consequence classes, 224, 232, 236–237, 255–264
construction works, 181–218
loads, 185–187, 198–199, 204–210
material density, 11, 29, 55–56
material storage, 205, 206, *207*
self-weight, 11, 31–33
thermal actions, 177–178
continental snow fall, 84
continuous floor slabs, 25–27
crushing, 41
cumulative distribution, 78, 79, 111–112
cylindrical roofs, 89, 95–99

dancing, 17–18
deflagration index, 281–282
demolition, 182, 217
density, 11, 14, 29, 55–56
snow loads, 68, 121–122
derailed rail traffic, 245–248
design situations
accidental actions, 229–237
actions during execution, 182, 189–195
density, self-weight and imposed loads, 19–27
snow loads, 73–75, 80, 85–86, 107
thermal actions, 165–166
wind actions, 131
design working lives, 6–7
directional factors, 128
displacement height, 144

disproportionate collapse, 224–225, 231, 257
divergent wind actions, 155
domestic area loads, 38–44
drifts, snow loads, 67–68, 71, 74–75, 84–86, 92–99, 101–103, 107
dust explosions, 281–283
dynamic characteristics of structures, 155, 156
dynamic design, impacts, 277–279
dynamic factors, 149–151, 155
dynamic loading, 17–18

earthquakes, 80, 211, 222, 229
EN 1990
  action classification/definitions, 4–6
  requirements affecting EN 1991-1-7, 222–223
EN 1991-1-1, 11–59
  action classification, 16–18
  action representation, 31–32, 35–37
  administration areas, 38–44
  angle of repose, 14
  Annex A, 55–56
  Annex B, 57–58
  arrangements of loads, 37–38
  beams, 23–24, 37–38
  building's imposed loads, 11–12, 17–18, 22–27, 31–33, 35–53
  cantilever beam against overturning, 23–24
  car parks, 49–50, 57–58
  characteristic value of actions, 20, 23, 32–33, 38–53
  classification of actions, 16–18
  columns, imposed loads, 38
  combination value, 23
  commercial area loads, 38–44
  construction material density, 11, 29, 55–56
  construction works, 11, 31–33
  continuous floor slabs, 25–27
  definitions, 13–14
  density, 11, 14, 29, 55–56, 68
  design situations, 19–27
  domestic area loads, 38–44
  dynamic loading, 17–18
  examples, 20, 23–25
  floors, 25–27, 35–38
  forklifts, 47, *48*
  free actions, 16, 17–18, *17*, 23–27
  frequent value, 23
  garages, 49–50, 57–58
  general, 11–14
  gross weight of vehicle, 14
  horizontal loads, 52–53
  imposed loads, 11–12, 17–18, 22–27, 31–53, 57–58
  industrial activity areas, 44–48
  lightweight structure loads, 18
  limit states, 19–20
  long-span structures, 18
  machinery dynamic loads, 18
  maintenance special devices, 48–49
  movable partitions, 14
  non-structural elements, 14
  notation, 14
  office area loads, 38–44
  parapets, 52–53, 57–58
  partitions, 14, 41–42, 52–53
  people, 35–38
  permanent actions, 15–16, 20
  permanent loads, 20–22
  quasi-permanent value, 33
  reduction factors, 42–44
  representation of actions, 31–32, 35–37
  residential area loads, 38–44
  roofs, 35, 36–38, 50–53
  self-weight, 11, 15–17, 19–27, 31–33
  serviceability limit states, 19–20
  shopping area loads, 38–44
  social area loads, 38–44
  special devices for maintenance, 48–49
  storage areas, 44–48
  stored material density, 11, 29, 55–56
  structural elements, 14
  symbols, 14
  synchronised rhythmical movement, 17–18
  traffic areas, 49–50
  transport vehicles, 47
  ultimate limit states, 19–20
  variable actions, 20, 22–27
  vehicles, 35, 47–50
  walls, 38
  water levels, 22
EN 1991-1-2, fire, 63
EN 1991-1-3
  snow loads, 67–123, 192–193, 203
    accidental actions, 71, 74–75, 93, 99
    accidental design situation, 74–75, 80, 85–86
    action classification, 71
    actions during execution, 192, 193, 203
    Annex A, 67–68, 107
    Annex B, 68, 109
    Annex C, 68, 77–78, 111–115
    Annex D, 68, 117–119
    Annex E, 68, 121–122
    arrangement, 83–89, 107
    bulk weight density, 68, 121–122
    characteristic values, 68, 73–74, 77–81, 111–115
    classification of actions, 71
    coefficients, 87–99, 117–119
    collapse of roofs, 86–87, *88*
    combination values, 73
    continental snow fall, 84
    cylindrical roofs, 89, 95–99
    definitions, 69–70
    density, 68, 121–122
    design assisted by testing, 69
    design situations, 73–75, 85–86, 107
    drifts, 67–68, 71, 74–75, 84–86, 92–99, 101–103, 107
    European load maps, 111–115

EN 1991-1-3 (*continued*)
snow loads
exceptional drifts, 67–68, 71, 74–75, 107
exceptional fall, 67–68, 71, 74, 75, 78–81, 107
exposure characteristics, 87–88
exposure coefficients, 87–88, 92, 94
fixed actions, 71
frequent value, 73, 80
general, 67–70
ground loads, 68, 77–81, 111–115, 117–119
imposed loads, 81
limit states, 73, 80
local effects, 101–105
maps, 68, 77–78, 111–115
maritime snow fall, 84–85
monopitch roofs, 89, 90–91
multi-span roofs, 89, 90, 93–95
non-exceptional drifts, 67–68, 74, 107
non-exceptional fall, 67–68, 74–75, 107
normal conditions, 67–68, 74
notation, 70
obstacles, 104–105
obstruction drifting, 101–103
overhang on roofs, 102–104
parapets, 102, *103*
permanent actions, 81
persistent design situation, 74–75, 85–86, 107
pitch angles, 90–92
pitched roofs, 89, 90, 91–93
projection drifting, 101–103
quasi-permanent value, 73, 80
rainfall, 86
return periods, 68, 117–119
roofs, 83–99, 102–105
seismic design, 80
self-weight, 81
serviceability limit states, 73, 80
shape coefficients, 89–99, 109
shapes on roofs, 83–99
snowguards, 104–105
symbols, 70
testing and design, 69
thermal characteristics, 88–89
thermal coefficients, 88–89, 92, 94
transient design situation, 74–75, 85–86, 107
ultimate limit states, 73
variable actions, 71, 73–74
wind, 69, 81, 83–84, 87–88, 91, 95–96
wind tunnel tests, 69, 83–84, 91, 95–96
EN 1991-1-4
wind actions, 126–156
actions during execution, 192–194
aerodynamic coefficients, 153–154
aeroelasticity, 155
Annex F, 155, 156
Annexes A to E, 155
background response factors, 127
buffeting, 151
coefficients, 128, 146–148, 153–154
definitions, 127–129
design situations, 131
directional factors, 128
displacement height, 144
divergence, 155
dynamic characteristics of structures, 155, 156
dynamic factors, 149–151, 155
end-effect factors, 147
exposure factors, 128, 139–143
external forces, 147
flutter, 155
forces, 128, 129, 145–148, 153–154
friction, 128, 147–148, 154
galloping, 155
general, 127–129
height displacement, 144
internal forces, 147
mean wind velocity, 128, 137–138
modelling, 133
National Annex, 138, 140–143, 149–151
notation, 127–129
orography factors, 128, 140–143
peak velocity pressure, 128, 135, 139–143
pressures, 128, 129, 133, 135–144, 145, *146*, 153–154
probability factors, 128
reduction factors, 146, 154
resonant factors, 128
roughness factors, 128, 137–138
season factors, 128
shapes force coefficients, 146–148
size factors, 143, 149–151, 155
structural factors, 128, 149–151, 155
surface wind pressures, 145, *146*
symbols, 127–129
terrain factors, 135–143, 155
turbulence intensity, 129, 138–139
velocity, 128, 133, 135–144
vortex shedding, 155
wake buffeting, 151
wind forces, 128, 129, 145–148, 153–154
wind pressures, 128, 129, 133, 135–144, 145, *146*, 153–154
wind velocity, 128, 133, 135–144
EN 1991-1-5
thermal actions, 159–178, 191–192, 202
action classification, 163–164
actions during execution, 191–192
Annex A, 175
Annex B, 177–178
architectural massing junctions, 160
bracing systems, 160
bridges, 175
buildings, 167–173, 177–178
characteristic values, 165–166
classification of actions, 163–164

coefficients of linear expansion, 160, 167–168, 175
combination values, 165–166
construction works, 177–178
definitions, 161
design situations, 165–166
framing systems, 160
frequency factors, 160
frequent value, 165–166
general, 159–161
geometry, 160, 167–168
materials, 160, 167–168
movement, 160
notation, 161
quasi-permanent value, 166
representation of actions, 167–168
restraints, 160
return periods, 159, 163–164
serviceability limit states, 165
shade air temperatures, 163, 167, 175
solar radiation, 163, 167, 172–173
strains, 160, 167–168
stresses, 167–168
symbols, 161
temperature, 160, 167–173, 177–178
ultimate limit states, 165
EN 1991-1-6
actions during execution, 181–218
accidental actions, 210–211
action classification, 181–182, 185–187
action representation, 182, 197–211
alterations, 182, 217
Annex A, 182, 213–215
Annex B, 182, 217
atmospheric icing, 204
auxiliary construction works, 182
bridges, 182, 215
buildings, 181–218
characteristic value, 190–194, 197–199
classification of actions, 181–182, 185–187
climatic action, 191–194
combination value, 197, 198
concrete construction work, 193, 200, 209–210
construction, 181–218
construction loads, 185–187, 198–199, 204–210
definitions, 183
demolition, 182, 217
design situations, 182, 189–195
general, 181–183
geotechnical actions, 200–201
handling factors, 199–200
horizontal actions, 214
impacts, 193
limit states, 182, 194–195, 213–214
notation, 183
predeformations, 201–202
prestressing, 201, *202*
quasi-permanent value, 197, 198
reconstructions, 182, 217
representation of actions, 182, 197–211
return periods, 190–191
seismic actions, 211
self-weight, 199–200
serviceability limit states, 182, 194–195, 213–214
supplementary rules, 182, 213–215
symbols, 183
temperature, 202
ultimate limit states, 194, 213
variable actions, 197–199
water, 203
EN 1991-1-7
accidental actions, 221–284
accepted risks, 232, *233*, 269–270
action classification, 227–228, 240–241
action representation, 240–241, 252–253
Annex A, 255–264
Annex B, 265–275
Annex C, 277–279
Annex D, 281–283
bomb attacks, 234
buildings, 222, 241–245, 255–264, 265–275
classification of actions, 227–228
collapse, 224–225, 231, 234–236, 257
collisions, 221–222, 240–249
consequence classes, 224, 232, 236–237, 255–264
damage/failure, 230–231, 232, 234–236, 255–264
definitions, 223–225
design principles, 253–254
design situations, 229–237
EN 1990 requirements, 222–223
explosions, 221–222, 251–254, 281–283
fire, 222, 254
forklift trucks, 239, 245
general, 221–225
helicopters, 239, 249
horizontal ties, 259–263
identified actions, 230–231, 232–234
impacts, 221–222, 240–249, 274–275, 277–279
internal explosions, 251–254, 281–283
limit states, 227
localised damage/failure, 230–231, 232, 234–236, 255–264
National Annexes, 242, 244–248, 258
notation, 225
rail traffic, 239, 245–248, 274–275, 283
representation of actions, 240–241, 252–253
risk analysis, 222, 265–275
road vehicles, 239, 241–245, 274, 279, 283
robustness, 224–225, 230, 233–234, 256–257
seismic action, 222, 229
ship traffic, 248–249, 274, 279
symbols, 225

EN 1991-1-7 (*continued*)
accidental actions
ties, 235–236, 259–264
trains, 239, 245–248, 274–275
unidentified actions, 230–231, 255–264
vertical ties, 263–264
end-effect factors, 147
equipment loads, *205*, 206, *207*
Eurocodes
benefits, 2–3
field of application, 3
implementation, 3–4
links between, 2
National Annexes, 3, 4
National Standard implementation, 3–4
Nationally Determined Parameters, 3–4
objectives, 1–2
programme background, 1–3
rules and contents, 3
status, 1–2, 3
exceptional snow drifts, 67–68, 71, 74–75, 107
exceptional snow fall, 67–68, 71, 74, 75, 78–81, 107
execution actions *see* actions during execution
explosions, 221–222, 251–254, 256–257, 281–283
exposure characteristics, 87–88
exposure coefficients, 87–88, 92, 94
exposure factors, 128, 139–143
external forces, wind actions, 147
external temperature factors, 171–173
extreme event risks, 272, 273–274

fatal accidental actions, 232
fire, 63, 222, 254
fixed actions, 71, 186, 187
floors, 25–27, 35–38
flutter, 155
force coefficients, wind, 128, 146–148, 153–154
forklifts, 47, *48*, 239, 245
framing systems, 160, 235–236, 259–264
free actions, 16–18, 23–27, 186, 187, 227
frequency of temperature measurements, 160
frequent value, 5–6, 23, 73, 80, 165–166
friction, 128, 147–148, 154

galloping, 155
garage loads, 49–50, 57–58
gas explosions, 251–252, 256–257, 283
geometry, thermal actions, 160, 167–168
geotechnical actions during execution, 200–201
gross weight of vehicle, 14
ground snow loads, 68, 77–81, 111–115, 117–119
Gumbel type cumulative distribution, 78, 79, 111–112

handling factors, 199–200
handtools, 205–206
hard impact, 278–279
heavy machinery/equipment loads, *205*, 206, *207*
height displacement, 144
helicopters, *51*, 52, 239, 249
horizontal actions during execution, 214
horizontal loads, 52–53
horizontal ties, 259–263
hydration effects, 202

icing, 204
see also snow loads
identified accidental actions, 230–231, 232–234
impacts
accidental actions, 221–222, 240–249, 274–275, 277–279
actions during execution, 193
imposed loads
EN 1991-1-1, 11–12, 17–18, 22–27, 31–53, 57–58
arrangements, 37–38
beams, 23–24, 37–38
buildings, 11–12, 17–18, 22–27, 31–33, 35–53, 57–58
car parks, 49–50, 57–58
characteristic values, 38–53
design situations, 22–27
floors, 25–27, 35–38
forklifts, 47, *48*
garages, 49–50, 57–58
industrial activity areas, 44–48
parapets, 52–53, 57–58
people, 35–38
reduction factors, 42–44
roofs, 35, 36–38, 50–53
storage areas, 44–48
traffic areas, 49–50
transport vehicles, 47
vehicles, 35, 47–50
snow loads, 81
industrial activity area loads, 44–48
inner environment temperatures, 171–172
internal explosions, 251–254, 281–283
internal forces, wind actions, 147
internal temperatures, 171–172
Italian snow loads, 78–79, 92

jumping, 17–18
junctions, architectural massing, 160

lightweight structure loads, 18
limit states, 19–20
accidental actions, 227
actions during execution, 182, 194–195, 213–214
snow loads, 73, 80
thermal actions, 165
load-bearing walls, 261–264
local effects, snow loads, 101–105
localised damage/failure, 230–231, 232, 234–236, 255–264
long-span structures, 18

machinery
see also equipment loads
dynamic loads, 18
maintenance special devices, 48–49
maps, snow loads, 68, 77–78, 111–115
maritime
shipping, 248–249, 274, 279
snow fall, 84–85
masonry structures, 32–33
materials
density, 11, 29, 55–56
storage, 205, 206, *207*
thermal actions, 160, 167–168
waste accumulation, *205*, 206–207, *208*
mean wind velocity, 128, 137–138
mechanical actions, 63
mitigation of risk, 269–271
modelling wind actions, 133
monopitch roofs, 89, 90–91
movable partitions, 14, 41–42
moveable item's storage, *205*, 206
movement, thermal actions, 160
multi-span roofs, 89, 90, 93–95

National Annexes, 3, 4, 7
accidental actions, 242, 244–248, 258
wind actions, 138, 140–143, 149–151
national standard implementation, 3–4
Nationally Determined Parameters (NDPs), 3–4
nominal section, load-bearing walls, 264
non-exceptional snow drifts, 67–68, 74, 107
non-exceptional snow fall, 67–68, 74–75, 107
non-permanent equipment, *205*, 206, *208*
non-structural elements, 14
non-structural members during handling, 199–200
non-structural risk analysis, 271–273–5
normal conditions, snow loads, 67–68, 74

obstacles on roofs, 104–105
obstruction drifting, 101–103
office area loads, 38–44
orography factors, wind actions, 128, 140–143
oscillations, 17–18
outer environment temperatures, 172–173
overhanging snow on roofs, 102–104

parapets, 52–53, 57–58, 102, *103*
partitions, 14, 41–42, 52–53
peak velocity pressure, 128, 135, 139–143
people, imposed loads, 35–38
permanent actions, 5, 15–16, 20, 81
permanent loads, 20–22
persistent design situation loads, 74–75, 85–86, 107
personnel, construction loads, 205–206
pitch angles, 90–92
pitched roofs, 89, 90, 91–93

predeformations, 201–202
pressure, 128–129, 133–146, 153–154, 253–254
prestressing, 201, *202*
probability factors, 128
projection drifting, 101–103
punching, 41

qualitative risk analysis, 267–268
quantitative risk analysis, 268–269
quasi-permanent value, 5, *6*
actions during execution, 197, 198
imposed loads, 33
snow loads, 73, 80
thermal actions, 166

rail traffic, 239, 245–248, 274–275, 283
rainfall, 86
reconstruction actions, 182, 217
reduction coefficients, 192–193
reduction factors, 42–44, 146, 154
repose, angle of, 14
representation of actions
accidental actions, 240–241, 252–253
actions during execution, 182, 197–211
imposed loads, 35–37
self-weight, 31–32
thermal actions, 167–168
residential area loads, 38–44
resonance, 17–18, 128
restraints, ties, 160, 235–236, 259–264
return periods
actions during execution, 190–191, 192–193
characteristic value coefficient k, 6, *7*
snow loads, 68, 117–119
thermal actions, 159, 163–164
risk
acceptance, 232, *233*, 269–270
definition, 266
mitigation, 269–271
risk analysis, 222, 265–275
buildings, 222, 265–275
communication, 271
definition, 266
methods, 267–269
road vehicles
accidental actions, 239, 241–245, 274, 279, 283
imposed loads, 47
robustness, 224–225, 230, 233–234, 256–257
Ronan Point collapse, 256–257
roofs
collapse, 86–87, *88*
imposed loads, 35, 36–38, 50–53
snow loads, 83–99, 102–105, 109
roughness factors, 128, 137–138

Scottish Highland snow loads, 93
season factors, 128
seismic actions, 80, 211, 222, 229

self-weight, 5
actions during execution, 199–200
EN 1991-1-1, 11, 15–17, 19–27, 31–33
snow loads, 81
serviceability limit states, 19–20, 22–23, 73, 80
actions during execution, 182, 194–195, 213–214
thermal actions, 165
shade air temperatures, 163, 167, 175
shape force coefficients, wind, 146–148
shapes on roofs, snow, 83–99
ship traffic, 248–249, 274, 279
shopping area loads, 38–44
shrinkage effects, 202
single action classification, 5
size factors, 143, 149–151, 155
snow loads
actions during execution, 192, 193, 203
EN 1991-1-3, 67–123, 192–193, 203
accidental actions, 71, 74–75, 80, 85–86, 93, 99
accidental design situation, 74–75, 80, 85–86
action classification, 71
actions during execution, 192, 193, 203
arrangement of loads, 83–89, 107
bulk weight density, 68, 121–122
characteristic value, 68, 73–74, 77–81, 111–115
classification of actions, 71
coefficients, 87–99, 109, 117–119
collapse of roofs, 86–87, *88*
combination value, 73
cylindrical roofs, 89, 95–99
density, 68, 121–122
design situations, 73–75, 80, 85–86, 107
drifts, 67–68, 71, 74–75, 84–86, 92–99, 101–103, 107
European load maps, 111–115
exceptional drifts, 67–68, 71, 74–75, 107
exceptional fall, 67–68, 71, 74, 75, 78–81, 107
exposure characteristics, 87–88
exposure coefficients, 87–88, 92, 94
fixed actions, 71
frequent value, 73, 80
ground loads, 68, 77–81, 111–115, 117–119
imposed loads, 81
local effects, 101–105
maps, 68, 77–78, 111–115
monopitch roofs, 89, 90–91
multi-span roofs, 89, 90, 93–95
non-exceptional drifts, 67–68, 74, 107
non-exceptional fall, 67–68, 74–75, 107
normal conditions, 67–68, 74
obstacles, 104–105
obstruction drifting, 101–103
overhang on roofs, 102–104
parapets, 102, *103*
permanent actions, 81
persistent design situation, 74–75, 85–86, 107
pitch angles, 90–92
projection drifting, 101–103
quasi-permanent value, 73, 80
rainfall, 86
return periods, 68, 117–119
roofs, 83–99, 109
seismic design situation, 80
self-weight, 81
serviceability limit states, 73, 80
shape coefficients, 89–99, 109
shapes on roofs, 83–99
snow overhang, 102–104
snowguards, 104–105
thermal characteristics, 88–89
thermal coefficients, 88–89, 92, 94
transient design situation, 74–75, 85–86, 107
ultimate limit states, 73
variable actions, 71, 73–74
variations (coefficients of), 117–119
wind, 69, 81, 83–84, 87–88, 91, 95–96
wind tunnel tests, 69, 83–84, 91, 95–96
social area loads, 38–44
soft impacts, 279
solar radiation, 163, 167, 172–173
steel structures, 33
storage area loads, 44–48
storage of construction loads, *205*, 206
stored material density, 11, 29, 55–56
strains, 160, 167–168
stresses, 167–168
structural elements, 14
structural factors, wind, 128, 149–151, 155
structural members during handling, 199–200
structural risk analysis, 271, 272–273
substructures, 225
superstructures, 225, 242–245
supplementary rules to EN, 182, 213–215, 1991–11–6
surface wind pressures, 145, *146*
surfacing depths, 175
Swedish snow loads, 93–94
synchronised rhythmical movement, 17–18

temperature
see also thermal
actions during execution, 202
changes, 160, 169–173
determination/range, 160, 169–173
profiles, 167–168, 170–173, 177–178
strains, 160
surfacing depths, 175
temporary state structure loads, *205*, 207–208, *209*
terrain factors, wind, 135–143, 155
thermal actions, 63, 159–178, 191–192, 202
actions during execution, 191–192, 202
EN 1991-1-5, 159–178, 191–192, 202

action classification, 163–164
architectural massing junctions, 160
bracing systems, 160
bridges, 175
buildings, 167–173, 177–178
characteristic value, 165–166
classification of actions, 163–164
coefficients of linear expansion, 160, 167–168, 175
combination value, 165–166
construction works, 177–178
design situations, 165–166
framing systems, 160
frequency factors, 160
frequent value, 165–166
geometry, 160, 167–168
materials, 160, 167–168
movement, 160
quasi-permanent value, 166
representation of actions, 167–168
restraints, 160
return periods, 159, 163–164
serviceability limit states, 165
shade air temperatures, 163, 167, 175
solar radiation, 163, 167, 172–173
strains, 160, 167–168
stresses, 167–168
temperature, 160, 167–173, 177–178
ultimate limit states, 165
thermal characteristics of snow loads, 88–89
thermal coefficients of snow loads, 88–89, 92, 94
three-span continuous floor slabs, 25–27
ties, 160, 235–236, 259–264
timber structure self weight values, 33
track end structures, 248
traffic
accidental actions, 239, 241–245, 274, 279, 283
area loads, 49–50
imposed loads, 47
trains, 239, 245–248, 274–275
transient design situations, 74–75, 85–86, 107, 190
transport vehicles
accidental actions, 239, 241–245, 274, 279, 283
imposed loads, 47
tunnel explosions, 283
turbulence intensity, 129, 138–139

ultimate limit states, 19–20
accidental actions, 227
actions during execution, 194, 212
snow loads, 73
thermal actions, 165
unidentified accidental actions, 230–231, 255–264

variable actions, 5, *6*, 20, 22–27
accidental actions, 227–228, 234
actions during execution, 197–199
snow loads, 71, 73–74
vehicles
accidental actions, 239, 241–245, 274, 279, 283
imposed loads, 35, 47–50
velocity, wind, 128, 133, 135–144
venting panels, 253–254
vertical ties, 263–264
vibrations, 17–18
vortex shedding, 155

wake buffeting, 151
walkways, 53
walls
accidental actions, 241–242, 261–264
imposed loads, 38
waste material accumulation, *205*, 206–207, *208*
water
actions during execution, 203
levels, 22
wind actions
actions during execution, 192–194, 202–203
EN 1991-1-3, snow loads, 69, 81, 83–84, 87–88, 91, 95–96
EN 1991-1-4, 126–156
actions during execution, 192–194
aerodynamic coefficients, 153–154
aeroelasticity, 155
background response factors, 127
buffeting, 151
coefficients, 128, 146–148, 153–154
design situations, 131
directional factors, 128
displacement height, 144
divergence, 155
dynamic factors, 149–151, 155, 156
end-effect factors, 147
exposure factors, 128, 139–143
external forces, 147
flutter, 155
forces, 128, 129, 145–148, 153–154
friction, 128, 147–148, 154
galloping, 155
height displacement, 144
internal forces, 147
mean wind velocity, 128, 137–138
modelling, 133
National Annex, 149–151
orography factors, 128, 140–143
peak velocity pressure, 128, 135, 139–143
pressures, 128, 129, 133, 135–144, 145, *146*, 153–154
probability factors, 128
reduction factors, 146, 154
resonant factors, 128
roughness factors, 128, 137–138
season factors, 128
shapes force coefficients, 146–148
size factors, 143, 149–151, 155

wind actions (*continued*)
EN 1991-1-4
structural factors, 128, 149–151, 155
surface wind pressures, 145, *146*
terrain factors, 135–143, 155
turbulence intensity, 129, 138–139
velocity, 128, 133, 135–144
vortex shedding, 155
wake buffeting, 151
wind forces, 128, 129, 145–148, 153–154
wind pressures, 128, 129, 133, 135–144, 145, *146*, 153–154
wind velocity, 128, 133, 135–144
snow loads, 69, 81, 83–84, 87–88, 91, 95–96
wind tunnel snow load tests, 69, 83–84, 91, 95–96